Buddhism and Ecology

Harvard University
Center for the Study of World Religions
Publications

General Editor: Lawrence E. Sullivan
Senior Editor: Kathryn Dodgson

Religions of the World and Ecology
Series Editors:
Mary Evelyn Tucker and John Grim

Cambridge, Massachusetts

Buddhism and Ecology
The Interconnection
of Dharma and Deeds

edited by
Mary Evelyn Tucker
and
Duncan Ryūken Williams

Distributed by Harvard University Press
for the
Harvard University Center for the Study of World Religions

Grateful acknowledgment is made for permission to reprint the following:

Rita M. Gross, "Buddhist Resources of Issues of Population, Consumption, and the Environment," in *Population, Consumption, and the Environment: Religious and Secular Responses*, edited by Harold Coward. By permission of the State University of New York Press. Copyright © 1995 by the State University of New York Press.

Paul O. Ingram, "The Jeweled Net of Nature," *Process Studies* 22, no. 3 (fall 1993).

John Daido Loori, "The Precepts and the Environment," *Mountain Record*, spring 1996.

Steve Odin, "The Japanese Concept of Nature in Relation to the Environmental Ethics and Conservation of Aldo Leopold," *Environmental Ethics* 13 (winter 1991):345–60.

Alan Sponberg, "Green Buddhism and the Hierarchy of Compassion," *Western Buddhist Review* 1 (December 1994):131–55.

Library of Congress Cataloging-in-Publication Data

Buddhism and ecology : the interconnection of dharma and deeds / edited
    by Mary Evelyn Tucker and Duncan Ryūken Williams.
        p.   cm. — (Religions of the world and ecology)
    Includes bibliographical references and index.
    ISBN 0-945454-13-9 (hard cover : alk. paper)
    ISBN 0-945454-14-7 (pbk. : alk. paper)
    1. Human ecology—Religious aspects—Buddhism. 2. Ecology—
Religious aspects—Buddhism. I. Tucker, Mary Evelyn. II. Williams,
Duncan Ryūken. III. Series.
BQ4570.E23B83 1997
294.3'378362—dc21   97-37528
                                                                              CIP

Acknowledgments

The series of conferences on religions of the world and ecology will take place from 1996 through 1998, with supervision at the Harvard University Center for the Study of World Religions by Don Kunkel and Malgorzata Radziszewska-Hedderick and with the assistance of Janey Bosch, Naomi Wilshire, and Lilli Leggio. Narges Moshiri, also at the Center, was indispensable in helping to arrange the first two conferences. A series of volumes developing the themes explored at the conferences will be published by the Center and distributed by Harvard University Press under the editorial direction of Kathryn Dodgson and with the skilled assistance of Eric Edstam.

These efforts have been generously supported by major funding from the V. Kann Rasmussen Foundation. The conference organizers appreciate also the support of the following institutions and individuals: Association of Shinto Shrines, Nathan Cummings Foundation, Dharam Hinduja Indic Research Center at Columbia University, Harvard Buddhist Studies Forum, Harvard Divinity School Center for the Study of Values in Public Life, Jain Academic Foundation of North America, Laurance Rockefeller, Sacharuna Foundation, and Theological Education to Meet the Environmental Challenge. The conferences were originally made possible by the Center for Respect of Life and Environment of the Humane Society of the United States, which continues to be a principal cosponsor. Bucknell University, also a cosponsor, has provided support in the form of leave time from teaching for conference coordinators Mary Evelyn Tucker and John Grim as well as the invaluable administrative assistance of Stephanie Snyder. Her thoughtful attention to critical details is legendary. President William Adams of Bucknell University and Vice-President for Academic Affairs Daniel Little have also granted travel funds for faculty and students to attend the conferences. Grateful acknowledgment is here made for the advice from key area specialists in shaping each conference and in editing the published volumes. Their generosity in time and talent has been indispensable at every step of the project. Finally, throughout this process, the support, advice, and encouragement from Martin S. Kaplan has been invaluable.

Contents

Preface
Lawrence E. Sullivan   xi

Series Foreword
Mary Evelyn Tucker and *John Grim*   xv

Introduction
Duncan Ryūken Williams   xxxv

Overview: Framing the Issues

Buddhism and Ecology: Collective Cultural
Perceptions
Lewis Lancaster   3

Theravāda Buddhism and Ecology: The Case of Thailand

The Hermeneutics of Buddhist Ecology in
Contemporary Thailand: Buddhadāsa and
Dhammapiṭaka
Donald K. Swearer   21

A Theoretical Analysis of the Potential
Contribution of the Monastic Community in
Promoting a Green Society in Thailand
Leslie E. Sponsel and *Poranee Natadecha-Sponsel*   45

Mahāyāna Buddhism and Ecology: The Case of Japan

The Jeweled Net of Nature
Paul O. Ingram   71

The Japanese Concept of Nature in Relation
to the Environmental Ethics and Conservation
Aesthetics of Aldo Leopold
Steve Odin   89

Voices of Mountains, Trees, and Rivers: Kūkai,
Dōgen, and a Deeper Ecology
Graham Parkes   111

Buddhism and Animals: India and Japan

Animals and Environment in the Buddhist
Birth Stories
Christopher Key Chapple   131

Animal Liberation, Death, and the State:
Rites to Release Animals in Medieval Japan
Duncan Ryūken Williams   149

Zen Buddhism: Problems and Prospects

Mountains and Rivers and the Great Earth:
Zen and Ecology
Ruben L. F. Habito   165

The Precepts and the Environment
John Daido Loori   177

American Buddhism: Creating Ecological Communities

Great Earth *Saṅgha*: Gary Snyder's View
of Nature as Community
David Landis Barnhill   187

American Buddhist Response to the Land:
Ecological Practice at Two West Coast
Retreat Centers
Stephanie Kaza   219

The Greening of Zen Mountain Center:
A Case Study
Jeff Yamauchi   249

Applications of Buddhist Ecological Worldviews

Nuclear Ecology and Engaged Buddhism
Kenneth Kraft   269

Buddhist Resources for Issues of Population,
Consumption, and the Environment
Rita M. Gross   291

Buddhism, Global Ethics, and the Earth Charter
Steven C. Rockefeller   313

**Theoretical and Methodological Issues in Buddhism
and Ecology**

Is There a Buddhist Philosophy of Nature?
Malcolm David Eckel   327

Green Buddhism and the Hierarchy of Compassion
Alan Sponberg   351

Buddhism and the Discourse of Environmental
Concern: Some Methodological Problems Considered
Ian Harris   377

Bibliography on Buddhism and Ecology
Duncan Ryūken Williams   403

Notes on Contributors   427

Index   433

Preface

Lawrence E. Sullivan

Religion distinguishes the human species from all others, just as human presence on earth distinguishes the ecology of our planet from other places in the known universe. Religious life and the earth's ecology are inextricably linked, organically related.

Human belief and practice mark the earth. One can hardly think of a natural system that has not been considerably altered, for better or worse, by human culture. "Nor is this the work of the industrial centuries," observes Simon Schama. "It is coeval with the entirety of our social existence. And it is this irreversibly modified world, from the polar caps to the equatorial forests, that is all the nature we have" (*Landscape and Memory* [New York: Vintage Books, 1996], 7). In Schama's examination even landscapes that appear to be most free of human culture turn out, on closer inspection, to be its product.

Human beliefs about the nature of ecology are the distinctive contribution of our species to the ecology itself. Religious beliefs—especially those concerning the nature of powers that create and animate—become an effective part of ecological systems. They attract the power of will and channel the forces of labor toward purposive transformations. Religious rituals model relations with material life and transmit habits of practice and attitudes of mind to succeeding generations.

This is not simply to say that religious thoughts occasionally touch the world and leave traces that accumulate over time. The matter is the other way around. From the point of view of environmental studies, religious worldviews propel communities into the world with fundamental predispositions toward it because such religious worldviews are primordial, all-encompassing, and unique. They are *primordial* because they probe behind secondary appearances and stray thoughts to rivet human attention on realities of the first order: life at its source, creativity in its fullest manifestation, death and destruction at their origin, renewal and salvation in their germ. The revelation of first things is compelling and moves communities to take creative action. Primordial ideas are prime movers.

Religious worldviews are *all-encompassing* because they fully absorb the natural world within them. They provide human beings both a view of the whole and at the same time a penetrating image of their own ironic position as the beings in the cosmos who possess the capacity for symbolic thought: the part that contains the whole—or at least a picture of the whole—within itself. As all-encompassing, therefore, religious ideas do not just contend with other ideas as equals; they frame the mind-set within which all sorts of ideas commingle in a cosmology. For this reason, their role in ecology must be better understood.

Religious worldviews are *unique* because they draw the world of nature into a wholly other kind of universe, one that appears only in the religious imagination. From the point of view of environmental studies, the risk of such religious views, on the one hand, is of disinterest in or disregard for the natural world. On the other hand, only in the religious world can nature be compared and contrasted to other kinds of being—the supernatural world or forms of power not always fully manifest in nature. Only then can nature be revealed as distinctive, set in a new light startlingly different from its own. That is to say, only religious perspectives enable human beings to evaluate the world of nature in terms distinct from all else. In this same step toward intelligibility, the natural world is evaluated in terms consonant with human beings' own distinctive (religious and imaginative) nature in the world, thus grounding a self-conscious relationship and a role with limits and responsibilities.

In the struggle to sustain the earth's environment as viable for future generations, environmental studies has thus far left the role of religion unprobed. This contrasts starkly with the emphasis given, for example, the role of science and technology in threatening or sustaining the ecology. Ignorance of religion prevents environmental studies from achieving its goals, however, for though science and technology share many important features of human culture with religion, they leave unexplored essential wellsprings of human motivation and concern that shape the world as we know it. No understanding of the environment is adequate without a grasp of the religious life that constitutes the human societies which saturate the natural environment.

A great deal of what we know about the religions of the world is new knowledge. As is the case for geology and astronomy, so too for religious studies: many new discoveries about the nature and function of religion are, in fact, clearer understandings of events and processes that began to unfold long ago. Much of what we are learning now about the religions of the world was previously not known outside of a circle of adepts. From the ancient history of traditions and from the ongoing creativity of the world's contemporary religions we are opening a treasury of motives, disciplines, and awarenesses.

A geology of the religious spirit of humankind can well serve our need to relate fruitfully to the earth and its myriad life-forms. Changing our habits of consumption and patterns of distribution, reevaluating modes of production, and reestablishing a strong sense of solidarity with the matrix of material life—these achievements will arrive along with spiritual modulations that unveil attractive new images of well-being and prosperity, respecting the limits of life in a sustainable world while revering life at its sources. Remarkable religious views are presented in this series—from the nature mysticism of Bashō in Japan or Saint Francis in Italy to the ecstatic physiologies and embryologies of shamanic healers, Taoist meditators, and Vedic practitioners; from indigenous people's ritual responses to projects funded by the World Bank, to religiously grounded criticisms of hazardous waste sites, deforestation, and environmental racism.

The power to modify the world is both frightening and fascinating and has been subjected to reflection, particularly religious reflection, from time immemorial to the present day. We will understand ecology better when we understand the religions that form the rich soil of memory and practice, belief and relationships where life on earth is rooted. Knowledge of these views will help us reappraise our ways and reorient ourselves toward the sources and resources of life.

This volume is one in a series that addresses the critical gap in our contemporary understanding of religion and ecology. The series results from research conducted at the Harvard University Center for the Study of World Religions over a three-year period. I wish especially to acknowledge President Neil L. Rudenstine of Harvard University for his leadership in instituting the environmental initiative at Harvard and thank him for his warm encouragement and characteristic support of our program. Mary Evelyn Tucker and John Grim of Bucknell University coordinated the research, involving the direct participation of some six hundred scholars, religious leaders, and environmental specialists brought to Harvard from around the world during the period of research and inquiry. Professors Tucker and Grim have brought great vision and energy to this enormous project, as has their team of conference convenors. The commitment and advice of Martin S. Kaplan of Hale and Dorr have been of great value. Our goals have been achieved for this research and publication program because of the extraordinary dedication and talents of Center for the Study of World Religions staff members Don Kunkel, Malgorzata Radziszewska-Hedderick, Kathryn Dodgson, Janey Bosch, Naomi Wilshire, Lilli Leggio, and Eric Edstam and with the unstinting help of Stephanie Snyder of Bucknell. To these individuals, and to all the sponsors and participants whose efforts made this series possible, go deepest thanks and appreciation.

Series Foreword

Mary Evelyn Tucker and John Grim

The Nature of the Environmental Crisis

Ours is a period when the human community is in search of new and sustaining relationships to the earth amidst an environmental crisis that threatens the very existence of all life-forms on the planet. While the particular causes and solutions of this crisis are being debated by scientists, economists, and policymakers, the facts of widespread destruction are causing alarm in many quarters. Indeed, from some perspectives the future of human life itself appears threatened. As Daniel Maguire has succinctly observed, "If current trends continue, we will not."[1] Thomas Berry, the former director of the Riverdale Center for Religious Research, has also raised the stark question, "Is the human a viable species on an endangered planet?"

From resource depletion and species extinction to pollution overload and toxic surplus, the planet is struggling against unprecedented assaults. This is aggravated by population explosion, industrial growth, technological manipulation, and military proliferation heretofore unknown by the human community. From many accounts the basic elements which sustain life—sufficient water, clean air, and arable land—are at risk. The challenges are formidable and well documented. The solutions, however, are more elusive and complex. Clearly, this crisis has economic, political, and social dimensions which require more detailed analysis than we can provide here. Suffice it to say, however, as did the *Global 2000 Report*: ". . .once such global environmental problems are in motion they are difficult to reverse. In fact few if any of the problems addressed in the *Global 2000 Report* are amenable to quick technological or policy fixes; rather, they are inextricably mixed with the world's most perplexing social and economic problems."[2]

Peter Raven, the director of the Missouri Botanical Garden, wrote in a paper titled "We Are Killing Our World" with a similar sense of urgency regarding the magnitude of the environmental crisis: "The world that provides our evolutionary and ecological context is in serious trouble, trouble of a kind that demands our urgent attention. By formulating adequate plans for dealing with these large-scale problems, we will be laying the foundation for peace and prosperity in the future; by ignoring them, drifting passively while attending to what may seem more urgent, personal priorities, we are courting disaster."

Rethinking Worldviews and Ethics

For many people an environmental crisis of this complexity and scope is not only the result of certain economic, political, and social factors. It is also a moral and spiritual crisis which, in order to be addressed, will require broader philosophical and religious under-standings of ourselves as creatures of nature, embedded in life cycles and dependent on ecosystems. Religions, thus, need to be re-examined in light of the current environmental crisis. This is because religions help to shape our attitudes toward nature in both conscious and unconscious ways. Religions provide basic interpretive stories of who we are, what nature is, where we have come from, and where we are going. This comprises a worldview of a society. Religions also suggest how we should treat other humans and how we should relate to nature. These values make up the ethical orientation of a society. Religions thus generate worldviews and ethics which underlie fundamental attitudes and values of different cultures and societies. As the historian Lynn White observed, "What people do about their ecology depends on what they think about themselves in relation to things around them. Human ecology is deeply conditioned by beliefs about our nature and destiny—that is, by religion."[3]

In trying to reorient ourselves in relation to the earth, it has become apparent that we have lost our appreciation for the intricate nature of matter and materiality. Our feeling of alienation in the modern period has extended beyond the human community and its patterns of material exchanges to our interaction with nature itself. Especially in technologically sophisticated urban societies, we have become removed from the recognition of our dependence on nature. We no longer know who we are as earthlings; we no longer see the earth as sacred.

Thomas Berry suggests that we have become autistic in our interactions with the natural world. In other words, we are unable to value the life and beauty of nature because we are locked in our own egocentric perspectives and shortsighted needs. He suggests that we need a new cosmology, cultural coding, and motivating energy to overcome this deprivation.[4] He observes that the magnitude of destructive industrial processes is so great that we must initiate a radical rethinking of the myth of progress and of humanity's role in the evolutionary process. Indeed, he speaks of evolution as a new story of the universe, namely, as a vast cosmological perspective that will resituate human meaning and direction in the context of four and a half billion years of earth history.[5]

For Berry and for many others an important component of the current environmental crisis is spiritual and ethical. It is here that the religions of the world may have a role to play in cooperation with other individuals, institutions, and initiatives that have been engaged with environmental issues for a considerable period of time. Despite their lateness in addressing the crisis, religions are beginning to respond in remarkably creative ways. They are not only rethinking their theologies but are also reorienting their sustainable practices and long-term environmental commitments. In so doing, the very nature of religion and of ethics is being challenged and changed. This is true because the reexamination of other worldviews created by religious beliefs and practices may be critical to our recovery of sufficiently comprehensive cosmologies, broad conceptual frameworks, and effective environmental ethics for the twenty-first century.

While in the past none of the religions of the world have had to face an environmental crisis such as we are now confronting, they remain key instruments in shaping attitudes toward nature. The unintended consequences of the modern industrial drive for unlimited economic growth and resource development have led us to an impasse regarding the survival of many life-forms and appropriate management of varied ecosystems. The religious traditions may indeed be critical in helping to reimagine the viable conditions and long-range strategies for fostering mutually enhancing human-earth relations.[6] Indeed, as E. N. Anderson has documented with impressive detail, "All traditional societies that have succeeded in managing resources well, over time, have done it in part through religious or ritual representation of resource management."[7]

It is in this context that a series of conferences and publications exploring the various religions of the world and their relation to ecology was initiated by the Center for the Study of World Religions at Harvard. Directed by Lawrence Sullivan and coordinated by Mary Evelyn Tucker and John Grim, the conferences will involve some six hundred scholars, graduate students, religious leaders, and environmental activists over a period of three years. The collaborative nature of the project is intentional. Such collaboration will maximize the opportunity for dialogical reflection on this issue of enormous complexity and will accentuate the diversity of local manifestations of ecologically sustainable alternatives.

The conferences and the volumes are intended to serve as initial explorations of the emerging field of religion and ecology while pointing toward areas for further research. We are not unaware of the difficulties of engaging in such a task, yet we are encouraged by the enthusiastic response to the conferences within the academic community, by the larger interest they have generated beyond academia, and by the probing examinations gathered in the volumes. We trust that this series and these volumes will be useful not only for scholars of religion but also for those shaping seminary education and institutional religious practices, as well as for those involved in public policy on environmental issues.

We see these conferences and publications as expanding the growing dialogue regarding the role of the world's religions as moral forces in stemming the environmental crisis. While, clearly, there are major methodological issues involved in utilizing traditional philosophical and religious ideas for contemporary concerns, there are also compelling reasons to support such efforts, however modest they may be. The world's religions in all their complexity and variety remain one of the principal resources for symbolic ideas, spiritual inspiration, and ethical principles. Indeed, despite their limitations, historically they have provided comprehensive cosmologies for interpretive direction, moral foundations for social cohesion, spiritual guidance for cultural expression, and ritual celebrations for meaningful life. In our search for more comprehensive ecological worldviews and more effective environmental ethics, it is inevitable that we will draw from the symbolic and conceptual resources of the religious traditions of the world. The effort to do this is not without precedent or problems, some of which will be signaled below. With this volume and with this series we hope the field of reflection and discussion regarding religion and ecology will begin to broaden, deepen, and complexify.

Qualifications and Goals

The Problems and Promise of Religions

These conferences and volumes, then, are built on the premise that the religions of the world may be instrumental in addressing the moral dilemmas created by the environmental crisis. At the same time we recognize the limitations of such efforts on the part of religions. We also acknowledge that the complexity of the problem requires interlocking approaches from such fields as science, economics, politics, health, and public policy. As the human community struggles to formulate different attitudes toward nature and to articulate broader conceptions of ethics embracing species and ecosystems, religions may thus be a necessary, though only contributing, part of this multidisciplinary approach.

It is becoming increasingly evident that abundant scientific knowledge of the crisis is available and numerous political and economic statements have been formulated. Yet we seem to lack the political, economic, and scientific leadership to make necessary changes. Moreover, what is still lacking is the religious commitment, moral imagination, and ethical engagement to transform the environmental crisis from an issue on paper to one of effective policy, from rhetoric in print to realism in action. Why, nearly fifty years after Fairfield Osborne's warning regarding *Our Plundered Planet* and more than thirty years since Rachel Carson's *Silent Spring*, are we still wondering, is it too late?[8]

It is important to ask where the religions have been on these issues and why they themselves have been so late in their involvement. Have issues of personal salvation superseded all others? Have divine-human relations been primary? Have anthropocentric ethics been all-consuming? Has the material world of nature been devalued by religion? Does the search for otherworldly rewards override commitment to this world? Did the religions simply surrender their natural theologies and concerns with exploring purpose in nature to positivistic scientific cosmologies? In beginning to address these questions, we still have not exhausted all the reasons for religions' lack of attention to the environmental crisis. The reasons may not be readily apparent, but clearly they require further exploration and explanation.

In discussing the involvement of religions in this issue, it is also appropriate to acknowledge the dark side of religion in both its institutional expressions and dogmatic forms. In addition to their oversight with regard to the environment, religions have been the source of enormous manipulation of power in fostering wars, in ignoring racial and social injustice, and in promoting unequal gender relations, to name only a few abuses. One does not want to underplay this shadow side or to claim too much for religions' potential for ethical persuasiveness. The problems are too vast and complex for unqualified optimism. Yet there is a growing consensus that religions may now have a significant role to play, just as in the past they have sustained individuals and cultures in the face of internal and external threats.

A final caveat is the inevitable gap that arises between theories and practices in religions. As has been noted, even societies with religious traditions which appear sympathetic to the environment have in the past often misused resources. While it is clear that religions may have some disjunction between the ideal and the real, this should not lessen our endeavor to identify resources from within the world's religions for a more ecologically sound cosmology and environmentally supportive ethics. This disjunction of theory and practice is present within all philosophies and religions and is frequently the source of disillusionment, skepticism, and cynicism. A more realistic observation might be made, however, that this disjunction should not automatically invalidate the complex world-views and rich cosmologies embedded in traditional religions. Rather, it is our task to explore these conceptual resources so as to broaden and expand our own perspectives in challenging and fruitful ways.

In summary, we recognize that religions have elements which are both prophetic and transformative as well as conservative and constraining. These elements are continually in tension, a condition which creates the great variety of thought and interpretation within religious traditions. To recognize these various tensions and limits, however, is not to lessen the urgency of the overall goals of this project. Rather, it is to circumscribe our efforts with healthy skepticism, cautious optimism, and modest ambitions. It is to suggest that this is a beginning in a new field of study which will affect both religion and ecology. On the one hand, this process of reflection will inevitably change how religions conceive of their own roles, missions, and identities, for such reflections demand a new sense of the sacred as not divorced from the earth itself. On the other hand, environmental studies can recognize that religions have helped to shape attitudes toward nature. Thus, as religions themselves evolve they may be indispensable in fostering a more expansive appreciation for the complexity and beauty of the natural world. At the same time as religions foster awe and reverence for nature, they may provide the transforming energies for ethical practices to protect endangered ecosystems, threatened species, and diminishing resources.

Methodological Concerns

It is important to acknowledge that there are, inevitably, challenging methodological issues involved in such a project as we are undertaking in this emerging field of religion and ecology.[9] Some of the key interpretive challenges we face in this project concern issues of time, place, space, and positionality. With regard to time, it is necessary to recognize the vast historical complexity of each religious tradition, which cannot be easily condensed in these conferences or volumes. With respect to place, we need to signal the diverse cultural contexts in which these religions have developed. With regard to space, we recognize the varied frameworks of institutions and traditions in which these religions unfold. Finally, with respect to positionality, we acknowledge our own historical situatedness at the end of the twentieth century with distinctive contemporary concerns.

Not only is each religious tradition historically complex and

culturally diverse, but its beliefs, scriptures, and institutions have
themselves been subject to vast commentaries and revisions over
time. Thus, we recognize the radical diversity that exists within and
among religious traditions which cannot be encompassed in any
single volume. We acknowledge also that distortions may arise as
we examine earlier historical traditions in light of contemporary
issues.

Nonetheless, the environmental ethics philosopher J. Baird
Callicott has suggested that scholars and others "mine the conceptual
resources" of the religious traditions as a means of creating a more
inclusive global environmental ethics.[10] As Callicott himself notes,
however, the notion of "mining" is problematic, for it conjures up
images of exploitation which may cause apprehension among certain
religious communities, especially those of indigenous peoples.
Moreover, we cannot simply expect to borrow or adopt ideas and
place them from one tradition directly into another. Even efforts to
formulate global environmental ethics need to be sensitive to cultural
particularity and diversity. We do not aim at creating a simple
bricolage or bland fusion of perspectives. Rather, these conferences
and volumes are an attempt to display before us a multiperspectival
cross section of the symbolic richness regarding attitudes toward
nature within the religions of the world. To do so will help to reveal
certain commonalities among traditions, as well as limitations within
traditions, as they begin to converge around this challenge presented
by the environmental crisis.

We need to identify our concerns, then, as embedded in the
constraints of our own perspectival limits at the same time as we
seek common ground. In describing various attitudes toward nature
historically, we are aiming at *critical understanding* of the complexity, contexts, and frameworks in which these religions articulate
such views. In addition, we are striving for *empathetic appreciation*
for the traditions without idealizing their ecological potential or
ignoring their environmental oversights. Finally, we are aiming at
the *creative revisioning* of mutually enhancing human-earth relations. This revisioning may be assisted by highlighting the multiperspectival attitudes toward nature which these traditions disclose.
The prismatic effect of examining such attitudes and relationships
may provide some necessary clarification and symbolic resources
for reimagining our own situation and shared concerns at the end

of the twentieth century. It will also be sharpened by identifying the multilayered symbol systems in world religions which have traditionally oriented humans in establishing relational resonances between the microcosm of the self and the macrocosm of the social and natural orders. In short, religious traditions may help to supply both creative resources of symbols, rituals, and texts as well as inspiring visions for reimagining ourselves as part of, not apart from, the natural world.

Aims

The methodological issues outlined above are implied in the overall goals of the conferences, which are described as follows:

1) To identify and evaluate the *distinctive ecological attitudes*, values, and practices of diverse religious traditions, making clear their links to intellectual, political, and other resources associated with these distinctive traditions.

2) To describe and analyze the *commonalities* that exist within and among religious traditions with respect to ecology.

3) To identify the *minimum common ground* on which to base constructive understanding, motivating discussion, and concerted action in diverse locations across the globe; and to highlight the specific religious resources that comprise such fertile ecological ground: within scripture, ritual, myth, symbol, cosmology, sacrament, and so on.

4) To articulate in clear and moving terms *a desirable mode of human presence with the earth;* in short, to highlight means of respecting and valuing nature, to note what has already been actualized, and to indicate how best to achieve what is desirable beyond these examples.

5) To outline the most significant areas, with regard to religion and ecology, in need of *further study*; to enumerate questions of highest priority within those areas and propose possible approaches to use in addressing them.

In these conferences and volumes, then, we are not intending to obliterate difference or ignore diversity. The aim is to celebrate plurality by raising to conscious awareness multiple perspectives regarding nature and human-earth relations as articulated in the religions of the world. The spectrum of cosmologies, myths, symbols, and rituals within the religious traditions will be instructive in resituating us within the rhythms and limits of nature.

We are not looking for a unified worldview or a single global ethic. We are, however, deeply sympathetic with the efforts toward formulating a global ethic made by individuals, such as the theologian, Hans Kung, or the environmental philosopher, J. Baird Callicott, and groups, such as Global Education Associates and United Religions. A minimum content of environmental ethics needs to be seriously considered. We are, then, keenly interested in the contribution this series might make to discussions of environmental policy in national and international arenas. Important intersections may be made with work in the field of development ethics.[11] In addition, the findings of the conferences have bearing on the ethical formulation of the Earth Charter that will be presented to the United Nations for adoption by the end of the century. Thus, we are seeking both the grounds for common concern and the constructive conceptual basis for rethinking our current situation of estrangement from the earth. In so doing we will be able to reconceive a means of creating the basis not just for sustainable development, but also for sustainable life on the planet.

As scientist Brian Swimme has suggested, we are currently making macrophase changes to the life systems of the planet with microphase wisdom. Clearly, we need to expand and deepen the wisdom base for human intervention with nature and other humans. This is particularly true as issues of genetic alteration of natural processes are already available and in use. If religions have traditionally concentrated on divine-human and human-human relations, the challenge is that they now explore more fully divine-human-earth relations. Without such further exploration, adequate environmental ethics may not emerge in a comprehensive context.

Resources: Environmental Ethics Found in the World's Religions

For many people, when challenges such as the environmental crisis are raised in relation to religion in the contemporary world, there frequently arises a sense of loss or a nostalgia for earlier, seemingly less complicated eras when the constant questioning of religious beliefs and practices was not so apparent. This is, no doubt, something of a reified reading of history. There is, however, a decidedly anxious tone to the questioning and soul-searching that appears to haunt many contemporary religious groups as they seek to find their particular role in the midst of rapid technological change and dominant secular values.

One of the greatest challenges, however, to contemporary religions remains how to respond to the environmental crisis, which many believe has been perpetuated because of the enormous inroads made by unrestrained materialism, secularization, and industrialization in contemporary societies, especially those societies arising in or influenced by the modern West. Indeed, some suggest that the very division of religion from secular life may be a major cause of the crisis.

Others, such as the medieval historian Lynn White, have cited religion's negative role in the crisis. White has suggested that the emphasis in Judaism and Christianity on the transcendence of God above nature and the dominion of humans over nature has led to a devaluing of the natural world and a subsequent destruction of its resources for utilitarian ends.[12] While the particulars of this argument have been vehemently debated, it is increasingly clear that the environmental crisis and its perpetuation due to industrialization, secularization, and ethical indifference present a serious challenge to the world's religions. This is especially true because many of these religions have traditionally been concerned with the path of personal salvation, which frequently emphasized otherworldly goals and rejected this world as corrupting. Thus, as we have noted, how to adapt religious teachings to this task of revaluing nature so as to prevent its destruction marks a significant new phase in religious thought. Indeed, as Thomas Berry has so aptly pointed out, what is necessary is a comprehensive reevaluation of human-earth relations if the human is to continue as a viable species on an increasingly degraded planet. This will require, in addition to major economic and political changes, examining worldviews and ethics among the world's religions that differ from those that have captured the imagination of contemporary industrialized societies which regard nature primarily as a commodity to be utilized. It should be noted that when we are searching for effective resources for formulating environmental ethics, each of the religious traditions have both positive and negative features.

For the most part, the worldviews associated with the Western Abrahamic traditions of Judaism, Christianity, and Islam have created a dominantly human-focused morality. Because these worldviews are largely anthropocentric, nature is viewed as being of secondary importance. This is reinforced by a strong sense of the transcendence of God above nature. On the other hand, there are rich resources for rethinking views of nature in the covenantal tradition of the Hebrew Bible, in sacramental theology, in incarnational Christology, and in the vice-regency (*khalifa Allah*) concept of the Qur'an. The covenantal tradition draws on the legal agreements of biblical thought which are extended to all of creation. Sacramental theology in Christianity underscores the sacred dimension of material reality, especially for ritual purposes.[13] Incarnational Christology proposes that because God became flesh in the person of Christ, the entire natural order can be viewed as sacred. The concept of humans as vice-regents of Allah on earth suggests that humans have particular privileges, responsibilities, and obligations to creation.[14]

In Hinduism, although there is a significant emphasis on performing one's *dharma*, or duty, in the world, there is also a strong pull toward *moksa*, or liberation, from the world of suffering, or *samsāra*. To heal this kind of suffering and alienation through spiritual discipline and meditation, one turns away from the world (*prakrti*) to a timeless world of spirit (*purusa*). Yet at the same time there are numerous traditions in Hinduism which affirm particular rivers, mountains, or forests as sacred. Moreover, in the concept of *līlā*, the creative play of the gods, Hindu theology engages the world as a creative manifestation of the divine. This same tension between withdrawal from the world and affirmation of it is present in Buddhism. Certain Theravāda schools of Buddhism emphasize withdrawing in meditation from the transient world of suffering (*samsāra*) to seek release in *nirvāna*. On the other hand, later Mahāyāna schools of Buddhism, such as Hua-yen, underscore the remarkable interconnection of reality in such images as the jeweled net of Indra, where each jewel reflects all the others in the universe. Likewise, the Zen gardens in East Asia express the fullness of the Buddha-nature (*tathāgatagarbha*) in the natural world. In recent years, socially engaged Buddhism has been active in protecting the environment in both Asia and the United States.

The East Asian traditions of Confucianism and Taoism remain, in certain ways, some of the most life-affirming in the spectrum of world religions.[15] The seamless interconnection between the divine, human, and natural worlds that characterizes these traditions has been described as an anthropocosmic worldview.[16] There is no emphasis on radical transcendence as there is in the Western traditions. Rather, there is a cosmology of a continuity of creation stressing the dynamic movements of nature through the seasons and the agricultural cycles. This organic cosmology is grounded in the philosophy of *ch'i* (material force), which provides a basis for appreciating the profound interconnection of matter and spirit. To be in harmony with nature and with other humans while being attentive to the movements of the *Tao* (Way) is the aim of personal cultivation in both Confucianism and Taoism. It should be noted, however, that this positive worldview has not prevented environmental degradation (such as deforestation) in parts of East Asia in both the premodern and modern period.

In a similar vein, indigenous peoples, while having ecological cosmologies have, in some instances, caused damage to local environments through such practices as slash-and-burn agriculture. Nonetheless, most indigenous peoples have environmental ethics embedded in their worldviews. This is evident in the complex reciprocal obligations surrounding life-taking and resource-gathering which mark a community's relations with the local bioregion. The religious views at the basis of indigenous lifeways involve respect for the sources of food, clothing, and shelter that nature provides. Gratitude to the creator and to the spiritual forces in creation is at the heart of most indigenous traditions. The ritual calendars of many indigenous peoples are carefully coordinated with seasonal events such as the sound of returning birds, the blooming of certain plants, the movements of the sun, and the changes of the moon.

The difficulty at present is that for the most part we have developed in the world's religions certain ethical prohibitions regarding homicide and restraints concerning genocide and suicide, but none for biocide or geocide. We are clearly in need of exploring such comprehensive cosmological perspectives and communitarian environmental ethics as the most compelling context for motivating change regarding the destruction of the natural world.

Responses of Religions to the Environmental Crisis

How to chart possible paths toward mutually enhancing human-earth relations remains, thus, one of the greatest challenges to the world's religions. It is with some encouragement, however, that we note the growing calls for the world's religions to participate in these efforts toward a more sustainable planetary future. There have been various appeals from environmental groups and from scientists and parliamentarians for religious leaders to respond to the environmental crisis. For example, in 1990 the Joint Appeal in Religion and Science was released highlighting the urgency of collaboration around the issue of the destruction of the environment. In 1992 the Union of Concerned Scientists issued a statement of "Warning to Humanity" signed by over 1,000 scientists from 70 countries, including 105 Nobel laureates, regarding the gravity of the environmental crisis. They specifically cited the need for a new ethic toward the earth.

Numerous national and international conferences have also been held on this subject and collaborative efforts have been established. Environmental groups such as World Wildlife Fund have sponsored interreligious meetings such as the one in Assisi in 1986. The Center for Respect of Life and Environment of the Humane Society of the United States has also held a series of conferences in Assisi on Spirituality and Sustainability and has helped to organize one at the World Bank. The United Nations Environmental Programme in North America has established an Environmental Sabbath, each year distributing thousands of packets of materials for use in congregations throughout North America. Similarly, the National Religious Partnership on the Environment at the Cathedral of St. John the Divine in New York City has promoted dialogue, distributed materials, and created a remarkable alliance of the various Jewish and Christian denominations in the United States around the issue of the environment. The Parliament of World Religions held in 1993 in Chicago and attended by some 8,000 people from all over the globe issued a statement of Global Ethics of Cooperation of Religions on Human and Environmental Issues. International meetings on the environment have been organized. One example of these, the Global Forum of Spiritual and Parliamentary Leaders held in Oxford in 1988, Moscow in 1990, Rio in 1992, and Kyoto in

1993, included world religious leaders, such as the Dalai Lama, and diplomats and heads of state, such as Mikhail Gorbachev. Indeed, Gorbachev hosted the Moscow conference and attended the Kyoto conference to set up a Green Cross International for environmental emergencies.

Since the United Nations Conference on Environment and Development (the Earth Summit) held in Rio in 1992, there have been concerted efforts intended to lead toward the adoption of an *Earth Charter* by the year 2000. This *Earth Charter* initiative is under way with the leadership of the Earth Council and Green Cross International, with support from the government of the Netherlands. Maurice Strong, Mikhail Gorbachev, Steven Rockefeller, and other members of the Earth Charter Project have been instrumental in this process. At the March 1997 Rio + 5 Conference a benchmark draft of the *Earth Charter* was issued. The time is thus propitious for further investigation of the potential contributions of particular religions toward mitigating the environmental crisis, especially by developing more comprehensive environmental ethics for the earth community.

Expanding the Dialogue of Religion and Ecology

More than two decades ago Thomas Berry anticipated such an exploration when he called for "creating a new consciousness of the multiform religious traditions of humankind" as a means toward renewal of the human spirit in addressing the urgent problems of contemporary society.[17] Tu Weiming has written of the need to go "Beyond the Enlightenment Mentality" in exploring the spiritual resources of the global community to meet the challenge of the ecological crisis.[18] While this exploration is also the intention of these conferences and volumes, other significant efforts have preceded our current endeavor.[19] Our discussion here highlights only the last decade.

In 1986 Eugene Hargrove edited a volume titled *Religion and Environmental Crisis*.[20] In 1991 Charlene Spretnak explored this topic in her book *States of Grace: The Recovery of Meaning in the Post-Modern Age*.[21] Her subtitle states her constructivist project clearly: "Reclaiming the Core Teachings and Practices of the Great

Wisdom Traditions for the Well-Being of the Earth Community."
In 1992 Steven Rockefeller and John Elder edited a book based on
a conference at Middlebury College titled *Spirit and Nature: Why
the Environment Is a Religious Issue.*[22] In the same year Peter
Marshall published *Nature's Web: Rethinking Our Place on Earth*,[23]
drawing on the resources of the world's traditions. An edited volume
on *Worldviews and Ecology*, compiled in 1993, contains articles
reflecting on views of nature from the world's religions and from
contemporary philosophies, such as process thought and deep
ecology.[24] In this same vein, in 1994 J. Baird Callicott published
Earth's Insights which examines the intellectual resources of the
world's religions for a more comprehensive global environmental
ethics.[25] This expands on his 1989 volumes, *Nature in Asian
Traditions of Thought* and *In Defense of the Land Ethic.*[26] In 1995
David Kinsley issued a book titled *Ecology and Religion: Ecological
Spirituality in a Cross-Cultural Perspective*[27] which draws on
traditional religions and contemporary movements, such as deep
ecology and ecospirituality. Seyyed Hossein Nasr wrote a comprehensive study of *Religion and the Order of Nature* in 1996.[28] Several
volumes of religious responses to a particular topic or theme have
also been published. For example, J. Ronald Engel and Joan Gibb
Engel compiled a monograph in 1990 on *Ethics of Environment and
Development: Global Challenge, International Response*[29] and in
1995 Harold Coward edited a volume on *Population, Consumption
and the Environment: Religious and Secular Responses.*[30] Roger
Gottlieb edited a useful source book, *This Sacred Earth: Religion,
Nature, Environment.*[31] Single volumes on the world's religions and
ecology were published by the Worldwide Fund for Nature.[32]

The conferences and volumes in the series Religions of the World
and Ecology are thus intended to expand the discussion already
under way in certain circles and to invite further collaboration on a
topic of common concern—the fate of the earth as a religious
responsibility. To broaden and deepen the reflective basis for mutual
collaboration has been an underlying aim of the conferences
themselves. While some might see this as a diversion from pressing
scientific or policy issues, it is with a sense of humility and yet
conviction that we enter into the arena of reflection and debate on
this issue. In the field of the study of world religions, we see this
as a timely challenge for scholars of religion to respond as engaged

intellectuals with deepening creative reflection. We hope that these conferences and volumes will be simply a beginning of further study of conceptual and symbolic resources, methodological concerns, and practical directions for meeting this environmental crisis.

Notes

1. He goes on to say, "And that is qualitatively and epochally true. If religion does not speak to [this], it is an obsolete distraction." Daniel Maguire, *The Moral Core of Judaism and Christianity: Reclaiming the Revolution* (Philadelphia: Fortress Press, 1993), 13.

2. Gerald Barney, *Global 2000 Report to the President of the United States*, (Washington. D.C.: Supt. of Docs. U.S. Government Printing Office, 1980–1981), 40.

3. Lynn White, Jr., "The Historical Roots of Our Ecologic Crisis," *Science* 155 (March 1967):1204.

4. Thomas Berry, *The Dream of the Earth* (San Francisco: Sierra Club Books, 1988).

5. Brian Swimme and Thomas Berry, *The Universe Story* (San Francisco: Harper San Francisco, 1992).

6. At the same time we recognize the limits to such a project, especially because ideas and action, theory and practice do not always occur in conjunction.

7. E. N. Anderson, *Ecologies of the Heart: Emotion, Belief, and the Environment* (New York and Oxford: Oxford University Press, 1996), 166. He qualifies this statement by saying, "The key point is not religion per se, but the use of emotionally powerful symbols to sell particular moral codes and management systems" (166). He notes, however, in various case studies how ecological wisdom is embedded in myths, symbols, and cosmologies of traditional societies.

8. *Is It Too Late?* is also the title of a book by John Cobb, first published in 1972 by Bruce and reissued in 1995 by Environmental Ethics Books.

9. Because we cannot identify here all of the methodological issues that need to be addressed, we invite further discussion by other engaged scholars.

10. See J. Baird Callicott, *Earth's Insights: A Survey of Ecological Ethics from the Mediterranean Basin to the Australian Outback* (Berkeley: University of California Press, 1994).

11. See, for example, *The Quality of Life*, ed. Martha C. Nussbaum and Amartya Sen, WIDER Studies in Development Economics (Oxford: Oxford University Press, 1993).

12. White, "The Historical Roots of Our Ecologic Crisis," 1203–7.

13. Process theology, creation-centered spirituality, and ecotheology have done much to promote these kinds of holistic perspectives within Christianity.

14. These are resources already being explored by theologians and biblical scholars.

15. While this is true theoretically, it should be noted that, like all ideologies, these traditions have at times been used for purposes of political power and social control. Moreover, they have not been able to prevent certain kinds of environmental destruction, such as deforestation in China.

16. The term "anthropocosmic" has been used by Tu Weiming in *Commonality and Centrality* (Albany: State University of New York Press, 1989).

17. Thomas Berry, "Religious Studies and the Global Human Community," unpublished manuscript.

18. Tu Weiming, "Beyond the Enlightenment Mentality," in *Worldviews and Ecology*, ed. Mary Evelyn Tucker and John Grim (Lewisburg, Penn.: Bucknell University Press, 1993; reissued, Maryknoll, N.Y.: Orbis Press, 1994).

19. This history has been described more fully by Roderick Nash in his chapter entitled "The Greening of Religion," in *The Rights of Nature: A History of Environmental Ethics* (Madison: University of Wisconsin, 1989).

20. *Religion and Environmental Crisis*, ed. Eugene Hargrove (Athens: University of Georgia Press, 1986).

21. Charlene Spretnak, *States of Grace: The Recovery of Meaning in the Post-Modern Age* (San Francisco: Harper San Francisco, 1991).

22. *Spirit and Nature: Why the Environment Is a Religious Issue*, ed. Steven Rockefeller and John Elder (Boston: Beacon Press, 1992).

23. Peter Marshall, *Nature's Web: Rethinking Our Place on Earth* (Armonk, N.Y.: M. E. Sharpe, 1992).

24. *Worldviews and Ecology*, ed. Mary Evelyn Tucker and John Grim (Lewisburg, Penn.: Bucknell University Press, 1993; reissued, Maryknoll, N.Y.: Orbis Books, 1994).

25. Callicott, *Earth's Insights*.

26. Both are State University of New York Press publications.

27. David Kinsley, *Ecology and Religion: Ecological Spirituality in a Cross-Cultural Perspective* (Englewood Cliffs, N.J.: Prentice Hall, 1995).

28. Seyyed Hossein Nasr, *Religion and the Order of Nature* (Oxford: Oxford University Press, 1996).

29. *Ethics of Environment and Development: Global Challenge, International Response*, ed. J. Ronald Engel and Joan Gibb Engel (Tucson: University of Arizona Press, 1990).

30. *Population, Consumption, and the Environment: Religious and Secular Responses*, ed. Harold Coward (Albany: State University of New York Press, 1995).

31. *This Sacred Earth: Religion, Nature, Environment*, ed. Roger S. Gottlieb (New York and London: Routledge, 1996).

32. These include volumes on Hinduism, Buddhism, Judaism, Christianity, and Islam.

Introduction

Duncan Ryūken Williams

Throughout the past several decades, Buddhist practitioners in both Asia and the West have engaged in a wide variety of efforts to protect the environment. A Buddhist priest led a recent campaign to save an ancient urban forest in Tokyo from being turned into an apartment complex; the priest erected a large sign near the grove stating that the trees have "Buddha-nature." Similar efforts in forest conservation from a Buddhist perspective have occurred in Thailand, where a number of environmentally minded monks have selectively "ordained" trees in the forests. Traditionally, a Thai Buddhist novice is ordained by the shaving of the monk's hair and by his acceptance of saffron robes. Thai monks have used this symbolic act of initiation to "ordain" the trees in the rain forest as "members of a Buddhist order" by tying strips of saffron cloth around them. This rather unique tactic has actually prevented the logging of quite a number of acres of forest. This creative adaptation of Buddhist concepts and practices for environmental concerns has been taking place since the early 1960s in three larger communities: the academic, the Buddhist, and the environmental.

In the academic community, scholars from a variety of disciplines have evaluated Buddhist perspectives on nature, ecological ethics, and actions taken by Buddhists for environmental causes. While Buddhologists have focused on Buddhist *sūtra*s and other textual sources, as well as on individual Buddhist thinkers' perspectives on nature, environmental philosophers have turned to Buddhism as a conceptual resource for a new ecological ethics. At the same time, scholars in the fields of anthropology and sociology have studied contemporary Buddhist movements and individuals who have been involved as "engaged Buddhists" in environmental activism.

Members of the second group, both ordained and lay members of the Buddhist community in Asia and the West, have been speaking, writing, and organizing activities leading toward a more active Buddhist role in addressing the environmental crisis. Well-known ordained leaders, such as the Dalai Lama, Buddhadāsa Bhikkhu, and Thich Nhat Hanh, have recognized the need to address such contemporary issues as ecology if Buddhism is to continue to be relevant to many members of the Buddhist community. Institutionally, such organizations as the International Network of Engaged Buddhists, the Buddhist Peace Fellowship, and Buddhists Concerned for Animals have served as vehicles for expressing particular Buddhist positions on ecological and peace concerns. Furthermore, the efforts of local temples and lay Buddhists in environmental education, activism, and conservation have been noteworthy. In Japan, for example, even without institutional backing, local temple priests have played key roles in protecting marine life in the Himeiji region, protesting nuclear power and waste in western Japan, and preserving the few ancient groves left in Tokyo. Perhaps even more remarkable have been key lay Buddhists in both Asia and the West, such as Sulak Sivaraksa, Yanase Giryō, Gary Snyder, and Joanna Macy, who, through their writings and activism grounded in a Buddhist perspective, have made a significant contribution to ecological awareness.

Finally, a number of environmentalists have found Buddhist doctrines, such as Buddha-nature, and Buddhist practices, such as meditation, to be extremely useful. Many environmentalists are familiar with the deep ecology movement, inspired and influenced in part by Buddhism, which espouses a nonanthropocentric worldview. Moreover, many environmentalists are familiar with the Council of All Beings, a ritual in which one places oneself in the position of another species, which was designed by Buddhists Joanna Macy and John Seed. Environmental activists who are drawn to Buddhism, but who are not officially Buddhists, might be what Thomas Tweed has called "Buddhist sympathizers," or persons who are positioned between adherents and nonadherents. Although Buddhism has certainly been influenced by the environmental movement, these last examples suggest ways in which Buddhism, in its worldview and practice, has penetrated the environmentalist community.

This volume was based on a three-day conference at the Harvard University Center for the Study of World Religions that brought together scholars of Buddhism and environmentally engaged Buddhists. While it reflects some of the juxtapositions of those two groups, the significance of this volume lies in the fact that it is primarily a scholarly one. Previous publications in this area have largely been written by practitioners and environmentalists. Moreover, the two previous scholarly books of note, Lambert Schmithausen's *Buddhism and Nature* and the collection *Nature in Asian Traditions of Thought*, edited by J. Baird Callicott and Roger T. Ames, are limited in both their scope and treatment of the range of Buddhist traditions. This volume, although scholarly in nature, is intended for undergraduate and graduate students as well as for an educated public with some basic knowledge of Buddhist teachings. Rather than being exhaustive, it should serve as a modest beginning so as to encourage further research on the topic of Buddhism and ecology.

The volume begins with an essay by Lewis Lancaster, an overview highlighting some of the key issues and complexities inherent in a study of this topic. One of these involves the problem of generalizing about the Buddhist tradition as a whole. Lancaster signals the need to be aware of the cultural and geographical diversity of Buddhism as well as of the historical contexts of particular Buddhist teachings and practices. Moreover, methodological issues, such as utilizing ideas from the past to inform contemporary issues, are also recognized as problematic in certain respects. Yet, the spirit of this volume is one that, while acknowledging these difficulties, also notes that traditions have always been changing in relation to present circumstances. In addition, it accepts the premise that views of nature are, to a large extent, conditioned by religious and cultural worldviews. Hence, it is important to probe these views historically so as to shed light not only on the past but also on present circumstances. The issue may be described as the coexistence of traditional ideas with modern conditions—or the adaptation of the former to the latter. While this may be an uneasy coexistence, it is not without historical precedent, given the manner in which traditions have adapted themselves to particular times, places, and situations.

The first five sections of the volume reflect cultural, thematic, and denominational approaches to the study of Buddhism in general and to the study of Buddhism and ecology in particular. The cultural areas represented in this volume include Southeast Asia, East Asia, and North America, and specific examples are drawn from Thailand, Japan, and the United States. The section on Thailand includes an essay by Donald Swearer on two key figures in the Theravāda Buddhist world—Buddhadāsa and Dhammapiṭaka—who have figured prominently in contemporary discussions of Buddhist ecological theories and practices in Thailand. The anthropological essay of Leslie Sponsel and Poranee Natadecha-Sponsel complements this with a discussion of how the Thai Buddhist monastic community is involved in promoting environmental awareness and action.

The following three chapters focus on particular Japanese Buddhist thinkers' views of nature as a starting point for a discussion of what the Japanese tradition offers in terms of environmental worldviews and ethics. Paul Ingram discusses the case of the medieval Shingon monk Kūkai and his *maṇḍala*-like world of interconnectedness that Ingram terms the "jeweled net of nature." Graham Parkes also begins with Kūkai's doctrine of this earth being the manifestation of Buddha-nature. He then moves on to discuss the similarly nonanthropocentric and nature-affirming worldview of the medieval Zen monk Dōgen. Parkes concludes with reflections on the philosophical and practical problems in the undifferentiated affirmation of all things "natural," including tuberculosis or toxic waste dumps. Steve Odin considers a wide range of sources to highlight an aesthetic and salvific aspect to a specifically Japanese concept of nature. He links this perspective to the environmental ethics and conservation aesthetics of Aldo Leopold to propose what he calls an "East-West Gaia theory of nature."

The third geographical and cultural area taken up in the volume is the United States where, despite its relatively brief history, Buddhism has played an important role in the formation of a Buddhist ecology and in the creation of environmentally friendly Buddhist temples. David Barnhill analyzes the work of the Buddhist poet and environmental activist Gary Snyder, who was one of the first Westerners to recognize the rich potential of the interface between Buddhism and ecology. In particular, Snyder articulates a

Buddhist-inspired bioregionalism and a Buddhist form of deep ecology. His concept of wildness and his shamanic/mythological orientation is drawn, Barnhill suggests, from his feelings for the dramatic landscape of the Pacific Northwest and his affinities with Native American views of community and land. Stephanie Kaza's essay focuses on two environmentally sustainable rural communities in Northern California, namely, Green Gulch Farm, a Zen meditation center, and Spirit Rock, a *vipassanā* meditation center. Kaza draws on Gary Snyder's ecological guidelines for reinhabitation of the land to evaluate the environmental stewardship and educational practices of these centers. Jeff Yamauchi's essay complements this discussion with another case study of the process of "greening" a Buddhist retreat center, the Zen Mountain Center in Southern California. He surveys efforts to protect the flora and fauna of the region and discusses fire prevention and management of the forest.

This volume also includes a thematic section on the place of animals in Buddhism, in which the particular cultural areas and traditions of India and Japan are examined. Christopher Chapple's essay deals with various images of animals as found in the early Indian Buddhist stories known as the Jātaka tales. Chapple suggests that the wise, compassionate, and foolish animals appearing in these narratives illustrate that Buddhists had a keen awareness of animals and their place in Buddhist cosmology. My essay takes up the Buddhist ritual of releasing animals for merit that has been practiced in both East and Southeast Asia. The study of this ritual in medieval Japan reveals the ironic relationship between the effort at "animal liberation" in the Buddhist tradition and the unintended consequence to this ritual of the loss of animal life.

Another approach to the study of Buddhism in general is to examine different traditions or denominations of Buddhism. In this volume, Ruben Habito and John Daido Loori look to the possibilities and limitations of what the Zen Buddhist tradition can offer to this discussion of environmental issues. Habito points to the experiential realization in Zen of nonseparation of oneself and the world as the starting point for embracing an ecologically engaged way of life. This affirms living in the present moment. However, Habito acknowledges another impulse in Zen that may promote detachment from this world and absorption in cultivating the inner life. Loori, the head abbot of Zen Mountain Monastery, gives a Zen interpretation of the Buddhist precepts as a map for an environmental ethic. In his article, originally delivered as a Zen *Dharma* talk at the monastery, he suggests interpreting the Buddhist precepts so as to develop a way of life that is in harmony with the natural world.

The last two sections of the volume focus on practical/policy-level contributions that Buddhism can make and on theoretical/methodological issues that ought to be considered for future research. The section on the practical application of Buddhism to environmental problems begins with Kenneth Kraft's chapter on the issue of nuclear waste. Kraft documents the background of Buddhist concern over this unresolved issue and reflects on the responsibilities of the scholar and the engaged Buddhist in facing this particular aspect of the environmental crisis. Rita Gross draws from the wide-ranging spectrum of Buddhist thought to construct a position that undercuts what she calls a pronatalist view toward population. Gross suggests that particular Buddhist teachings on desires and sexuality could help to moderate the more polemical discussions of population and consumption. She thus points toward a middle way between irresolvable extremes on these two issues. In his essay, Steven Rockefeller outlines the core elements of a Buddhist contribution to an emerging global ethics. He focuses particularly on the Earth Charter, which is expected to be submitted to the United Nations General Assembly by the year 2000. The Charter is intended to function as a "soft law" document to undergird efforts at sustainable development in the international community. The practical problems and initiatives discussed by these three authors provide models for future considerations of ways in which Buddhist values can be applied to environmental issues.

The final section of the volume focuses on broader theoretical and methodological questions regarding the interface between Buddhism and ecology. David Eckel and Ian Harris both question facile assumptions that Asian, and particularly Buddhist, worldviews are inherently environmentally friendly. Indeed, they ask when and why Buddhism came to be seen as ecofriendly. They both argue that this conception is relatively recent and that the term "nature" is itself a complex and somewhat problematic term in Buddhist history. Eckel proposes a means of circumventing the complexity of Buddhist views of nature, while Harris advocates continued vigilance in translating Western environmental discourse into a

Buddhist setting. Alan Sponberg also observes that there are limits to what he calls "Green Buddhism." In particular, he questions the view that Buddhism advocates a notion of interrelatedness between all beings that is entirely egalitarian. Sponberg suggests, instead, the need to assess traditional Buddhism more accurately, first, by noting that Buddhism often advocated a hierarchical conception of the human and natural world, and second, by recognizing the usefulness of what he calls the "hierarchy of compassion" in contributing to a specifically Buddhist approach to environmental ethics.

The essays in this volume, then, span a wide range of possible approaches to the study of Buddhism and ecology. The chapters adopt various methodological perspectives, including anthropology, sociology, textual analysis, historical studies, and philosophical or theological approaches. The essays also share tensions between a descriptive and a critical perspective on the one hand and a more interpretive and engaged perspective on the other. In his response at the conference, Charles Hallisey identified this tension as one between the historical and the prophetic. This may be a fruitful tension between an approach that descriptively historicizes certain Buddhist views of nature, or particular examples of Buddhist engagement with environmental issues, and an approach that reinterprets and advocates, with a prophetic voice, Buddhist involvement with particular issues. This volume represents the full spectrum of these orientations and suggests that various approaches are necessary for an adequate understanding of Buddhist views on ecology.

There has never been any one Buddhist perspective on nature or ecology that might be considered definitive. There have been Indian, Tibetan, American, Thai, or Japanese Buddhist perspectives on the natural world, and they differ considerably according to each one's place and time in history. There is no core "Buddhistic" element to each cultural worldview but rather a diversity of perspectives that might all legitimately be identified as Buddhist.

The essays in this volume may, however, begin to reveal some general orientations that would elicit what might be a more Buddhist than, say, a Christian approach to ecology. Or, as it is a religious tradition, perhaps we can see a Buddhist perspective in contra-distinction to a secular one. It is hoped that this volume might spark a continuing inquiry, both to further a more diverse understanding of Buddhist views on ecology (for example, in underresearched areas, such as Tibetan and Chinese Buddhism) as well to help ascertain common Buddhist themes that might be offered as resources for a new religious contribution to environmental problems.

In conclusion, the editors would like to acknowledge the contributions made by several scholars and engaged Buddhists who participated in the conference upon which this volume is based. These include Joe Franke, Larry Gross, Joan Halifax, Charles Hallisey, Joanna Handlin-Smith, Jeffrey Hopkins, Leslie Kawamura, William LaFleur, Susan Murcott, Marty Peale, Christopher Queen, and David Shaner. The editors are particularly grateful for the assistance of Donald Swearer and Kenneth Kraft in shaping this volume. They also wish to acknowledge the initiative of Masatoshi Nagatomi in teaching a course on Buddhist views of nature at Harvard University for several years before his retirement in 1996. His opening address and his presence throughout the conference was a source of inspiration for the participants.

Overview:
Framing the Issues

Buddhism and Ecology:
Collective Cultural Perceptions

Lewis Lancaster

This is a significant moment in the field of Buddhist studies, a time of reappraisal of methods and sources, and the topic of Buddhism and ecology requires thoughtful and considered dialogue. There is a need for innovative ways of exploring the data, and there is a hope that we can provide direction for future contributions of Buddhism to the problems of our contemporary society.

We have an expression in our lore—"preaching to the choir." It is probable that the readers of a book such as this one are fully aware of the ecological issues and need not be reminded of the dilemmas and the dangers of our present situation. You are in that sense a "choir," already converted to a position of support for the preservation of our ecosystem. Rather than recite a litany of the past and current events that have created this crisis, I would like to reflect on the challenge we face as researchers looking at the Buddhist approach.

By preparing a study on this topic, we have already assumed that Buddhism is a tradition which can offer help to a world undergoing rapid and sometimes destructive changes in its fragile ecosystem. It is a great responsiblity to define that help and to bring to the attention of all concerned the Buddhist solutions, as contrasted to other possible systems. Let me be frank in stating that I am fearful that in our publications we will fail in our attempt to give an adequate definition of the "Buddhist solution." It is not so easy to make these determinations about the Buddhist traditions, and we may run the risk of using the collective perceptions of our Western heritage as a template for defining the principles that we attribute to Buddhism. We may seek only to find expressions and practices in Buddhism that can be interpreted as supportive of ethical norms and values established in our modern and postmodern era.

I believe, however, that the range of essays in this volume is not limited to a narrow definition of Buddhism and its practices. The authors have not succumbed to the situation which Clifford Geertz has called "scripturalism," in which can be found only "a collection of strained apologies."[1] Nor do the essays indicate that we are avoiding the difficult issues. There is, for example, little doubt that population pressure is the most crucial aspect of ecological matters, and I am pleased Rita Gross's essay directly addresses this.

It is reasonable that the Buddhist tradition should be considered when we discuss the problems facing us in the world. Buddhism is one of the largest religions in terms of the number of participants. The place of this institution in Asia is well established. Until recent times, it was the only religious or social tradition to be found throughout all the cultural spheres of Asia. Therefore, its impact on the character of the history and the current state of cultural and social development in Asia is enormous. Buddhism also offers us a model for an era of international contacts between people of different backgrounds and cultural histories. In this regard, we can say that Buddhism was the first world religion; it was the first to transcend boundaries of language, kinship patterns, political structures, cultural areas, geography. In its history, Buddhism has already proven that it can move from one culture to another without a loss of power. It is this portable Buddhism, able to move from the Ganges across the deserts of Inner Asia into the East Asian kingdoms, that may hold within it some patterns which can be used in our current global interchange. For some centuries, Buddhism had become a fixed religion, tied to the various kingdoms and cultural areas, and we study the history of Chinese, Korean, or Tibetan Buddhism. But over the last century Buddhism has once again become portable and, driven by the energies of contemporary Asian societies, it is finding new homes in places as remote from one another as Brazil, Western and Eastern Europe, Australia, and the United States. The essays included in this volume are themselves an indication of this portable Buddhism, and we read here of Thai, Japanese, and American groups operating throughout the world.

The main issue before us is the question of whether in Buddhism we can find a unique way of dealing with ecological issues. What approach might we take that will allow the Buddhist answers to be made plain for all to see? How can we avoid the inclination to use our own collective perceptions and cultural backgrounds as a guide for scanning Buddhist practice and literature? Will our discoveries be tainted by our method of search? I have used the expression "collective cultural perceptions," implying that we have ideas and values that are held collectively. These collective perceptions are pervasive and so widely accepted that they are not always seen as perceptions but are considered as simply "the way things are." Only by looking over long periods of time can we begin to identify these perceptions and some of the ways in which they influence our lives. Unless we have a basic understanding of our collective perceptions, we are apt to fall into the easy approach of extracting supportive fragments from the Buddhist texts for our own existing views. The full force of the uniqueness and power of the Buddhist tradition may well be deflected if we are unaware of our perceptions and our cultural history.

Let me give you an example of what I think to be a collective perception of western European culture and, by extension, a part of American life. This may seem a bit far removed from ecological issues, but I will try to use my example to suggest the problems which face us when we consider any issue and turn to Buddhism for answers.

We in the West have a pervasive collective perception that people who are good, moral, ethical, and worthy give help to the poor and oppressed. Aid to the poor and oppressed constitutes a major dimension of our cultural development, and we have ample proof of it in our churches and synagogues, our confraternities and consororities, such as Lions, Exchange, and Masons, our aid projects abroad, hospitals, orphanages, homes for the aged, food banks, public housing, shelters, and hospices. These are enterprises that seek to meet the pressing needs of a society where there are inequities in the resources necessary for existence itself. The poor—especially the deserving poor, such as widows, orphans, and the physically handicapped—and the oppressed must be helped. We seldom stop to think that this type of activity has a history, that it is susceptible to study.

Many of us do not realize that this strong focus on the poor and oppressed, while present in the Hebrew Bible,[2] did not reach its high point in Western European life until the time of the sale of indulgences. That is to say, helping the poor and oppressed reached a new level of social importance when the church declared that one did not have to seek the help of the transfer of merit from a monk to secure passage of the dead from purgatory to heaven. In place of the ascetic merit, an indulgence could be used to effect the movement of the dead into paradise. For the indulgence to work, one had to confess and then perform some act of charity toward the poor or oppressed. Gradually confraternities arose for just this purpose[3]— and we still have them operating at noon in most cities of the United States, with groups of men and women performing acts of help for the poor and oppressed. When carrying out these acts was tied to moving the dead into paradise, the poor and oppressed became important. There is nothing like tying a practice to the dead to bring it into prominence.[4]

Our perception of aid for the deserving poor and oppressed has undergone cultural transformations which have shifted and changed through the centuries. As wonderful as it is to have such help, there is a dark side to it. Today, psychologists and social-welfare workers report that programs based on the identification of people as poor or oppressed create deep anger. In other words, giving to those whom we call poor and oppressed can be patronizing: by the very act of giving to an identified group, we are saying, "You are inferior to me in resources, education, power, health." The gift becomes one which transfers not only merit but also shame. The call for Buddhism to become more involved in giving aid to the poor and oppressed is in one sense an attempt to have Buddhism mimic a practice which has deep roots in our European heritage. I am suggesting that in this case, where we have a collective agreement on the importance of helping the poor, our perception has a potential downside. If we attempt to project our practice onto another culture, the problems as well as the benefits must be considered.

I have tried to look into this issue with Buddhist leaders and teachers, to ask about the problems of the poor and oppressed and to see what particular and unique solutions might be offered by the Buddhist tradition as opposed to those being attempted in the West.

The answers I have received are summarized in the following series of statements:

> The first Noble Truth is the recognition of the reality of suffering. . . . Suffering is universal; we all suffer, there is no distinction between the rich and the poor. . . . Help should be given to all. We are all brother and sisters in suffering. . . . No one escapes suffering. . . . We who suffer greatly offer help to others because we understand and identify with their suffering. . . . If we acknowledge our own suffering, how much more is our compassion for others who suffer.

These comments from the leaders of the tradition suggest that it is possible to give support to others without there being a sense of patronizing. Perhaps Buddhists can offer the world more by finding solutions from within their own tradition than by being forced into a model composed of approaches taken from the West.

I began to search for an example of Buddhist assistance for the poor which might be a reflection of the responses I received when talking to teachers and spiritual leaders. This search led me to observe the work of the Taiwanese nun Jen-yen. She had built a hospital in Hwalin, on the eastern side of Taiwan, an area that had little in the way of hospital care. Her goal was to build one of the finest medical facilities in Taiwan, complete with a medical college, which, when it was constructed, would be open to everyone. Her idea was that the hospital should be the best of its kind and hence should serve the entire population. When it came to payment, each patient was asked to pay voluntarily whatever he or she could afford. Without forms or interviews, without shame, each paid according to individual resources. Since then, this nun has sent out medical staff to the Middle East, to Africa, and even to Los Angeles following the recent riots there. In every area, the services are available universally. While Jen-yen serves the poor, she also serves others. Many middle-class Americans appreciated the portable showers she erected after the earthquake in southern California cut off the normal water supply and gas lines. Her approach of helping even the affluent residents of Los Angeles is suggestive of a different attitude toward "doing good." She is perhaps the best example of a way in which Buddhist service is expressed. Few charities have equaled the amount of money she raises through these voluntary services.

Following an academic conference I attended in Taiwan a few years ago, we were taken on a tour and stopped to visit with the nun Jen-yen. She so moved these hardhearted professors that the group decided to donate all honoraria received for giving papers at the conference. There is something appealing about this woman, who made an elitist group of academics feel that they would be served with as much compassion in that hospital as the poorest beggar. It was powerful to experience that sensation. I do not say that this approach would necessarily work within our environment, but it is one worth considering.

There are possible negative results when we impose our cultural perceptions onto others. The most current example of this has been the critique of Western influence on Sri Lankan Buddhism. Gananath Obeyesekere coined the term "Protestant Buddhism" to define the events associated with the so-called revival of Buddhism in Sri Lanka under the influence of Colonel Henry Steel Olcott and the Theosophical Society.[5] When Olcott, operating from a Victorian view of religion, rejected many of the practices of the Buddhist population of Sri Lanka while accepting the tradition as he found it in the Pali texts, he made a profound impact on the tradition. In a recent article, Stephen Prothero tells us that the form of Buddhism put forward by Olcott was characterized by the idea that the normative and essential aspects of the religion are to be found in the texts—and, by implication, not in the practices. Second, this Protestant Buddhism is defined as an ethical system rather than as a spiritual one.[6] Stanley Tambiah and Gananath Obeyesekere have stated that this was a betrayal of Buddhism; it was a rejection of the rituals and practices dealing with matters such as health, fortune, births, and deaths that had been at the heart of Sri Lankan Buddhism. As Christopher Queen states in his excellent article in the new volume *Engaged Buddhism*, Tambiah, Obeyesekere, and Walpola Rahula describe "a tragic picture of the long descent of Sinhala Buddhism first into passivity and finally into sectarian, ethnic and political violence."[7] If these scholars are correct in their assessment, then we should be wary of the rejection of popular religious practices in the name of reaching for the pure essence of the Buddhist tradition.

These examples, I hope, offer some indication of what I mean by collective cultural perceptions which control many of our actions and modes of thinking. When we come to the matter of Buddhism and ecology, we will need to try to identify some of these perceptions from both the Western and the Buddhist traditions and then see how they can be of help. One of the important elements in ecological discussions is the role of industry, transnational corporations, and commerce in all of its forms. It is impossible to discuss the cutting of rain forests, the pollution of water, toxic emissions into the atmosphere, and the expansion of agricultural land into the natural habitat of animals without dealing with capital and mercantile activity. Here, too, we run into a collective perception in the West. From biblical sources onward, wealth—and its companion, mercantile activity—has often been denigrated in the Western sources. We see this reflected in Marxism, where the evils of mercantile life become nearly demonic. The workings of transnational corporations and other mercantile activities then become, by the very nature of our perception, an evil. When we begin to face ecological problems, we are immediately able to summon up our perception regarding bankers, money changers, merchants. They are seen as greedy, uncaring, the source of much woe, the chargers of usurious interest. Therefore, no small part of our ecological discourse deals with an attack against the mercantile.

When I turn to Buddhist history and texts and current practice, I find quite a different picture. Buddhism has been the religion of merchants from its earliest days, and the spread of Buddhism has been accomplished by the mercantile community. The Buddha talked to kings and secured large donations from merchants who were close to him. One of his most important supporters was the money changer Anāthapiṇḍika, who provided the Buddha and his disciples with a grove of trees inside a walled area.[8] From his name, we know that Anāthapiṇḍika was "one who gave support to those who were without protectors." This indicates that we have a very different perception regarding wealth and merchants in Buddhism than we do in Western European cultural systems. Buddhists, depending on the merchants and holding them in high esteem, directed much of their teaching toward this lay group.[9] Even more importantly, some merchants and kings came to have an understanding of Buddhism that allowed them to teach and convert. We have the report of a merchant converting a king to Buddhism; this would be incomprehensible in the antimerchant environment of ancient Palestine and the Greek Testament. Many of the supporters of the Venerable Jen-yen—of whom I spoke earlier—are merchants and corporate officials. The role of the merchant layman in Buddhism today is as strong as it was in ancient India. If ecological discourse assumes a rejection of this particular group, then one of the pillars of the Buddhist community will be under strong attack. Perhaps we can learn from Buddhism in this regard. We need to seek out the merchants and the corporate leaders, include them in our conferences, urge them to be active partners in the search for answers to the ecological crisis. I would hope that in some future conference on this topic, we could have lay Buddhists from banks and business offices speaking and responding to the problem. After all, many of the brightest and best of our society are involved in corporate life; a rejection of this group may do a great disservice to our search for solutions.

In the West, one of our views of nature is that of the Garden of Eden, when nature was without pain. I believe there are forms of Buddhism that would not subscribe to this view of nature. We find in many places texts which speak of the terrible and frightening forest, of the wilderness infested with robbers,[10] vermin, beasts of prey,[11] and flesh-eating ghouls,[12] of areas swarming with snakes,[13] where there is neither food nor water.[14] A man emerging from a huge wild forest and seeing indications of a town or some other inhabited place will feel happier. And, in Buddhism, one way that we know things are getting worse is when the realm of animal rebirth becomes crowded, for the overpopulation of the animals indicates just how many beings have sunk to a lower level of birth.

From the Prajñāpāramitā (Perfection of Wisdom) literature we find that the *bodhisattva*, exemplar of practice, living in the wilderness and suffering from all the ills caused by insects, by lack of food and water, will, on account of the horrors of the forest, have compassion. This *bodhisattva* believes that by experiencing the agonies of the jungle, one can fully have compassion for those who are forced to live lives that face these dangers daily. The *bodhisattva* takes a vow that, in the Buddha land that will be brought into existence at the time of his elevation to Buddhahood, there will be no animals, the inhabitants will eat only divine food, there will be plenty of water—it will be like a pleasure grove near a great city.[15]

As we turn our attention to the *Jātakamāla*,[16] or birth stories, however, we find that one of the reasons to follow the Buddha is the fact that for lifetime after lifetime he has expressed his compassion for animals and other beings. This, then, is the challenge before us: how do we adequately express the variety of teachings and practices found in Buddhism? I can no more claim that the view of the wilderness as horrible is the essence of Buddhism than can one who selects the *Jātakamāla* as solely representational of the tradition.

Western perceptions have also influenced our views of the life of Śākyamuni—though our resources for reconstructing his activities are slim. The popular idea of the Buddha as a young Luther, a reformer, one who spoke out against the establishment of his time, a rebel, an individualist, has great appeal to us. But there is little evidence to support these claims. Noritoshi Aramaki of Kyoto has worked diligently to try to determine the most ancient sayings of the Buddha using the technique of finding passages from the ancient Buddhist text, the *Suttanipāta*, that are echoed in the oldest layers of the Upaniṣads and Jain literature. He reasons that the collective perceptions of those times can best be discerned from words that are found in these three early texts of three religious streams of India. What emerges from the sayings which Aramaki finds in the *Suttanipāta* and in the writings from the other two traditions is a picture of a person in despair over the endless cycle of rebirth and continual move from birth to death and back to birth. In his anxiety to escape from this endless round, Śākyamuni turned to the ascetic solution, leaving home and seeking an enlightened state in a homeless unattached life. But this, Aramaki claims, was a perception generally held by society throughout the Gangetic plain. Śākyamuni chose the solution, but he did not do so as a reformer. As Richard Gombrich has pointed out, the Buddha's "concern was to reform individuals and help them to leave society forever, not to reform the world."[17] Śākyamuni was then a person of his time, and while his tradition brought forth many innovations, the ascetic solution, his chosen life-style, was not one of them.[18]

We strive also to learn what the situation was for the region where Śākyamuni lived and taught. Most scholars now agree it was a time when deforestation of the Ganges region was taking place, population growth was sizable, urban centers were the important hubs—urban islands in a sea of rain forests. This urbanization was characterized by a growth of mercantile activity, long-range trading, and travel between the population centers. Buddhism came into existence not as a tradition that was limited to the wilderness but as part of the growing urban movement; it found major supporters among merchants, bankers, and kings, as well as among a population where job specialization was rapidly growing, with increasing numbers of barbers, carpenters, jewelers, grain merchants, ferryboat operators, and even robbers who preyed on the trade system between the urban centers. The Buddha was constantly in contact with all these people, and many of his earliest sermons were directed at one layman or another belonging to any one of a wide range of occupations. In other words, we would do well to reconstruct carefully the ancient history of Buddhism and to try to see it in its complexity and within its social context. While Śākyamuni was a wandering ascetic, he nonetheless taught and lived in the precincts of the growing urban world of his time. His childhood was lived within a city environment and I suspect his view of the forests was very close to what I have described above—as a place of danger and suffering.

Although I have suggested that the imposition of one culture's collective perception onto another may sometimes result in problems, let me give an example of how Buddhism moved from one culture to another and promoted entirely opposite perceptions in the two environments. In the Ganges Valley, Buddhism reported the perception that the forest was a source of pain, danger, and struggle. When it moved into the Han cultural area, a quite different perception of nature was held by that society.

While India of the time of the Buddha was composed of urban islands in the sea of the forest, by the time of the arrival of Buddhism among the Chinese, nature was beginning to be seen to consist of islands of mountains within a sea of cultivated fields. The sages of China who sought nature as a way of renewing their humanity left the cultivated areas and repaired to the remaining islands of untamed nature in the mountains. When Buddhism came from India with a perception of nature that was quite different from that of the ancient sages of China, it still was able to provide the

Chinese with an important approach to their physical environment. John Jorgenson, in his groundbreaking dissertation from Australian National University,[19] points out that while the Chinese had long held to the importance of contact with nature, and knew that this contact was healing and supportive, they had no explanation for it. They could only affirm that this was so. In Buddhism they found a way of explaining this close connection between man and nature— of this relationship which went deep into the very essence of the human experience.

The great contribution of Buddhism to this collective perception about nature in China was the concept of Buddha-nature. Teaching that everything has Buddha-nature was a revolutionary development in China. Every person has Buddha-nature, but what was of such importance to the Chinese was the teaching that insentient objects also have it. The rocks, trees, lotuses, streams, mountains—all have Buddha-nature. Therefore, one's mind, which has Buddha-nature as its essence, shares a common aspect with every part of insentient nature, which also possesses this same Buddha-nature. With this introduction of the idea that the mind and the natural objects had the same Buddha-nature, the Chinese at last had an explanation for the power of nature. Buddhist poetry written by Ch'an meditation monks was not limited to words of doctrine; it was about nature and the references were to snow and falling leaves and water running over rocks, for this was an expression of Buddha-nature. If the artist painting a lotus or a rock could capture its essence, this was to depict Buddha-nature and was a valid way of dealing with this important essence of the religion. Here, suggests Jorgenson, was a happy meeting of a Buddhist concept with the prevailing collective perceptions of the Han people. As a result Chinese literature was enriched and art found a firm place within the religious system. This is echoed in the essays in this volume that discuss Zen Buddhism and ecology.

This perception of everything having a Buddha-nature resonates with us in North America and Europe. If we look to Buddhism to support our views of the wonder of nature, it is probable that we will rely more on East Asian than Indian forms of the tradition. This tells us something about Buddhism: it was able to move into a cultural sphere quite different from that of its origins and was able to supply a doctrine of great value to the new region. Our challenge now, in looking at ecology, is to find what aspect of Buddhist teaching can provide us with the greatest help.

Claude Lévi-Strauss has made an interesting study of cultures, which he identifies according to the terms used by different societies: raw–cooked, fresh–rotten.[20] From this he infers that in some cases there is a transformation of objects in ways that are not found in nature. Cooked food is the most basic example of cultural transformation. By comparison, fresh–rotten implies cultural patterns where food is gathered fresh, not transformed by heat, and lasts only until it rots in the natural process. Our perceptions of nature are involved in this distinction. We may wish to have "fresh–rotten," leaving nature to follow its own processes without intervention. But such a course is nearly impossible, since the agriculture methods, planting seeds, building permanent shelter—all of the ways in which we live—are predominantly "raw–cooked." And, in the transformations, in the "cooking," we create problems because this is an intrusion into the natural process. We may long for "fresh–rotten" but we are living in a "raw–cooked" society. We travel by car and airplane, we eat food cooked in ovens and stoves, we shower with water that is heated and pressurized. While there is a tendency to glorify the "noble savage," to seek to return to the Garden of Eden, where everything would be "fresh–rotten," it is unrealistic to hope that the billions of people now living on this planet could possibly achieve such a state. If Buddhism has something to offer our ecological process, it must be within the "raw–cooked" sphere.

I have tried to indicate some of the difficulties, challenges, and complexities, the hidden but powerful perceptions which exist as a collective view, matters that impinge on our discussion of Buddhism as a religious system. Perhaps Buddhists offer us aspects which we did not expect. Buddhists sometimes have taught us to be fearful of nature. They have suggested that the world is an endless net of causality where every event sends ripples throughout the whole fabric of the universe. This may be a healthy lesson, demonstrating that we need to be more fearful of the consequences of what we do with regard to nature. There are dangers to our very existence, just as there were dangers for those who entered the ancient forests of India. Perhaps it is better for us to have a respect for nature and its power. We must not fall into the trap of seeing nature as the poor and oppressed and ourselves as the powerful rescuer of the "victim."

Nature, with its microbes, its fierce rays which pierce through damaged ozone, is awesome. Any belief that we can conquer it or defeat it or heal it is naïve and arrogant. Our ploys are successful only to the degree that they imitate nature. Buddhism teaches us that all is in flux. Whatever is in flux will never exist in a permanent state. We yearn for all of the germs and viruses to remain in an unchanged state so that we might have the luxury of time to invent instruments targeted to destroy them. In the Buddhist texts and teachings we hear the hard truth that none of these perceived dangers will remain unchanged or permanent, and we must learn how to survive in a natural state of constant change.

From China we see the other side of nature, the healing and pleasant one. But we should remember that this view of nature grew out of the period following the deforestation of the entire kingdom. From views of the ancient and modern landscape, Chinese culture appears to be anti-tree. That nature which the sages sought centuries ago was even then the fragile remnants of the primeval wilderness of ancient times. To say merely that the sages' support for the natural process and their love of nature is an accurate description of the Chinese approach to nature misses the reality of a situation where the real appreciation of that time was for the ploughed field.

Buddhism may also lead us to reevaluate the role of the business community in this struggle. There have been far too many books which depict Buddhism as otherworldly, and it sometimes comes as a shock to think of it as having a partnership with merchants.

Śākyamuni followed and advocated the ascetic solution. It is possible that we need a "new asceticism" for our times, an asceticism that involves using less of the resources and that most certainly means control of population growth. Recently, at a lecture in Berkeley, the Dalai Lama spoke about population. One solution, he suggested, was for all the thousands gathered in the Greek Theatre to become monks and nuns. With a twinkle in his eye, he mused that probably most wouldn't want to do that. He then said, "many people consider abortion to be an act of violence, so for those who do not wish to have violence, the practice of birth control must be used." Is it not the case that practices such as birth control, using less, saving, recycling, changing our diet, forgoing convenience in favor of conservation are all forms of a modern asceticism? Maybe the ancient solution of Śākyamuni is still an important and viable one in the present world and we can construct a new asceticism. This asceticism should be more than prescriptive; it should be fulfilling and life-affirming, perhaps even playful. If it espouses the dull drabness of a puritanical approach, it will fail to recruit supporters.

What we discuss in this volume is of importance and it is urgent. My plea is that we should let Buddhism in all its complex forms be represented in our discussion and that we should seek to be aware of our cultural perceptions with their potential to blind us to other solutions. Buddhism may give some answers to our questions. These answers may surprise us. However, if we look with care and awareness at these varied and changing positions of Buddhism, we will find ourselves open to the possibilities of discovering innovative approaches to the problems facing our environment.

Notes

1. See Christopher Queen's introduction to *Engaged Buddhism: Buddhist Liberation Movements in Asia*, ed. Christopher S. Queen and Sallie B. King (Albany: State University of New York Press, 1996), 115.

2. See Exod. 23:11 and Deut. 15:4.

3. Ruth Murphy, *Saint François de Sales et la civilité chrétienne* (Paris: Nizet, 1964), provides a suggestion of the way in which social codes have developed.

4. A description of the Canon Law regarding indulgences can be found in the *Catholic Encyclopedia*, ed. Charles Harbermann et al. (New York: Gilmary, 1957), 783–88. See also Conrad Boerma, *Rich Man, Poor Man—and the Bible*, trans. John Bowden (London: SCM Press, 1979).

5. Gananath Obeyesekere, "Religious Symbolism and Political Change in Ceylon," *Modern Ceylon Studies* 1 (1970):43–63.

6. Stephen Prothero, "Henry Steel Olcott and 'Protestant Buddhism,'" *Journal of the American Academy of Religion* 63, no. 2 (summer 1995):281–302.

7. Queen, introduction to *Engaged Buddhism*, 5.

8. The story appears in many places. See *Manorathapūraṇī, Aṅguttara Commentary*, 1:208; *Dhammapadatthakatha*, 1:128.

9. In former lives, the *bodhisattva* was a merchant (*Jātaka*, 1:405) and a money changer (*Sāratthappakāsinī, Saṃyutta Commentary*, 1:240). Elders within the monastic group were identified as the sons or daughters of money changers (*Theragāthā Commentary*, 1:312), merchants (*Therīgāthā Commentary*, 260), and caravan leaders (*Theragāthā Commentary*, 1:238). On the negative side some merchants were described as misers and suffered accordingly (*Jātaka*, 1:349).

10. *Jātaka*, 1:332, *Dhammapadatthakatha*, 2:254, and *Jātaka*, 2:335, give stories about robbers and thieves.

11. We read of fierce Nagas (*Manorathapūraṇī, Aṅguttara Commentary*, 1:165); enormous spiders (*Jātaka*, 5:469–70); snakes (*Udāna Commentary*, 60; *Jātaka*, 2:145); wild elephants (*Apadāna* 1:198).

12. Ogres of both sexes are said to devour people (*Sumaṅgala Vilasini*, 2:483; *Dhammapadatthakatha*, 1:37).

13. Snakes and snake bites are popular themes (*Udāna Commentary*, 60; *Jātaka*, 2:145).

14. Thirst is a common problem for those who travel (*Petavatthu Commentary*, 141; *Petavatthu*, 28).

15. Edward Conze, trans., *Astasahasrika Prajnaparamita* (Calcutta: Asiatic Society, 1958), 139.

16. See H. Kern, ed., *The Jatakamala*, Harvard Oriental Series, vol. 1 (Cambridge, Mass.: Harvard University Press, 1943). The famous tale of the *bodhisattva* giving his life for the hungry tigress (*Vyāghrijātaka*) is found in the *Mūlasarvāstivādavinaya* and in the Avadānas (51 and 95). The English translation

of *The Jatakamala* is by J. S. Speyer (Delhi: Motilal Banarsidass, 1971; reprint of Sacred Books of the Buddhists, vol. 1).

17. Richard Gombrich, *Theravada Buddhism: A Social History from Ancient Benares to Modern Colombo* (London: Routledge and Kegan Paul, 1988), 30.

18. Richard Gombrich, "The Fundamental Truth of Buddhism: *Pratitya-samutpada*—Conditioned Becoming and Conditionless Being," *Machikaneyama Ronso* (Osaka) 22 (1988):28–29.

19. John Jorgenson, "Sensibility of the Insensible: The Genealogy of a Ch'an Aesthetic and the Passionate Dream of Poetic Creation" (Ph.D. diss., Australia National University, 1989).

20. Claude Lévi-Strauss, *Le cru et le cuit* (Paris: Plon, 1964).

Theravāda Buddhism and Ecology:
The Case of Thailand

The Hermeneutics of Buddhist Ecology in Contemporary Thailand: Buddhadāsa and Dhammapiṭaka[1]

Donald K. Swearer

The world's environmental crisis has prompted religiously committed, socially concerned people throughout the world to search their traditions for resources to address its root causes and its symptoms. Buddhists are no exception. The compatibility between the Buddhist worldview of interdependence and an "environmentally friendly" way of living in the world, the values of compassion and nonviolence, and the example of the Buddha's life-style and the early *saṅgha* are cited as important contributions to the dialogue on ways to live in an increasingly threatened world. This essay seeks to interject a particular insight into this discussion through an examination of selected writings of Buddhadāsa Bhikkhu and Phra Prayudh Payutto (current monastic title, Dhammapiṭaka), the Thai *saṅgha*'s most highly regarded interpreters of the *buddhadhamma*.[2] In particular, I propose to explore their distinctive ecological hermeneutics, that is to say, the particular environmental lessons each draws from the texts and traditions of Thai Buddhism. In conclusion, I shall briefly assess the recent critical evaluation of Buddhist environmentalism by Ian Harris[3] from the perspective of my construction of the ecological hermeneutics of Buddhadāsa Bhikkhu and Phra Prayudh.

Introduction

During the past half century, economic and social configurations have changed dramatically throughout the world as a consequence of population increases, urbanization, industrialization, and technical achievement. These changes have, to a certain extent, created a common economic culture determined by the necessities of the modern nation-state and the business interests of multinational corporations. This economic culture is primarily "materialistic" in nature in the sense that human well-being tends to be defined in terms of the production and consumption of goods. It is commonplace, for example, to measure the wealth of a nation in terms of its GNP (gross national product).

The consequences of the development of an economically defined modern culture are manifold. For example, it has led to a general increase in life expectancy among most populations of the world as a consequence of improved health services, more adequate housing, and so forth. In short, in respect to material aspects of life more people share in the benefits of the increased production and use of various kinds of goods. Yet even from an economic perspective, the increase in the production and use of goods has been a mixed blessing. In general, even though by GNP measurements the world has seen a significant increase in the amount of material wealth, critics are quick to point out the gross disparity between the rich and the poor, not only in "developing" countries, such as Thailand, but also in "developed" countries, such as the United States. For instance, in Thailand conflicts that began in 1988 over water use between the wealthier industrial/urban sector and the poorer agricultural/rural sector have prompted numerous farmer protests over low water supplies that came to a head in the drought year 1993.[4] Internationally, it can also be pointed out that despite improvements in agricultural technology hunger has emerged as a persistent and pervasive worldwide problem. The capital-intensive green revolution, with its dependence on chemical fertilizers and pesticides, has produced more systemic, long-range problems than it has solved, and biotechnology may raise even more questions about the consequences of genetic engineering.[5]

Developments in many different kinds of technologies have led to dramatic breakthroughs in everything from space exploration to microscopic laser surgery. At the same time, however, technological advancement has contributed to the sense of hopelessness and prevalent violence experienced by modern society, as evidenced by the plague of drug addiction, increasing levels of armed violence, or the seemingly insurmountable problem of waste disposal, especially the threat of the widespread nuclear waste contamination and the toxic contamination of water and food supplies.

Our modern economic culture has also had a generally dele-terious effect on classical moral values and religious worldviews and on traditional ways of understanding human existence and what constitutes the good or happy life. In the face of a perceived threat to traditional ways of being by modern economic culture, some seek a return to the verities of a simpler era believed to be embodied in an earlier historical age or represented by an idealized, mythic time of primal beginnings. Religious fundamentalisms, whether Christian, Jewish, Muslim, Hindu, or Buddhist, may be interpreted as a retreat from the confusions and threats of the modern world to the truths and values of an earlier age. But there are other, more creative and constructive religious responses to modernity than today's various fundamentalisms Thoughtful religious adherents throughout the world are seeking to understand and interpret their traditions in ways that preserve the lasting insights and values of their faith, while at the same time engaging the realities of existence in today's world rather than retreating from them.

In the past several years the media in Thailand has devoted considerable attention to the conflicts between the goals of national and commercial development, the well-being of the majority of the Thai people (especially the rural, farming populations), and the health of the environment. In particular, the Seventh National Development Plan has been criticized for following in the footsteps of its predecessors by emphasizing material growth at the expense of a more balanced development and an equitable distribution of wealth. Dr. Ananda Kanchanapan of the Faculty of the Social Sciences at Chiang Mai University observes that development in Thailand has emphasized the GNP and in doing so has undermined the moral and spiritual integration between the social and natural environment.[6] An article in the *Matichon* newspaper representative of this point of view charges that development in Thailand has benefited the elites at the expense of the environment and proposes a reformist Buddhist perspective that would challenge selfishness and greed and the excessive lifestyle that has resulted from "too much wealth, too much power, too much to eat and drink, too many cars and mistresses."[7]

Buddhadāsa Bhikkhu: Nature as *Dhamma*

Like Thomas Merton, the late American Trappist monk and peace activist, Buddhadāsa exemplifies the truth that thoughtful spiritual engagement with the world requires a degree of contemplative distance.[8] In much the same way as Merton, Buddhadāsa spent most of his active career living and teaching in a forest hermitage (Wat Suan Mokkhabalārāma [Thai, Mōkh], Chaiya, south Thailand). Like Merton, he was also extraordinarily responsive to the issues of his time. Although known in Thailand primarily as a teacher or a "monk of wisdom" (Thai, *phra paññā*), Buddhadāsa used the doctrinal tenets of non-attachment, dependent co-arising, and emptiness as the bases for addressing an exceptionally broad range of issues, problems, and concerns, from meditation, monastic discipline, and ritual observances to work, politics, women in Buddhism, and the environment.

The core of Buddhadāsa's ecological hermeneutic is found in his identification of the *dhamma* with nature (Thai, *thamachāt*; Pali, *dhammajāti*). It was his sense of the liberating power of nature-as-*dhamma* that inspired Buddhadāsa in 1932 to found Wat Suan Mōkh as a center for both teaching and practice in a forest near the small town of Chaiya in Surat Thani Province, rather than pursue a monastic career in Bangkok. For Buddhadāsa the natural surroundings of his forest monastery were nothing less than a medium for personal transformation.[9]

> Trees, rocks, sand, even dirt and insects can speak. This doesn't mean, as some people believe, that they are spirits [Thai, *phī*] or gods [Pali, *devatā*]. Rather, if we reside in nature near trees and rocks we'll discover feelings and thoughts arising that are truly out of the ordinary. At first we'll feel a sense of peace and quiet [Thai, *sangopyen*=quiet-cool] which may eventually move beyond that feeling to a transcendence of self. The deep sense of calm that nature

provides through separation [Pali, *viveka*] from the troubles and anxieties that plague us in the day-to-day world functions to protect heart and mind. Indeed, the lessons nature teaches us lead to a new birth beyond the suffering [Pali, *dukkha*] that results from attachment to self. Trees and rocks, then, can talk to us. They help us understand what it means to cool down from the heat of our confusion, despair, anxiety, and suffering.[10]

Buddhadāsa's identification of nature and *dhamma* prompts him to read nature as a text. Indeed, because experiencing nature involves not just the mind but all of the bodily senses, to listen to the "shouts of nature" is potentially more liberating (read *nibbāna*) than studying the Pali scriptures. Buddhadāsa, moreover, makes the extraordinarily strong claim that nature is a much more appropriate context or environment in which to pursue liberation than sitting at a desk: "If we don't spend time in places like this [Wat Suan Mōkh], it will be virtually impossible for us to experience peace and quiet. It is only by being in nature that the trees, rocks, earth, sand, animals, birds, and insects can teach us the lesson of self-forgetting."[11] In Buddhadāsa's spiritual biocentric view, being attuned to the lessons of nature is tantamount to at-one-ment with the *dhamma*. By inference, the destruction of nature implies the destruction of the *dhamma*.

Cynics could argue that Buddhadāsa's ecological hermeneutic is self-serving. After all, his essay *Shouts from Nature* (*Siang Takǫn Jāk Thamachāt*) was a Visākhā Pūja sermon at Wat Suan Mōkh, so could not his teaching be interpreted as a clever strategy to promote interest in and support of his forest ashram? Such an argument can be summarily dismissed in the face of Buddhadāsa's exemplary integrity over a monastic career of sixty-five years. Two additional, more serious criticisms might be made, however: 1) while his message is not gauged to promote Wat Suan Mōkh, it might be argued that it constructs Buddhist practice as a retreat to the forest rather than engagement with the world; 2) from a deep ecology perspective Buddhadāsa appears to be more anthropocentric than biocentric; that is to say, the forest is valued simply as a place for spiritual practice rather than for its inherent value. Although both criticisms are not without merit, I propose to challenge these two views.

Toward the end of his life the destruction of the natural environment became a matter of great concern for Buddhadāsa. One of his informal talks at Wat Suan Mōkh in 1990, three years before his death, was titled "Buddhists and the Care of Nature" (*Buddhasāsanik Kap Kān Anurak Thamachāt*). This essay provides insight into both the biocentric and ethical dimensions of Buddhadāsa's ecological hermeneutic.[12] Let us begin by exploring the essay's two central terms—"care" (Thai, *anurak*; Pali, *anurakkhā*) and "nature" (Thai, *thamachāt*; Pali, *dhammajāti*).[13]

Within the context of the worldwide concern for environmental destruction, the Thai term *anurak* is often translated into English as "conservation." In fact, the dozens of Thai monks involved in efforts to stop the exploitation of forests in their districts and provinces have been labeled *phra kānanurak pā*, or "forest conservation monks." *Anurak*, as embodied in the life and work of Buddhadāsa, however, conveys a richer, more nuanced meaning closer to its Pali roots, namely, to be imbued with the quality of protecting, sheltering, or caring for. By the term *anurak*, Buddhadāsa intends this deeper, dhammic sense of *anurakkhā*, an intrinsic, active "caring for" that issues forth from the very nature of our being. In this sense, to care for nature is linked with a pervasive feeling of human empathy (Pali, *anukampā*)[14] for all of our surroundings. If you will, caring is the active expression of empathy.

One cares for the forest because one empathizes with the forest just as one cares for people, including oneself, because one has become empathetic. *Anurak*, the active expression of a state of empathy, is fundamentally linked to non-attachment or liberation from preoccupation with self, which is at the very core of Buddhadāsa's thought. He develops this theme using various Thai and Pali terms, including *mai hen kae tua* (not being selfish),[15] *cit wāng* (non-attachment or having a liberated heart-mind), *anattā* (not-self), *suññatā* (emptiness). In a talk to the Dhamma Study Group at Sirirāt Hospital in Bangkok in 1961, he stated unequivocally the centrality of non-attachment to Buddhist spirituality: "This is the heart of the Buddhist Teachings, of all Dhamma: nothing whatsoever should be clung to."[16] It is just such non-attachment or self-forgetting—the heart of the *dhamma*—that we learn from nature.

We truly care for our total environment, including our fellow human beings, only when we have overcome selfishness and those qualities which empower it: desire, greed, hatred. Buddhadāsa's profound commitment to this truth can be seen in "Overcoming Selfishness Is Essential to a Political System" (*Khwām Mai Hen Kae Tua Jampen Samrap Rabop Kanmuang Khong Lōk* [1989]); "Serving Others Makes the World Peaceful" (*Kān Rapchai Phūœn Tham Hai Lōk Santi* [1960]); "Working with a Liberated Heart and Mind for the Good of Society" (*Kān Tham Ngān Duœ Cit Wāng Phū'a Sangkhom* [1975]). Note the persistent linkage between non-attachment, selflessness, and the capacity to be truly other-regarding. Caring in Buddhadāsa's dhammic sense, therefore, is the active expression of our empathetic identification with all life-forms: sentient and nonsentient, human beings and nature.

Caring in this deeper sense of the meaning of *anurak* goes beyond the well-publicized strategies to protect and conserve the forest, such as ordaining trees, implemented by the conservation monks, as important as these strategies have become in Thailand. This is where the second term, *thamachāt*, enters the picture. The Thai term *thamachāt* is usually translated as "nature." Its Pali root, however, denotes everything that is linked to *dhamma* or that is *dhamma* originated (*jāti*). That is to say, *thamachāt* includes all things in their true, natural state, a condition that Buddhadāsa refers to as "norm-al" or "norm-ative" (*pakati*), that is, the way things are in the true, dhammic condition. To conserve (*anurak*) nature (*thamachāt*), therefore, translates as having at the core of one's very being the quality of empathetic caring for all things in the world in their natural conditions; that is to say, to care for them as they really are rather than as I might benefit from them or as I might like them to be. Indeed, *anurak thamachāt* implies that the "I" is not over against nature but interactively co-dependent with it. In other words, the moral/spiritual quality of non-attachment or self-forgetfulness necessarily implies the ontological realization of interdependent co-arising.

From an ethical perspective this means that our care for nature derives from an ingrained selfless, empathetic response. It is not motivated by a need to satisfy our own pleasures as, say, in the maintenance of a beautiful garden or even by the admirable goal of conserving nature for our own physical and spiritual well-being or for the benefit of future generations. To care for nature in these pragmatic, functional terms has immense value, to be sure. I think that Buddhadāsa would not dispute this fact. A carefully tended garden is both meaningful to the gardener and inspirational to the viewer; furthermore, human survival may depend on whether or not we are able to conserve our dwindling natural resources and solve the problems of our increasingly polluted natural environment. Laudable as these two senses of conserving nature are, they lack the profound transformational or spiritual sense of what Buddhadāsa means by *anurak thamachāt*. I propose that Buddhadāsa's identification of nature and *dhamma* makes his view inherently biocentric. That is, listening to nature and caring for nature are both forms of dhammic self-forgetting, not merely instrumental to human flourishing.

The concept of active caring for other human beings needs little explication.[17] The word itself evokes numerous examples from our own experience: the parent who cares for a child, the mutual caring among friends, the responsible caring of citizens for the well-being of the state. But what does Buddhadāsa mean by caring for nature, *thamachāt*? By *thamachāt* Buddhadāsa does not have in mind either a metaphysical or a romantic concept of nature. Quite the contrary. For Buddhadāsa, things in their natural, true state are characterized by their dynamic, interdependent nature (*idappaccayatā, paṭicca samuppāda*). Everything is linked in a process of interdependent co-arising, or as Buddhadāsa often says, "We are mutual friends inextricably bound together in the same process of birth, old age, suffering, and death."[18] In other words, the world is a conjoint, interdynamic, cooperative whole (Thai, *sahakorn*; Pali, *saha+karaṇa*), not a collection of disparate, oppositional parts.[19] In the deepest sense, therefore, to care for nature means participation in this state of inter-becoming, not just human beings preserving nature for the sake of human beings.

While human linkages are self-evident to us, as in our relationships with family and friends, the interdependence of human beings and nature has been less self-evident. Only in recent years has it been commonly understood that the destruction of the Brazilian rain forest or the ocean dumping of toxic waste affects the entire world ecosystem; or, in more immediate and personal terms, that whether I personally conserve water, electricity, gasoline, and so on affects not only my utility bills but the health of the entire cosmos. To care for (*anurak*) nature (*thamachāt*), therefore, stems from a realization that I do not and cannot exist independently of my total environment. I am not "an island unto myself"; or, in Buddhadāsa's terminology, I do not and cannot exist unto myself (Pali, *atta*; Thai, *tua kū khong kū*) because to do so contravenes the very laws of nature (*dhammajāti=idappaccayatā*).

Buddhadāsa's sense of a cooperative society (*sahakorn*), therefore, extends to the broadest reaches of the cosmos.

> The entire cosmos is a cooperative. The sun, the moon, and the stars live together as a cooperative. The same is true for humans and animals, trees and the earth. Our bodily parts function as a cooperative. When we realize that the world is a mutual, interdependent, cooperative enterprise, that human beings are all mutual friends in the process of birth, old age, suffering, and death, then we can build a noble, even a heavenly environment. If our lives are not based on this truth then we'll all perish.[20]

My own personal well-being is inextricably dependent on the well-being of everything and everyone else, and vice versa. In Buddhadāsa's view this is an incontrovertible, absolute truth (*saccadhamma*). To go against this truth is to suffer the consequences. Today, we are suffering the consequences. As Buddhadāsa expressed it in terms approaching an apocalyptic vision:

> The greedy and selfish are destroying nature. . . . Our whole environment has been poisoned—prisons everywhere, hospitals filled with the physically ill, and we can't build enough facilities to take care of all the mentally ill. This is the consequence of utter selfishness [Thai, *khwām hen kae tua*]. . . . And in the face of all of this our greed and selfishness continues to increase. Is there no end to this madness?[21]

In Buddhadāsa's view, caring for *thamachāt* necessarily means not only that we care for other human beings and for nature, but also that we care for ourselves. Outwardly, *thamachāt* means physical nature. But the inner truth of nature is *dhammadhātu*, the essential or fundamental nature of *dhamma*, namely, the interdependent co-arising nature of things (*paṭicca samuppāda, idappaccayatā*). "When we realize this truth, the truth of *dhammadhātu*, when this law of the very nature of things is firmly in our hearts and minds, then we will overcome selfishness and greed. By caring for this inner truth we are then able to truly care for nature."[22]

Buddhadāsa's environmental philosophy can be characterized as a spiritual biocentrism based on the identification of nature and *dhamma*. The simplicity of his life-style amidst the natural surroundings of Suan Mōkh, furthermore, provides a compelling testimony to the possibility of putting these teachings into practice. By basing his ecological hermeneutic on the identification of nature and *dhamma*, Buddhadāsa challenges the criticisms that his environmental philosophy is either too otherworldly or too anthropocentric. Another kind of criticism, that Buddhadāsa fails to take sufficient account of Theravāda historical traditions to justify his ecological hermeneutic, brings us to a consideration of Phra Prayudh Payutto.

Dhammapiṭaka: Nature and the Pursuit of Enlightenment

Grant A. Olson's introduction to Dhammapiṭaka's (Phra Prayudh Payutto) *Buddhadhamma* provides a sketch of his life. Phra Prayudh was born in 1939, seven years after Buddhadāsa founded Suan Mōkh. His monastic career has followed a very different trajectory from that of Buddhadāsa. He passed the ninth and highest level of Pali studies in Thailand on the way to being acknowledged as the finest Pali scholar in the Thai *saṅgha*. His scholarly work includes two Pali dictionaries, editorial leadership in the newest edition of the Thai Pali *tipiṭaka* and the Mahidol University CD-ROM Pali canon, as well as his magnum opus of doctrinal interpretation, *Buddhadhamma: Natural Laws and Values for Life*.[23] Although in recent years Phra Prayudh has dedicated himself to scholarly work, from the mid-1960s to the mid-1970s he was actively involved in institutional leadership roles as the abbot of Phra Phirain Monastery in Bangkok and the deputy secretary-general of Mahāchulalongkorn University for Buddhist monks. He has also been awarded several honorary doctorates and in 1994 received the UNESCO Prize for Peace Education.

While Buddhadāsa's fame rests largely on his innovative, creative interpretation of the *dhamma*, Phra Prayudh's teachings are more systematic in nature and more consistently grounded in Pali texts and Theravāda historical traditions. These differences reflect, in part, their distinctive career patterns. Whereas Buddhadāsa built a monastic life-style essentially outside the normal structures and regimes of the Thai *sangha*, Phra Prayudh has chosen to work within them as educator and scholar. Perhaps even more importantly, he wrote *Buddhadhamma* as an objective presentation of the teachings of the Buddha free from subjective bias.[24] Buddhadāsa's teachings, in contrast, are grounded in certain fundamental themes—non-attachment, not-self, interdependent co-arising—which he orchestrates around various contextual issues with little concern for textual or "objective" historical reference. Buddhadāsa does not ignore the Pali canon, especially the *sutta*s; however, scriptural references are not definitive for his philosophical musings.

Buddhadāsa and Phra Prayudh use the resources of both Pali text and tradition to address environmental problems, but they do so employing distinctive hermeneutical techniques which reflect their differing histories, backgrounds, and relationships to the Thai *sangha*. In his recent monograph *Khon Thai Kap Pā* (Thais and the forest), Phra Prayudh delineates several doctrinal principles relevant to a Buddhist environmental ethic. Although these principles resonate with Buddhadāsa's interpretation, Phra Prayudh's hermeneutical strategy differs from Buddhadāsa's in several ways, in particular by extensive references to Pali texts, a topical use of Pali terms rather than Thai, and a more systematic organization and development. In other words, Phra Prayudh's writings, including those about the environment, reflect the concerns of a textual scholar and a systematically organized writer. Buddhadāsa, by contrast, is primarily a philosopher oriented more to an oral rather than a written medium.[25]

Phra Prayudh organizes *Thais and the Forest* around three chronological perspectives: past, present, and future. In regard to the present, he attributes environmental destruction to a Western worldview flawed by three erroneous beliefs: that humankind is separated from nature, that human beings are masters of nature, and that happiness results from the acquisition of material goods.[26] In his essay prepared for the 1993 World Parliament of Religions, Phra Prayudh develops the same position but from a more general, less polemical perspective. He identifies the three erroneous beliefs as wrong attitudes toward nature, fellow human beings, and personal life objective.[27] All three constitute a wrong view (*micchadiṭṭhi*) that must be transformed if environmentally destructive attitudes and actions are to be curbed. Phra Prayudh holds the conventional Theravāda position that right views lead to right action.[28] In agreement with Buddhadāsa and other environmental philosophers, he argues that until the right view prevails and human beings are seen as part of nature, the worldwide trend toward environmental devastation will continue unchecked.

In contrast to Buddhadāsa's dhammic biocentrism grounded in the identification of nature and *dhamma*, Phra Prayudh stresses the centrality of Buddhist ethical values for an environmental philosophy. He emphasizes three Buddhist moral values that promote a positive, beneficial attitude toward the environment, including plants, animals, and fellow human beings: *kataññū* (gratitude), *mettā* (loving-kindness), and *sukha* (happiness). His discussion of gratitude begins with a passage from the *Khuddaka Nikāya* (Collection of minor dialogues): "A person who sits or sleeps in the shade of a tree should not cut off a tree branch. One who injures such a friend is evil."[29] Phra Prayudh observes:

> This maxim reminds us that the shade of a tree we enjoy is enjoyed by others as well. A tree is like a friend which we have no reason to injure. To injure a tree is like hurting a friend. Such a virtuous inner attitude toward nature will prevent us from destructive behavior, on the one hand, and will prompt helpful actions, on the other.[30]

Phra Prayudh links together the moral values of gratitude and loving-kindness (*mettā*). The latter arises from the recognition that according to the law of nature (Thai, *kotthamachāt*) humans and all other sentient beings are bound together in a universal process of birth, old age, suffering, and death. This sense of mutuality, Phra Prayudh argues, promotes cooperative and helpful feelings and actions toward everything around us rather than competitive and hostile ones.[31] He suggests that the recognition of a common enemy, the King of Death (*maccurāja*) or Māra, serves to engender *mettā*. From this recognition he draws the causally framed ecological lesson that "Our use of plants and animals must be thought out carefully and rationally and not carelessly without contemplating the consequences of our actions,"[32] the implication being that with right understanding we will not willfully add to the balance of suffering in the natural and human world. In contrast to Buddhadāsa's more intuitive, ontologically oriented perspective, Phra Prayudh's approach to the environment is seen as rational and ethical. He emphasizes the karmic side of the mutual interdependence of all life-forms, noting that we need to weigh carefully the *consequences* of our actions so that we do not willingly increase the suffering of sentient and nonsentient beings.

For the third ecologically relevant moral value, Phra Prayudh looks to the Buddhist teaching that human happiness (*sukha*) is dependent on our natural surroundings in two ways: 1) simply living within a natural setting engenders a greater sense of happiness and well-being; and 2) nature serves as a teacher of both mind and spirit. Nature trains us not only in moral virtue but also in mental concentration and attentiveness. He argues that for this reason the forest was the context in which Buddhism arose. Monks pursued their vocation in the forest. The forest is the ideal location for training the body and mind to overcome defilements (*kilesa*) that hinder the attainment of mental freedom.[33] Here again Phra Prayudh's approach to nature, that is, to the forest, contrasts with Buddhadāsa's. Wild nature—the forest, mountains, caves—is the best context in which to overcome the defilements that hinder the attainment of *nibbāna*. This view is more anthropocentric and instrumentalist than Buddhadāsa's view of the intrinsic dhammic value of nature.

Phra Prayudh's ecological hermeneutic focuses on the life of the Buddha and the *sangha* as exemplifications of the Buddhist attitude toward nature, in particular toward the forest: "The history of Buddhism as found in various Pali texts clearly indicates that monks saw the forest as a place to practice the *dhamma* and to achieve a feeling of well-being, a happy state of mind, and eventually higher states of mental consciousness."[34] Specifically in regard to the life of the Buddha, Phra Prayudh, in concert with other Thai voices of "green Buddhism," such as Chatsumarn Kabilsingh,[35] observes:

> From the time the Buddha left his palace Buddhism has been associated with forests. The Buddha's quest for the truth (*saccadhamma*) took place in the forest. It was in the forest that for six years he sought to overcome suffering and it was under the

Bodhi tree that he attained enlightenment. Throughout his life the Lord Buddha was involved with forests, from his birth in the forest garden of Lumbini under the shade of a Sāl tree to his *parinibbāna* under the same kind of tree. Thus, Buddhism has been associated with the forest from the time of the life of its founder.[36]

Beyond general references to the example of the Buddha and the early *saṅgha*, however, Phra Prayudh cites specific passages from the Pali *suttas* to justify his views. For example, he notes that the Buddha spoke of nature as the best environment in which to seek enlightenment (*bodhiñāṇa*): "O monks, in search of the good (*kusala*), the best place is a rural area such as Uruvelā. There you will find a refreshing environment of trees and fields, a cool flowing river, pleasant landings with homes to go for alms (*gocaragāma*). Such delightful surroundings are suitable for monks to pursue their religious practice."[37] Phra Prayudh also cites stories of forest-dwelling disciples of the Buddha, such as Vanavaccha Thera, Citta Thera, and Cūla Thera, who praised mountains, birds, and insects as well as forests. He also mentions the Venerable Mahākassapa, who advised monks to dwell in caves and mountains situated in beautiful natural surroundings with forests, animals, and birds.[38]

Phra Prayudh grounds his argument for the value of nature for religious practice in stories of the Buddha and the early disciples found in Pali texts. Buddhadāsa also links nature and religious practice to spiritual realization but does so by using Suan Mōkh as his primary illustration rather than citing specific passages in canon and commentary. Phra Prayudh, furthermore, makes a strong appeal to reason. Unlike some Thai Buddhist environmentalists who encourage such practices as ordaining trees or the promotion of a tree deity cult to preserve a stand of trees, Phra Prayudh believes that modern Buddhists need to go beyond appealing to Buddhist values, such as gratitude and loving-kindness, and citing scripturally grounded stories of the Buddha and the early *saṅgha* and should utilize scientific evidence to address global problems, such as pollution and environmental preservation.

Phra Prayudh's response to the case of Phra Prajak Kuttajitto, a much publicized activist monk from Buriram Province in northeast Thailand, is instructive. Phra Prajak, who has returned to lay life, was twice arrested in 1991 for his efforts in forest conservation, first, for trespassing on National Forest Reserve land and establishing a meditation center there and, second, for organizing villagers in Korat Province. In both cases, he led a protest opposing the government's program to remove villagers from National Forest Reserves. Phra Prajak questioned the legality of the removal of villagers from the lands and also objected to the proposed replacement of natural, diversified forests with trees, principally eucalyptus, grown for commercial purposes.

In response to Phra Prajak's controversial activities Phra Prayudh delivered a talk on 2 October 1991, later printed under the title *Phra Kap Pā: Mī Panhā Arai?* (Monks and the forest: Is there a problem?). He began his remarks with the comment that he did not intend to speak to the Phra Prajak case per se, in particular whether or not he had acted correctly or had broken the law. Rather, his concern was for the possible detrimental impact on Thai Buddhism:

> We need to look at the case from the Buddhist perspective. For example, there's a rumor that the government may enact a law forbidding monks to enter forests. I don't know if this is true or false, but if such a law were to be enacted then we would need to examine it carefully from the perspective of Buddhism, especially the relationship between the *sangha* and the forest. If we understand the principles of this relationship then we'll act appropriately.[39]

Rather than taking sides on this politically sensitive issue, Phra Prayudh advocates a rational approach grounded in the texts and traditions of Theravāda Buddhism.

After observing that the Buddha's birth, enlightenment, and death all took place under trees, Phra Prayudh notes that many of the major monasteries donated to the *sangha* were in forest groves: Veluvana (donated by Bimbisāra), Jetavana (donated by Lord Jetam), Jīvakamphavana (given by the physician Jīvaka), and many others, such as the Mahāvana monastery where the Buddha resided when he visited Kapilavattu, the capital of the Sakyas. Although the Buddha advised monks to dwell in forests—"O, Ānanda, when a *bhikkhu* enters the Order he should be encouraged to practice the *dhamma*, to follow the *pāṭimokkha*, to limit conversation, and to live in a tranquil place, *if possible a forest*"[40]—and extolled the forest as a good environment to practice the *dhamma*, Phra Prayudh argues against a naïve, simplistic identification of Buddhism with nature. The *principle* behind the Buddha's advocacy of a forest as a monastic retreat was its appropriateness as a place for the pursuit of monastic training, not that forest dwelling was a necessary and sufficient condition of the monastic life. On the contrary, because a monk's responsibility extends not only to the pursuit of enlightenment but also to other members of the *sangha* and to lay society, the Buddha stipulated that monasteries were to be located not too far from or too near a town. This is the second principle that needs to be kept in mind. The monastery "should be a quiet place, appropriately isolated, not disorderly and noisy. Too close a proximity to a town tends to make a monastery too busy and noisy but being too far away may jeopardize the work of the monks."[41]

Monks have a responsibility toward one another. They are required to assemble twice monthly for formal business meetings (*sanghakamma*). Furthermore, monks are forbidden by *vinaya* rules to support themselves. Because monks depend on the laity for food, they cannot live in isolation from society. The first of these rules joins monks or nuns together as a community; the second links them to laypeople. Therefore, even though the Buddha praised forest dwelling, this did not suggest following the withdrawn, isolated life of an ascetic. Indeed, one finds in early Buddhism ambivalent feelings toward forest-dwelling ascetics, as suggested by the following fivefold classification of *dhutanga* monks: those who are thickheaded and stupid, those who seek fame and praise, those who are deranged, those who follow the praiseworthy example of the Buddha, and those who seek solitude and quiet in order to practice the *dhamma*.[42] Thus, although Phra Prayudh notes the importance of the forest in the experience of the Buddha and the early *sangha* as the best environment in which to pursue spiritual practice, he also suggests that early Buddhism considered the forest with some misgivings. Furthermore, he suggests that wild nature at a far remove from human habitation is problematic for monastic practice because monks are dependent upon the laity for food and other material necessities.

Phra Prayudh bases his ecological hermeneutic on a close reading of the life of the Buddha and the early *sangha* in the Pali scriptures and the primary intentionality of the *dhamma* to overcome suffering and realize personal liberation. He finds within the Buddhist worldview of mutual cooperation an alternative to Western dualism and materialism, which he holds responsible for many forms of global exploitation. Phra Prayudh, however, does not construct a theory of Buddhist *gaia* or biocentric ecology, nor does he identify nature and *dhamma* in the manner of Buddhadāsa or paint a romantic portrait of the Buddha and his disciples holding forth in shaded glens. He warns:

> The Buddha shouldn't be revered because he lived near trees or because he taught that one should eat only enough food to get by for one day. Rather, he should be respected as one who realized the *dhamma* and then taught it. The Buddha advocated a life of simplicity and sufficiency not as an end in itself but as the context for the development of knowledge of the cause and effect of all actions. The Buddha praised monks who lived in the forest such as Mahākassapa. . .[but he] said that whether or not one lived in the forest was a matter of individual intent.[43]

Buddhadāsa Bhikkhu and Phra Prayudh represent two distinctive, complementary approaches to the environment within the context of contemporary Thai Buddhism. Buddhadāsa's intuitive, ontologically oriented view of nature as *dhamma* and the ethic of caring-for-nature (*anurak thamachāt*) that flows from it finds a greater commonality with what Ian Harris terms "ecoBuddhism" than does the ethical approach of Phra Prayudh, which is grounded primarily in reason, texts, and historical tradition. Buddhism—as well as the other great world religions—is complex, variegated, and dynamic and defies general, facile characterizations. As these two examples from Thai Buddhism illustrate, even within a single contemporary cultural tradition there is no univocal Buddhist ecological hermeneutic.

Counterpoint:
Buddhist Environmentalism—Critics in the Forest

The effort of Buddhists and students of Buddhism to construct a Buddhist environmental ethic has encountered several disclaimers. Among the strongest critics of the ecoBuddhism project are Noriaki Hakamaya, Lambert Schmithausen, and Ian Harris.[44] This brief postscript cannot examine these criticisms in depth; rather, it is intended only to suggest the nature of this critical assessment in the light of this study of Buddhadāsa and Dhammapiṭaka.

In the view of Ian Harris, recent writings in the area of Buddhism and environmental ethics can be divided into four broad categories: 1) a full endorsement of Buddhist environmental ethics by traditional guardians of doxic truth, for example, His Holiness, the Dalai Lama; 2) a similar literature by Japanese and North American scholar-activists that seeks to identify the doctrinal bases for an environmental ethic, represented by Joanna Macy; 3) critical studies which nonetheless argue for an authentic Buddhist response to environmental problems, such as those by Lambert Schmithausen; and 4) an outright rejection of the possibility of Buddhist environmental ethics on the grounds of its otherworldliness, as put forth by Noriaki Hakamaya.[45] Harris identifies himself with the fourth position, although he admits that he is more sympathetic toward the third. This makes him a particularly strong critic of what he terms ecoBuddhism and also causes him to be suspicious of attempts to ground Buddhist environmental ethics in classical doctrines such as causality. Harris develops his critique in a series of articles published in *Religion* and the new electronic *Journal of Buddhist Ethics*. It is not my intent to give Harris's analysis the attention it deserves but rather to suggest the direction of his interpretation.

In his initial foray into this field, Harris established the critical stance he has continued to develop in subsequent articles. In contrast to the "ecospirituality," "ecojustice," and "ecotraditionalists" he cites,[46] Harris argues that the primacy of the spiritual quest in the Buddhist tradition privileges humans over the realms of animals and of nature. He points out, for example, that although the interconnected destinies of human beings and animals might suggest that humans should feel some solidarity with animals, in fact animals are regarded as particularly unfortunate. They cannot grow in the *dhamma* and *vinaya* nor can they be ordained as monks.[47] Furthermore, while animals may appear to be beings destined for final enlightenment, they have no intrinsic value in their animal form. Indeed, claims Harris, "The texts leave one with the impression that the animal kingdom was viewed. . .with a mixture of fear and bewilderment."[48] The plant world does not fare much better in Harris's analysis. He summarizes the canonical view of nature as being either something to be improved or cultivated or something to be confronted in a therapeutic encounter.[49]

In his study of ecoBuddhism as a contemporary American attempt to articulate an authentically Buddhist response to present environmental problems, he argues that this movement represents a teleological transformation of traditional Buddhist cosmogony.[50] In an earlier article which surveys Pali, Sarvāstivāda, Sautrāntika, Mādhyamika, and Yogācāra positions, Harris focuses his critique even more substantially on what he characterizes as the teleological transformation of Buddhist causality. There he argues, first, that a Buddhist action guide in regard to the natural world should be "specifically authorized by the Buddha," and, second, that the dysteleological nature of Buddhist thought does not lend itself to an environmental ethic in regard to such broadly contested issues as global warming or biodiversity.[51]

For the purposes of this essay, Harris's view of the problematic of a Buddhist environmental ethic serves primarily as a counterpoint to the views of Buddhadāsa and Phra Prayudh and to the general tenor of the essays on Buddhism and ecology in this volume. Although the ecological hermeneutics of Buddhadāsa and Phra Prayudh differ in some significant respects, both are at odds with Harris's critique of Buddhist eco-apologetics. Buddhadāsa and Phra Prayudh would, I believe, object to Harris's view on at least three general grounds: 1) His position is founded on too narrow a construction of the Buddhist view of nature and animals based on a selective reading of particular texts and traditions. Harris might have nuanced his claims about the Buddhist attitude toward animals had he included an analysis of selected Jātaka narratives, for example. 2) It is debatable whether or not a theory of causality (or conditionality) must be teleological in order to be environmentally viable. For instance, Buddhadāsa's biocentric ontology can be interpreted deontologically, or, as Buddhadāsa phrases it, nature implies certain moral maxims or duties. 3) Although the *buddhavacanam* is authoritative in the Theravāda tradition, moral action guides do not need to be authorized by the Buddha in a literal sense.

Although Phra Prayudh seems to agree with Harris that the primary positive view of nature in Buddhism is a context for spiritual development, that is, primarily for its therapeutic value,

Buddhadāsa's more biocentric perspective goes beyond such an instrumental understanding of nature as the ideal context for the pursuit of the ultimate goal of human flourishing. For Buddhadāsa nature has an inherent, dhammic value, not one merely instrumental to the monastic pursuit of spiritual transformation. In reacting against what he understands to be a well-intended but problematical interpretation of Buddhist thought by eco-apologists, Harris's normative standard of Buddhist orthodoxy judges Buddhadāsa's ecological hermeneutic to be inauthentically Buddhist or merely "accorded authenticity" by virtue of the fact that Buddhadāsa is a "high profile Buddhist" associated with "reformist circles" in Thai Buddhism.[52]

Harris's critical typology of Buddhist environmental ethics would evaluate Phra Prayudh's ecological hermeneutic more favorably than Buddhadāsa's because Phra Prayudh adheres more closely to Theravāda doctrinal orthodoxy. Phra Prayudh's position would be closest to Harris's type three, namely, an environmental ethic based on a critical reading of the tradition by a Buddhist monk. Buddhadāsa, in Harris's assessment, would be included in type one as an ecoBuddhist apologist of doxic truth. Buddhadāsa would probably not object to being associated with the Dalai Lama as a type one ecoBuddhist, although it is doubtful that he would consider himself to be a guardian of doxic Theravāda truth.

Notes

1. Dhammapiṭaka is the ecclesiastical title conferred in 1993 on Phra Prayudh Payutto, whose previous titles were Sivisuddhimoli, Rājavaramunī, and Debvedī. Published works by Phra Prayudh Payutto appear under all of these names. Here I use Dhammapiṭaka in the article title but in the text I use Phra Prayudh, following the convention established by Grant A. Olson in his translation of *Buddhadhamma*.

2. For introductions to the thought of Buddhadāsa Bhikkhu and Phra Prayudh Payutto, see Buddhadāsa Bhikkhu, *Me and Mine: Selected Essays of Bhikkhu Buddhadāsa*, ed. and with an introduction by Donald K. Swearer (Albany: State University of New York Press, 1989); and Phra Prayudh Payutto, *Buddhadhamma: Natural Laws and Values for Life*, trans. and with an introduction by Grant A. Olson (Albany: State University of New York Press, 1995). See also Grant A. Olson, "From Buddhadasa Bhikkhu to Phra Debvedi: Two Monks of Wisdom," in *Radical Conservatism: Buddhism in the Contemporary World* (Bangkok: Sathirakoses-Nagapradipa Foundation, 1990); and Santikaro Bhikkhu, "Buddhadasa Bhikkhu: Life and Society through the Natural Eyes of Voidness," in *Engaged Buddhism: Buddhist Liberation Movements in Asia*, ed. Christopher S. Queen and Sallie B. King (Albany: State University of New York Press, 1996). Essays in Thai consulted for this essay include Buddhadāsa, *Buddhasāsanik Kap Kān Anurak Thamachāt* (Buddhists and the care of nature) (Bangkok: Kōmol Khīmthọng Foundation, 1990); Buddhadāsa, *Siang Takọn Jāk Thamachāt* (Shouts from nature) (Bangkok: Sublime Life Mission, 1971); Debvedī (Phra Prayudh Payutto), *Phra Kap Pā: Mī Panhā Arai?* (Monks and the forest: Is there a problem?) (Bangkok: Vanāphidak Project, 1992); Dhammapiṭaka (Phra Prayudh Payutto), *Khon Thai Kap Pā* (Thais and the forest) (Bangkok: Association for Agriculture and Biology, 1994). For general essays in English on Thai culture and the natural environment, see *Culture and Environment in Thailand: A Symposium of the Siam Society* (Bangkok: Siam Society, 1989), *Man and Nature: A Cross-Cultural Perspective* (Bangkok: Chulalongkorn University Press, 1993). Transliteration of Thai terms follows the Library of Congress with some modifications, in particular "j" rather than "čh."

3. Ian Harris, "How Environmentalist Is Buddhism?" *Religion* 21 (April 1991):101–14; Harris, "Buddhist Environmental Ethics and Detraditionalization: The Case of EcoBuddhism," *Religion* 25, no. 3 (July 1995):199–211; Harris, "Causation and 'Telos': The Problem of Buddhist Environmental Ethics," *Journal of Buddhist Ethics* 1 (1994):45–57; Harris, "Getting to Grips with Buddhist Environmentalism: A Provisional Typology," *Journal of Buddhist Ethics* 2 (1995):173–90; and Harris's contribution in this volume.

4. "EGAT Warns of Low Water Level in Dams," *Bangkok Post*, Monday, 13 November 1989, pp. 1 and 3.

5. For example, see Francesca Bray, "Agriculture for Developing Nations," *Scientific American*, July 1994, 30–37; D. Pimentel et al., "Benefits and Risks of Genetic Engineering in Agriculture," *Bioscience* 39, no. 10 (1989):606–14.

6. Paraphrased from a lecture delivered at the McGilvary Theological Faculty of Payap University on 27 October 1989, entitled "Quam Khawjai Kiewkap Sangkhom Thai: Khabuankan Chai Amnāt lae Kanyaek Chīwit Ok Pen Suan" (Understanding Thai society: Violence and alienation).

7. It is interesting to observe that the first issue of *Generation* (October 1989), an expensive, elitist magazine, contained a lead article, "Namtatthakhot: Anicca Buddhasāsana nai Muang Thai" (The Buddha's tears: The decline of Buddhism in Thailand), 39–55. In the article some of the more important voices for reform of the Thai *saṅgha* and Thai society are mentioned, including Buddhadāsa Bhikkhu and Sulak Sivaraksa.

8. Buddhadāsa died 3 July 1993.

9. Buddhadāsa, *Siang Takǫn Jāk Thamachāt*. For essays on the relationship between ecoBuddhism and deep ecology, see *Dharma Gaia: A Harvest of Essays in Buddhism and Ecology*, ed. Allan Hunt Badiner (Berkeley: Parallax Press, 1990).

10. Buddhadāsa, *Siang Takǫn Jāk Thamachāt*, 5–7; translation mine.

11. Ibid., 7.

12. For an ethical critique of biocentrism, see Luc Ferry, *The New Ecological Order*, trans. Carol Volk (Chicago and London: University of Chicago Press, 1995).

13. Selections of my discussion of *Buddhasāsanik Kap Kān Anurak Thamachāt* appeared in "Buddhadāsa on Caring for Nature," *Seeds of Peace* 10, no. 2 (September-December 1994):36–38.

14. Western students of Buddhism often translate *anukampā* as "sympathy." In my view "empathy" is a more apt translation. I have in mind the image or metaphor of a tuning fork that resonates empathetically with its environment. See Harvey B. Aronson, *Love and Sympathy in Theravāda Buddhism* (Delhi: Motilal Banarsidass, 1980).

15. One of nine booklets published by the Dhamma Saphā, a group formed to disseminate Buddhadāsa's teaching, is *Kan Tham Lāi Khwām Hen Kae Tua* (Rooting out selfishness) (Bangkok: Dhamma Saphā, n.d.).

16. Buddhadāsa Bhikkhu, *Heartwood from the Bo Tree* (Bangkok: United States Overseas Mission Foundation, 1985), 13. Those who criticize Buddhadāsa for being a modernist, eclectic thinker should keep in mind that he never relinquished the centrality of the concept of non-attachment. While this notion is certainly pan-Buddhist and figures prominently in the ethical emphasis of modern Buddhist apologists, the concept of non-attachment is also fundamental to classical Theravāda *sīla-dhamma*.

17. For example, see Nel Noddings, *Caring: A Feminine Approach to Ethics and Moral Education* (Berkeley: University of California Press, 1984).

18. Buddhadāsa frequently used this phrase in his talks. See, for example, *Buddhasāsanik Kap Kān Anurak Thamachāt*, 34.

19. Buddhadāsa, *Buddhasāsanik Kap Kān Anurak Thamachāt*, 34–35.

20. Ibid., 35; translation mine. The similarity between Buddhadāsa's vision and comparable ecological visions in other religious traditions is striking. For example, see Ernesto Cardenal, "To Live Is to Love," in *Silent Fire: An Invitation to Western Mysticism*, ed. Walter Holden Capps and Wendy M. Wright (New York: Harper and Row, 1978).

21. Buddhadāsa, *Buddhasāsanik Kap Kān Anurak Thamachāt*, 15–16. I have given a free rendering of the Thai in order to convey my understanding of Buddhadāsa's meaning.

22. Ibid., 12–13.

23. Grant A. Olson translated the first edition (*Phutatham: Kotthamachāt læ Kham Samrup Chīwit* [Buddhadhamma: Natural laws and values for life] [Bangkok: Samnakphim Sukhaphāp, 1971]). The second edition is being translated in Thailand by Bruce G. Evans. Currently, Phra Prayudh's English monographs include a wide range of topics, e.g., *Thai Buddhism in the Buddhist World* (Bangkok: Amarin, 1984); *Looking to America to Solve Thailand's Problems* (Bangkok: Sathirakoses-Nagapradipa Foundation, 1987); *Toward a Sustainable Science* (Bangkok: Buddhadhamma Foundation, 1993); *Good, Evil, and Beyond: Kamma in the Buddha's Teaching* (Bangkok: Buddhadhamma Foundation, 1993); *A Buddhist Solution for the Twenty-First Century*, 2nd ed. (Bangkok: Sahathammik, 1993); *Buddhist Economics: A Middle Way for the Market Place*, 2nd ed. (Bangkok: Buddhadhamma Foundation, 1994).

24. Olson, introduction to Phra Prayudh Payutto, *Buddhadhamma*, 26–27.

25. This distinction between written and oral/aural mediums should not be drawn too sharply. Phra Prayudh gives many lectures; however, in contrast to Buddhadāsa, whose fame stems largely from his transcribed, published talks, Prayudh continues to be more oriented to the written word and is steeped in Pali canon and commentary.

26. Dhammapiṭaka (Phra Prayudh Payutto), *Khon Thai Kap Pā*, especially 43–68.

27. Phra Debvedī (Phra Prayudh Payutto), *A Buddhist Solution for the Twenty-First Century*, 7.

28. See Phra Prayudh Payutto, *Buddhadhamma*, pt. 2. This claim does not address the philosophical debate within Buddhism between those who argue for "no view" over "right view."

29. Dhammapiṭaka (Phra Prayudh Payutto), *Khon Thai Kap Pā*, 22; translation mine.

30. Ibid., 22–23.

31. Ibid., 24.

32. Ibid.; translation mine.

33. Ibid., 26.

34. Ibid., 27; translation mine.

35. For example, see Chatsumarn Kabilsingh, "Buddhist Monks and Forest Conservation," in *Radical Conservatism: Buddhism in the Contemporary World* (Bangkok: Sathirakoses-Nagapradipa Foundation, 1990), 301–11.

36. Debvedī (Phra Prayudh Payutto), *Phra Kap Pā*, 4; translation mine.

37. Dhammapiṭaka (Phra Prayudh Payutto), *Khon Thai Kap Pā*, 28; translation mine.

38. Ibid., 29–33.

39. Debvedī (Phra Prayudh Payutto), *Phra Kap Pā*, 3.

40. Ibid., 10; translation and italics mine.

41. Ibid., 11; translation mine.

42. Ibid., 15.

43. Ibid., 17.

44. Lambert Schmithausen, *Buddhism and Nature*, Studia Philologica Buddhica, Occasional Paper Series 7 (Tokyo: International Institute for Buddhist Studies, 1990); *The Problem of the Sentience of Plants*, Studia Philologica Buddhica, Occasional Paper Series 8 (Tokyo: International Institute for Buddhist Studies, 1991); and "The Early Buddhist Tradition and Ecological Ethics," *Journal of Buddhist Ethics* 4 (1997):1–42; and Noriaki Hakamaya, "Shizen-hihan to-shite no Bukkyo" (Buddhism as a criticism of physis/natura), *Komazawa Daigaku Bukkhogakubu Ronshu* (1990):380–403.

45. Harris, "Getting to Grips with Buddhist Environmentalism," 177. I have omitted Harris's category of engaged Buddhist activists. Doctrinally, they can be linked to his first type.

46. Ibid. In one sense Harris's typology represents forms of what he labels "eco-apologetics."

47. Harris, "How Environmentalist Is Buddhism?" 105.

48. Ibid., 107.

49. Ibid., 108.

50. Harris, "Buddhist Environmental Ethics and Detraditionalization." Harris focuses his critique on the transformation of the theory of causality in "Causation and 'Telos.'"

51. Harris, "Causation and 'Telos,'" 54.

52. Harris, "Getting to Grips with Buddhist Environmentalism," 177.

A Theoretical Analysis of the Potential Contribution of the Monastic Community in Promoting a Green Society in Thailand[1]

Leslie E. Sponsel and Poranee Natadecha-Sponsel

Multiple Interconnected Crises

In recent decades, Thailand has increasingly become an environmental disaster, largely as a result of the nearly wholesale acceptance of Westernization, including industrialism, urbanism, materialism, and consumerism.[2] As Dhira Phantumvanit and Khunying Suthawan Sathirathai observe: "For several decades, Thailand has indulged in the abundance of its natural resources without considering their long-term sustainability. As a result there are now ample signs of ecological stresses facing the nation."[3] For example, one symptom of the growing environmental crisis is deforestation; prior to World War II up to 75 percent of Thailand was still forested, whereas today less than 15 percent remains forested—and the latter figure is even an optimistic estimate.[4]

In the benchmark 1989 Siam Society symposium volume *Culture and Environment in Thailand*, a common underlying theme was the connection between the environmental crisis and the decline of adherence to Buddhism. In the concluding chapter, which summarizes the symposium, anthropologist Peter Kunstadter[5] records that the participants (most of whom were Thai) subscribed to a "theory of a moral collapse" as the cause of the growing ecological disequilibrium in Thailand (see figure 1).[6] In another context, Lily de Silva even goes so far as to view the environmental crisis as including the pollution, through Westernization, of mind and culture as well as of the environment.[7]

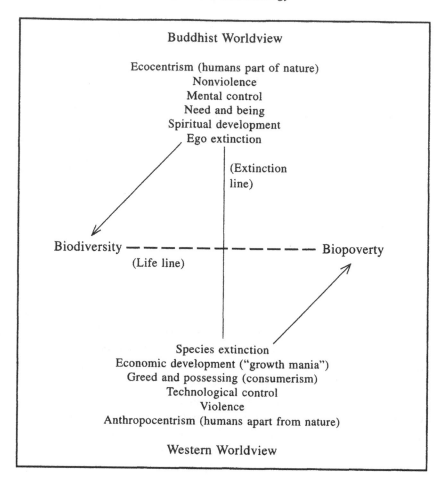

FIGURE 1
From Leslie E. Sponsel and Poranee Natadecha-Sponsel, "The Relevance
of Buddhism for the Development of an Environmental Ethic for the
Conservation of Biodiversity," in *Ethics, Religion, and Biodiversity:
Relations between Conservation and Cultural Values*, ed. Lawrence S.
Hamilton (Cambridge: White Horse Press, 1993), 87.

In previous publications we explored the relevance of the *dharma* (teaching of the Buddha) for resolving the environmental crisis in Thailand and developing a more ecologically appropriate (green) society, especially in relation to forests and deforestation. We noted that Buddhism is particularly relevant for coping with the environmental crisis in Thailand for four principal reasons: 1) About 95 percent of Thai people are Theravāda Buddhists, and Buddhism and culture in Thailand are intimately interconnected. 2) Some of the basic principles of Buddhism parallel those of ecology, although they are not identical. 3) Some of the fundamental principles of Buddhism can provide the basis for the construction of a green environmental philosophy and ethics. 4) Buddhism has a long history of mutualistic relationships with the forest, as illustrated by the lives of the Buddha and forest monks. Forests are optimum contexts for meditation, and they are needed by monks who choose to go to them for a period of ascetic practice (*dhutaṇga*). Indeed, deforestation, one of the most serious environmental problems in Thailand, is sacrilegious for Buddhism.[8] Furthermore, as Achan Pongsak Techathamamoo, a Thai monk who is an environmental activist (*phra nak anurak pa*), says:

> Dharma, the Buddhist word for truth and the teachings, is also the word for nature. That is because they are the same. Nature is the manifestation of truth and of the teachings. When we destroy nature we destroy the truth and the teachings. When we protect nature, we protect the truth and the teachings.[9]

Bhikkhu Bodhi nicely summarizes the relevance of the *dharma* for the development of an environmental ethic:

> With its philosophic insight into the interconnectedness and thoroughgoing interdependence of all conditioned things, with its thesis that happiness is to be found through the restraint of desire, with its goal of enlightenment through renunciation and contemplation and its ethic of non-injury and boundless loving-kindness for all beings, Buddhism provides all the essential elements for a relationship to the natural world characterized by respect, care and compassion.[10]

As a specific illustration of an environmentally friendly aspect of Buddhism,[11] consider the original reason for the rainy season retreat (*pansa* or *vassa*), a period when "nomadic" monks who usually wander the forests and countryside are largely confined to the temple (*wat*). The Buddha established this custom because the rainy season is a period when new life, including young crops, abounds. He wanted to minimize the destruction of life by the trampling feet of wandering monks.[12]

Within the context of the ecocrisis in Thailand and the ecological wisdom of Buddhism in principle, we explore *in theory* the *potential* of the local monastic community to contribute to the resolution or reduction of environmental and related problems. Specifically, we advance four propositions: 1) Ideally the monastic community exhibits attributes which are similar, and in some instances even identical, to many of the characteristics of a green society. 2) There is a tremendous contrast between these ideal principles and the behavior of lay communities and society as a whole in Thailand. 3) The monastic community has extraordinary status and power to transform Thailand into a more ecologically appropriate society by virtue of its antistructural and liminal social and moral roles. 4) By drawing on the ecological wisdom of the *dharma*, the local monastic communities have significant potential to contribute to the environmental education of the populace and thereby to help create a greener society.[13]

The Monastic Community as a Green Society

It is remarkable that three disparate and independent sources—political scientist Andrew Dobson, anthropologist John Bennett, and philosopher Philip Drengson[14]—nearly coincide in their prescriptions for an ecologically appropriate society, which they label respectively as green, equilibrium, and pernetarian. Our first hypothesis is that the local monastic communities of Thailand have the potential to serve as working models of a green society and that some actually do so. Eight of the more important characteristics which the *ideal* Buddhist monastic community in Thailand shares with these other societies are listed, with a key word for each, in table 1.[15] However, there is a tremendous contrast between these

TABLE 1: ECOLOGICALLY APPROPRIATE ATTRIBUTES
OF AN IDEAL MONASTIC COMMUNITY

1. POPULATION: small and controlled population
2. COMMUNALITY: egalitarian communal life based on mutual respect and cooperation
3. RESOURCES: sufficiency and sustainability by limiting resource consumption to satisfying basic needs and by self-restraint in wants and desires
4. ECONOMY: cooperative rather than competitive economy based on reciprocity and redistribution
5. ENVIRONMENT: limit environmental impact and practice stewardship with nature including the temple and vicinity as sacred space
6. PHILOSOPHY: holistic (systems), organic (ecology), and monistic (unity of humans and nature) worldview based on enhancing quality of life rather than accumulating quantity of material things (being rather than having)
7. VALUES: reverence (inherent worth), compassion or loving-kindness (*mettā*), and nonviolence (*ahiṃsā*) toward all life to promote harmony within society and between society and nature
8. SELF: "deep self" including self-examination, self-realization, self-fulfillment, and self-spirituality through meditation and eventually extinction of self (*anattā*)

TABLE 2: LIMINAL ATTRIBUTES OF AN IDEAL MONASTIC COMMUNITY

- totality
- homogeneity
- communitas
- equality
- anonymity
- absence of property
- uniform clothing
- sexual continence
- minimization of sexual distinction
- absence of rank
- humility
- disregard for personal appearance
- no distinctions of wealth
- unselfishness
- total obedience
- sacredness
- sacred instruction
- suspension of kinship rights and obligations
- continuous reference to mystical powers
- simplicity
- acceptance of pain and suffering

ideal principles for an ecologically appropriate society and the usual
practices of the lay communities and society as a whole in Thailand.
From this point of view, the monastic community can be considered
antistructural; that is, it stands in opposition to the structure of the
larger society, a point to be developed shortly. However, it should
be noted that the *saṅgha* (Pali, *sangha*; monastic community as a
whole), like Buddhism in general in Thailand, is far from uniform;
thus some individuals and communities are much closer to a green
society than others.[16]

We are not the first to make this observation; it is noteworthy
that the antistructural role of the monastic community has also been
recognized, albeit in other terms, by a Thai Buddhist monk, Phra
Phaisan Visalo:

> The lives of forest monks and the pattern of relationships in the
> Sangha convey to people in the larger society certain "messages,"
> some of which deny or resist the prevailing values. Such messages
> point to the true value of life, indicating that development of
> inwardness is much more important than wealth and power, that the
> life of tranquility and material simplicity is more rewarding and
> fulfilling. Such messages provide both hints and warnings which
> enable people to stop and reflect upon their lives, leading them to
> seek themselves rather than material gain and glory. Such messages
> are especially revolutionary for a society blindly obsessed by
> impoverished values. To have forest monasteries amidst, or, to put
> it more correctly, elevated above the laysociety, is to have *com-
> munities of resistance* that, by their nature and very existence,
> *question the validity of popular values*.
>
> These were the values and functions of forest monasteries in
> traditional Thai society. Nowadays these values and functions still
> exist; indeed, they have become *more important than ever*, because
> modern Thai society is increasingly influenced by degraded values
> and obsessed with material growth [emphasis added].[17]

The Monastic Community as Indefinite Liminality

We hypothesize that the monastic community has extraordinary
status and power to help transform Thailand into a more ecologically
appropriate society by virtue of its antistructural and liminal social

and moral roles. Here we must briefly digress to explain some anthropological theory.

Arnold van Gennep[18] recognized that as individuals enter a new status in any human society they go through a rite of passage. Furthermore, he observed that in every culture these rites have three distinct stages: separation, marginality (liminality), and reincorporation. In other words, the individual who is undergoing a ritual transition which formally changes his or her status is first separated from the community, then exists in an extraordinary, ambiguous, and even dangerous state, and then is reintegrated into the community with a new status and role. In the new status the individual possesses not only new rights but also new obligations, and he or she must act according to new norms as well. This transition is a socially significant event; indeed, frequently the liminal stage is even likened to death.[19]

Victor Turner elaborated on the ritual and symbolism of liminality. He noted that liminality often involves characteristics which stand in sharp contrast to society, even in opposition—something he refers to as antistructure. Turner also identified the characteristics of liminality in an elaborate list.[20]

Obviously, when an individual enters (*naga*) or leaves the monkhood he undergoes a rite of passage which includes a liminal stage.[21] However, Turner hinted that a monastic community may possess attributes of liminality.[22] Thus, we hypothesize that monkhood itself is in essence an indefinite liminality, which provides its primary source of power. This power in turn has great potential for contributing to the reduction and resolution of some of the environmental and other problems in Thailand. It helps explain the significant influence which environmental-activist monks are having on many people in Thailand.

The liminal attributes of the ideal monastic community are listed in table 2.[23] The monastic community is a totality in the sense that it is a social system in which the members participate with a shared sense and spirit of community (*communitas*). It is based on egalitarian principles with very limited distinctions by rank—the head monk, senior and junior monks, and novices. The extinction of the ego is a major objective; thus, members are anonymous in the sense that individuality is unimportant, as exemplified by the uniform clothing, disregard for personal appearance in the sense of distinguishing oneself physically, and absence of the accumulation of personal property and of distinctions of wealth. Sexual distinction is minimized by the type of clothing worn and the shaving of the hair on the head and eyebrows, while sexual continence is practiced. There is supposed to be total obedience in the form of allegiance to the Buddha, his teachings (*dharma*), and the monastic community (*saṅgha*). The individual's customary kinship rights and obligations are suspended. The teachings, which are considered to be sacred, emphasize humility, simplicity, unselfishness, nonviolence, compassion, and meditation. Pain and suffering are considered to be part of existence but reducible, if unavoidable, by minimizing ignorance, desires, selfishness, and greed. The comportment of monks in these and other matters is guided by no less than 227 rules (*vinaya* or *pāṭimokkha*).[24]

The spatial, sociocultural, religious, demographic, and ecological significance of monastic communities in Thailand is also important to consider here.[25] The temple compound (*wat*) usually includes buildings for worship by laypersons (*vihara*) and by monks (*bot*), meetings (*sala*), and residence (dormitory and/or huts) (*kuti*), as well as a cemetery with stupas (*chedi*), the bell-shaped stone monuments containing the cremated remains of deceased persons. There may also be a school for the lay community in or adjacent to the monastic compound. Commonly there are also trees in the temple yard, and typically these include species associated with the life of the Buddha, such as the bodhi (*Ficus religiosa*), banyan (*Ficus bengalensis*), and asoke (*Saraca indica*). Temples are often surrounded by groves of trees or even forests, which are also usually considered sacred places.[26]

Most temples are within walking distance of a village since the monks are usually dependent upon villagers for their daily food (*pindapata*). Likewise, in traditional Thai society the temple and monks were a pivotal component in the daily lives of most laypeople in the neighboring communities. Traditionally, the focus of the lay community is the local temple, where religious and sociocultural functions are integrated in many ways.[27] Indeed, the temple has been the most important institution beyond the family in the life of people in rural Thailand.[28] The rites of passage of lay individuals (including birth, puberty, marriage, and death) are usually marked by some community recognition through a ceremony at the temple. Rites of intensification, such as the Buddhist new year and "lent," are also marked by community activities at the temple. Lay individuals gain merit (*bun, punna*) by providing food for the bowls the monks (*bhikkhu,* or almsperson) carry on their daily early-morning walk through the community. Merit may also be gained by planting trees and by performing other more mundane activities in the temple yard. The temples have traditionally been the educational centers for children. Thus, the Thai temple is not a monastery in a Western sense of monks secluded from the larger community; rather, in Thailand the monastic and lay communities are interdependent and interact on a daily basis.[29]

In Thailand in 1992 there were about 63,358 villages, 29,002 temples, 288,637 monks, and 123,643 novices.[30] During the Buddhist rainy season retreat (*pansa*), it is customary for individuals to become monks and novices for a temporary period of days, weeks, or longer. Thus, in 1990, for example, approximately 106,500 monks and 26,800 novices were added to the temple population for the rainy season retreat.[31] It is important to realize that the majority of Thai males become novices (between the ages of eight and twenty years) or monks (twenty years and up) for up to three months, usually during the rainy season.[32]

Phra Phaisan Visalo of Sukato Forest Monastery in Chaiyaphum aptly describes the role of the forest monastery:

> Here we can see something of the contribution the forest monastery can render to society, since it is able to preserve the traditional wisdom so badly needed by Thailand and the modern world. This wisdom is not only found in the scriptures or expressed through words. It is manifested in living communities existing in the context of contemporary society, and is there to be perceived. Such wisdom cannot be apprehended, however, unless we perceive the forest monastery as a *system of relationships between the individual, society, and nature.* The Sangha in the forest monastery is a society aiming for human development amidst the natural environment. In this system of relationships we can see the wisdom which stresses the interrelatedness and interdependence of persons, society, and nature [emphasis added].[33]

We propose that by drawing on the environmental wisdom of the *dharma,* by serving as a model of a green society, and through the power afforded by their liminal status, local monastic communities have significant potential to contribute to the environmental awareness, information, and ethics of the populace, including daily visitors as well as participants in the rainy season retreat. This in turn could contribute toward a greening of society in Thailand. After all, environmental problems are one source of suffering (*dukkha*), one of the central concerns of Buddhism.[34]

Many examples could be cited of individual monks who are effective environmental activists. However, one must suffice here. Abbot Somneuk Natho resides at the forest monastery of Wat Plak Mai Lai in Nakhom Pathom Province, about ninety minutes from Bangkok. Within about ten years he restored one hundred *rai* (forty acres) of empty grassland to forest. He allowed the land, through its own resilience and natural processes of restoration, to return to forest, although he helped by planting some saplings of forest trees. He also maintains numerous medicinal plants. This successful initiative is in sharp contrast to the Royal Forestry Department's usual style of reforestation by clear-cutting areas of natural forest in order to establish monocrop commercial tree plantations, such as eucalyptus. The forest monastery is surrounded by fields of sugar, corn, and vegetable crops. It is noteworthy that Abbot Somneuk Natho became a monk despite resistance from his family and the fact that his father is a millionaire.[35] Given Abbot Somneuk Natho's example, we can better appreciate the statement by naturalists Mark Graham and Philip Round that monks are the "custodians of nature" in Thailand.[36]

Limitations of the Monastic Community

In Thailand, where more than 95 percent of the population is Buddhist, it is natural to explore the contributions Buddhism might make toward the reduction or resolution of environmental problems and the creation of a greener society. Indeed, while environmentalism may be originally a Western concept, in the Thai context it cannot be understood apart from Buddhism and the *sangha*. There are, of course, limitations as well as possibilities in the relevance of Buddhism and the *sangha* to the ecocrisis in Thailand, and it is appropriate to mention some of the limitations here, even though this is not the place to provide any elaborate analysis of the case against Buddhism.[37]

Some of the tenets of Buddhism may contribute more to the problem than to the solution. For instance, as Ruben Habito points out in his essay in this volume, there are adherents who interpret Buddhism as emphasizing individual self-examination and the present moment (being rather than doing), thus discouraging activism concerned with current social problems, which are viewed as ephemeral according to the principle of impermanence (*anicca*).[38]

Politics within the *sangha* can cause obstacles to the realization of the potential of Buddhism. Just as with any social institution, both the *sangha* and the state are subject to abuse and corruption. The *sangha* as a whole is hierarchical, its upper levels are conservative, and it has been closely allied with the state since the First Sangha Act in 1903.[39] For example, when the Council of Elders meets, its agenda is set by the Department of Religious Affairs of the Ministry of Education.[40] The upper levels of the monastic hierarchy as well as the state have opposed monks who have become environmental activists. Some of these monks have been threatened and harassed by the police, military, and others. Automatic weapons have even been fired into a temple.[41] On the other hand, such opposition to activist monks (*phra nak anurak pa*) like Phra Prachak Kuttachitto and Achan Pongsak Techathamamoo indicates that their efforts in challenging forest destruction and addressing other environmental concerns have met with some success. Yet there are conservative monks who oppose such activism and even label activist monks as renegade monks or spiritual outlaws. Thus, Santikaro, an American who has long been a monk in Thailand, says:

> Those of us who are thinkers in this network feel that the current monastic sangha is more likely to fall apart than to solve its crisis. I don't think the sangha is capable of solving its problems and reforming itself according to the current structure, a structure forced on them by dictatorial governments.[42]

In Thailand only a minority of monks are environmental activists, although the number has been growing rapidly in recent years. Of about 288,637 monks, only a few hundred may be environmental activists. Nevertheless, they have had a significant impact on raising

the environmental awareness and concern of many people throughout the country. From another perspective, Mehden[43] mentions that
in Asia Marxists have often viewed monks in general as something
like parasites on society, with their unproductive activities and
traditionalist views, and as undesirable as role models for the
advancement of society.

In Thailand another problem with Buddhism is the discrimination
against women. Only men may be ordained in Theravāda Buddhism.
There is no genuine institution of the nun in Thailand, although
some women (*mae chii*) renounce the world, shave their heads, wear
white robes, and undertake the eight precepts.[44] Yet one of the
concomitants of a green society is gender equity. Ecofeminists argue
that there is a direct connection between human domination and
violence against nature on the one hand and male domination and
violence against women on the other.[45] This aspect of the *saṅgha*
and society in Thailand will have to change if a green society is to
be realized to any extent.

Discussion

While the aforementioned limitations are serious, in our opinion the
greatest obstacle Buddhism presents in its own contribution to the
reduction or resolution of environmental problems is the discrepancy
between the ideals of Buddhism and the practices of Buddhists.
Despite the great potential of Buddhism, and the *Dharma* and
Saṅgha in particular, in reality Thailand is no ecotopia; it is
increasingly becoming an environmental disaster. Many Thai and
others, including the present authors, relate the environmental,
social, and moral crises of contemporary Thailand to the increasing
acceptance of Western worldviews and a correlated decline in
adherence to Buddhist worldviews.[46] In turn, this shift in worldview
results in the wide discrepancy between the theory and practice of
Buddhism in society—that is, between the ideals and actions of
Buddhists.[47] However, it should be understood that this discrepancy
represents not a failure of Buddhism per se but a failure of individual
Buddhists, who are, after all, only human. (This should not be
surprising, since, to some degree, there are internal discrepancies and contradictions in any individual, society, or religion.)

Nevertheless, while television, movie theaters, shopping malls, and other distractions have intruded into the daily lives of many Thai, Buddhism, the temple, and monks still retain some significance for most individuals. In particular, forest monasteries remain attractive for retreats by laypersons because the "naturalness" and peacefulness of the forests render them optimal sites for meditation by Buddhists.[48]

These discrepancies between ideals and actions are found in different forms and varying degrees, not only among laypersons but also among monks. Some monks, like the public, are trapped by materialism and consumerism, despite the professed allegiance to Buddhism by laity and monks alike.[49] This reflects the tremendously powerful social, economic, and political influences present in Thai society and the global community, including other nations of Asia as well as the West.[50]

Despite these difficulties, there are clear indications surfacing which demonstrate a growing disillusionment with the wholesale pursuit of Westernization and its associated phenomena.[51] For instance, there are revitalization movements of various kinds, including many environmental and sustainable-development nongovernmental organizations which seek to revive some of the traditional religion, culture, and ecology in order to create a greater degree of ecological and social equilibrium.[52] (Revitalization movements have occurred in diverse societies in response to the stresses and dissatisfactions accompanying rapid and profound cultural change.)[53] Thus, Mehden writes:

> The Thais have long viewed Buddhism as the core to their personal and national identity. To the vast majority, to be Thai is to be Buddhist. However, it was the recent Thai kings and postwar military-political leaders who were to systematically foster the ideology of king, Buddhism, and country as a means of reinforcing national integration. In the process, missionary efforts have brought formerly isolated peoples in the Kingdom into the fold in a planned effort to make Buddhism a tool of national ideology. However, the apparent growth of a sense of Thai-Buddhist identity in recent years has also been a reflection of negative reactions to what are considered to be undesirable aspects of modernization. The very

intrusion of material goods and values has activated a resurgence
of interest in traditional religious tenets as a means of retaining
national and personal identity.[54]

Among other things, the environmental and other crises in
Thailand signal a dire need to increase substantially the information
and awareness about the potential of Buddhism to contribute to
solving problems. There is also a critical need to increase information and awareness regarding maladaptation—the short-term and
long-term negative environmental (and economic) consequences of
the irreversible depletion of natural resources and the degradation
of the natural environment (including pollution).[55] In other words,
paraphrasing Kenneth Kraft's remarks in his essay included in this
volume, an important question Buddhist individuals must learn to
ask and answer is: What is my karmic responsibility for my
environmental legacy? As Donald Swearer observes: "In the final
analysis environmental problems are going to be solved when a
sufficient number of people understand the nature of the problem,
have the moral courage to do something about it, and offer an
alternative vision to challenge the status quo."[56]

Fortunately, there are extraordinary individuals, laypersons and
monks, such as Abbot Somneuk Natho, who provide role models
of people who are trying to restore a modicum of "ecosanity" in
Thailand. Many of these persons have been described in the superb
writings of Sanitsuda Ekachai.[57] Such exemplars deserve greater
recognition, consideration, and emulation if Thailand as a nation is
to cope successfully with the environmental and other crises it faces.

Nonviolent strategies of social change have proven remarkably
effective in numerous and diverse situations in recent history: the
rejection of British colonialism in India led by Mohandas K. Gandhi;
the advances against racism during the civil rights movement in the
United States led by Martin Luther King, Jr.; the movement to
overthrow the Marcos regime in the Philippines led by Benigno
Aquino; and the revolutionary changes in South Africa, resulting
in the abolition of the apartheid system, led by Nelson Mandela.
The political and economic struggles associated with the environmental movements in Thailand must be nonviolent if they are to be
effective and succeed; and Buddhism, by its very nature, can help
promote nonviolent strategies and actions.[58]

Future field research on the hypotheses, propositions, and other ideas developed in this article is sorely needed, at the levels of both broad national and regional surveys and intensive investigation at a representative sample of specific sites and events, in the latter applying the standard ethnographic methods of participant observation and interviews.[59] There is also a need for comparative research in other Theravāda Buddhist countries, especially neighboring Myanmar, Laos, and Kampuchea, since they are somewhat less Westernized than Thailand. This analysis has been primarily etic—that is, based on Western scientific analysis and interpretation—even though one of the authors is Thai and we have cited statements by Thai writers. More attention needs to be given to the emic side, with insights from members of the local communities through intensive interviews. From an ecological perspective, it would also be interesting to analyze temple compounds and their surroundings as ecosystems and in terms of biodiversity conservation.[60]

Conclusions

In recent decades Thailand has faced increasingly frequent environmental, social, and moral crises which endanger the vitality, quality of life, and future of the nation. While there are many resources which may contribute toward the resolution of these crises—education, science, technology, economics, politics, government, nongovernmental organizations, and even business and industry—we have argued that surely one of the most important resources is Buddhism, essentially because Buddhism has the potential to penetrate deeply to the very roots of the problems and to find lasting solutions rather than merely treat superficial symptoms and single issues. The history of Buddhism involves a mutualism between monks and forests; latent in this philosophy and religion are parallels to ecology and the basic principles for developing a green environmental philosophy and ethics; and Buddhism and culture are mutually reinforcing in Thailand, where the overwhelming majority of people are Buddhists.

In this essay we have focused on the ideal attributes which the local monastic community may have as an approximation of a green society. We have argued that by virtue of its indefinite liminality the local monastic community may have special status and power as a counter to maladaptive environmental and social trends. Of course, some will be skeptical of such notions, but then, as Chai Podhisita pointed out to us, there must even have been skeptics of the validity and utility of the Buddha's ideas during his time.

The fact that Buddhism has not prevented these crises from developing in the first place points not to a failure of Buddhism per se but to the discrepancies between the ideals of Buddhism and the actions of individual Buddhists. At the same time, there may be some ways in which Buddhism has been part of the problem rather than the solution. That thesis we have recognized in this paper, but it needs to be systematically and critically analyzed, a task that must be left to a subsequent publication.

In this essay we have focused on a theoretical analysis of the potential contribution of some monastic communities in helping to resolve the growing ecocrisis in Thailand. This ecocrisis is part of a multitude of complex and difficult problems associated with Westernization. Buddhism has survived, unlike the modern industrial-materialist-consumer society, for more than twenty-five hundred years because it has proven meaningful in numerous diverse contexts. While the ecocrisis is an unprecedented challenge, we join the scholars and activists who affirm the continuing relevance of engaged Buddhism in coping with individual and social problems in Thailand. As before, this relevance will depend on the adherents of Buddhism interpreting its principles in ways which they find meaningful, given the problems and challenges of their historical and sociopolitical contexts—and *without* distorting those principles—in the spirit of the radical conservativism of Buddhadāsa.[61]

Notes

1. We are most grateful to Duncan Williams and Mary Evelyn Tucker for the invitation to participate in the Consultation on Buddhism and Ecology as well as for the wonderful job they did in organizing and implementing the conference. As in the case of the Consultation, we would appreciate any critical commentary from other scholars and activists. We are most appreciative to Chai Podhisita (Mahidol University) and Decha Tangseefa (University of Hawaii), who offered comments on a draft of this paper based on their previous experiences in the monkhood in Thailand. However, any errors or deficiencies in this paper are the sole responsibility of the authors.

2. See Chalardchai Ramittanond, "Notes on the Role and Future of Thailand's Environmental Movement," in *Man and Nature: A Cross-Cultural Perspective* (Bangkok: Chulalongkorn University Press, 1993), 91–117; Anthony Reid, "Humans and Forests in Pre-colonial Southeast Asia," *Environment and History* 1 (1995):93–110; Jonathan Rigg, ed., *Counting the Costs: Economic Growth and Environmental Change in Thailand* (Singapore: Institute of Southeast Asian Studies, 1995); Santikaro Bhikkhu, "Planting Rice Together: Socially Engaged Monks in Thailand," *Turning Wheel* (summer 1996):16–20; Leslie E. Sponsel, "The Historical Ecology of Thailand: Increasing Thresholds of Human Environmental Impact from Prehistory to the Present," in *Advances in Historical Ecology*, ed. William Bale (New York: Columbia University Press, 1997); and Phra Phaisan Visalo, "The Forest Monastery and Its Relevance to Modern Thai Society," in *Radical Conservatism: Buddhism in the Contemporary World*, ed. Sulak Sivaraksa et al. (Bangkok: Thai Inter-Religious Commission for Development and International Network of Engaged Buddhists, 1990), 288–30. Buddhist environmentalists are not the only activists who are critical of industrialism, materialism, consumerism, and associated phenomena; such criticisms are central to radical ecology (Andrew McLaughlin, *Regarding Nature: Industrialism and Deep Ecology* [Albany: State University of New York Press, 1993]; Carolyn Merchant, *Radical Ecology: The Search for a Livable World* [New York: Routledge, 1992], and *Ecology* [New York: Routledge, 1994]), social ecology (Murray Bookchin, *The Ecology of Freedom* [Palo Alto, Calif.: Cheshire Books, 1982], and *The Philosophy of Social Ecology* [Montreal: Black Rose Books, 1990]), and the green movement (Andrew Dobson, ed., *The Green Reader* [San Francisco: Mercury House, 1990], and *Green Political Thought* [New York: Routledge, 1995]).

3. Dhira Phantumvanit and Khunying Suthawan Sathirathai, "Thailand: Degradation and Development in a Resource-Rich Land," *Environment* 30, no. 1 (1988):10–15.

4. Pinkaew Leungaramsri and Noel Rajesh, *The Future of People and Forests in Thailand after the Logging Ban* (Bangkok: Project for Ecological Recovery, 1992).

5. Peter Kunstadter, "The End of the Frontier: Culture and Environment Interactions in Thailand," in *Culture and Environment in Thailand*, (Bangkok: Siam Society, 1989), 548–50.

6. *Man and Nature: A Cross-Cultural Perspective* (Bangkok: Chulalongkorn University Press, 1993); and Ann Danaiya Usher, "After the Forest: AIDS as Ecological Collapse in Thailand," *Thai Development Newsletter* 26 (1994):20–32.

7. Lily de Silva, "The Buddhist Attitude toward Nature," in *Buddhist Perspectives on the Ecocrisis*, ed. Klas Sandell (Kandy, Sri Lanka: Buddhist Publication Society, 1987), 16.

8. Leslie E. Sponsel and Poranee Natadecha-Sponsel, "Buddhism, Ecology, and Forests in Thailand: Past, Present, and Future," in *Changing Tropical Forests: Historical Perspectives on Today's Challenges in Asia, Australasia, and Oceania*, ed. John Dargavel, Kay Dixon, and Noel Semple (Canberra, Australia: Centre for Resource and Environmental Studies, 1988), 305–25; "Nonviolent Ecology: The Possibilities of Buddhism," in *Buddhism and Nonviolent Global Problem-Solving*, ed. Glenn D. Paige and Sarah Gilliatt (Honolulu: Center for Global Nonviolence and Spark M. Matsunaga Institute for Peace, 1991), 139–50; "The Relevance of Buddhism for the Development of an Environmental Ethic for the Conservation of Biodiversity," in *Ethics, Religion, and Biodiversity: Relations between Conservation and Cultural Values*, ed. Lawrence S. Hamilton (Cambridge: White Horse Press, 1993), 75–97; and "The Role of Buddhism for Creating a More Sustainable Society in Thailand," in *Counting the Costs: Economic Growth and Environmental Change in Thailand*, ed. Jonathan Rigg (Singapore: Institute of Southeast Asian Studies, 1995), 27–46.

9. Kerry Brown, "In the Water There Were Fish and the Fields Were Full of Rice: Reawakening the Lost Harmony of Thailand," in *Buddhism and Ecology*, ed. Martine Batchelor and Kerry Brown (New York: Cassell Publishers, 1992), 87–99.

10. Bhikkhu Bodhi, foreword to *Buddhist Perspectives on the Ecocrisis*, ed. Klas Sandell (Kandy, Sri Lanka: Buddhist Publication Society, 1987), vii. Cf. Ian Harris, "How Environmentalist Is Buddhism?" *Religion* 21 (April 1991):101–14; Poul Pedersen, "The Study of Perceptions of Nature: Towards a Sociology of Knowledge about Nature," in *Asian Perceptions of Nature*, ed. Ole Bruun and Arne Kalland, Nordic Institute of Asian Studies Proceedings, no. 3 (Copenhagen: Nordic Institute of Asian Studies, 1992), 148–58.

11. Shann Davies, *Tree of Life: Buddhism and the Protection of Nature* (Bangkok: Buddhist Perception of Nature Project, 1987); Chatsumarn Kabilsingh, *A Cry from the Forest* (Bangkok: Wildlife Fund Thailand, 1987).

12. Marie Beuzeville Byles, *Footprints of Gautama the Buddha* (Wheaton, Ill.: Theosophical Publishing House, 1957), 116.

13. From the outset we should like to make clear that we do not consider Buddhism to be the only resource for reducing or resolving environmental and

social problems in Thailand. Science, technology, education, government, business, industry, nongovernmental organizations, and other resources are also important. However, it would be unrealistic to ignore the potential contribution of Buddhism in a country where more than 95 percent of the population is Buddhist.

Thai religion is synergistic, a dynamic and creative combination of elements from Buddhism, Hinduism, and animism. Chinese Thai may also include elements of Confucianism and Taoism. In the southernmost provinces of the peninsula, where the majority of people are of Malay heritage, the predominant religion is Islam (Charles F. Keyes, "Thai Religion," in *The Encyclopedia of Religion*, ed. Mircea Eliade [New York: Free Press, 1987], vol. 14, 416–21). Ideally, all of these religious influences should be considered, but space limitations and the theme of this volume force us to focus exclusively on Buddhism.

There are pros and cons for the involvement of religion in environmental problems and issues. Individuals committed to exclusively scientific and technological approaches for the resolution of environmental problems may dismiss religion as merely emotional, irrational, superstitious, and antiscientific (see Fred R. von der Mehden, "The Impact of Modernization on Religion," in his *Religion and Modernization in Southeast Asia* [Syracuse, N.Y.: Syracuse University Press, 1986], 16–17). Organized religion and its adherents may be criticized as hypocritical, given the discrepancies between ideals and actions. Indeed, many evils have been perpetrated in the name of some religion. Religious sects can be divisive and even lead to conflict and violence. Furthermore, some writers have even viewed certain religions as the ultimate cause of environmental problems (Lynn White, Jr., "The Historical Roots of Our Ecologic Crisis," *Science* 155 [March 1967]:1203–7; Eugene C. Hargrove, ed., *Religion and Environmental Crisis* [Athens: University of Georgia Press, 1986]).

On the other hand, religion is a cross-cultural universal; anthropologists have not found a society without religion. Moreover, religion can be a powerful and integrative force in individual behavior and in society. It is usually the ultimate source of one's worldview and values. Also, religion has been seen as a (or the) source for the solution to the ecocrisis, especially for constructing a viable environmental ethic. (For some exceptionally good studies of the relationship between religion and environment, see Roger S. Gottlieb, *This Sacred Earth: Religion, Nature, Environment* [New York: Routledge, 1996]; Lawrence S. Hamilton, ed., *Ethics, Religion, and Biodiversity: Relations between Conservation and Cultural Values* [Cambridge: White Horse Press, 1993]; Steven C. Rockefeller and John C. Elder, eds., *Spirit and Nature: Why the Environment Is a Religious Issue* [Boston: Beacon Press, 1992]; Rupert Sheldrake, *The Rebirth of Nature: The Greening of Science and God* [Rochester, Vt.: Park Street Press, 1991]; Henryk Skolimowski, *A Sacred Place to Dwell: Living with Reverence upon the Earth* [Rockport, Mass.: Element, 1993]; and Mary Evelyn Tucker and John A. Grim, eds., *Worldviews and Ecology* [Philadelphia: Bucknell University Press, 1993].)

14. Dobson, *The Green Reader*; John W. Bennett, *The Ecological Transition: Cultural Anthropology and Human Adaptation* (New York: Pergamon Press, 1976), 13; Alan R. Drengson, *Beyond Environmental Crisis: From Technocrat to Planetary Person* (New York: Peter Lang, 1989).

15. Also see P. A. Payutto, *Buddhist Economics: A Middle Way for the Market Place* (Bangkok: Buddhadhamma Foundation, 1994).

16. Donald K. Swearer, *The Buddhist World of Southeast Asia* (Albany: State University of New York Press, 1995).

17. Visalo, "The Forest Monastery," 293–94. See Swearer, *The Buddhist World of Southeast Asia*, 146.

18. Arnold van Gennep, *The Rites of Passage* (London: Routledge and Kegan Paul, 1909).

19. Victor Turner, *The Ritual Process: Structure and Anti-Structure* (New York: Aldine de Gruyter, 1969), 94–95.

20. Ibid., 106–7.

21. See Christopher Lamb, "Buddhism," in *Rites of Passage*, ed. Jean Holm and John Bowker (London: Pinter Publishers, 1994), 10–40.

22. Turner, *The Ritual Process*, 107.

23. Of course, the monastic community is composed of celibate males, and they are dependent for their daily food on the surrounding lay community. The *sangha* also has a bureaucratic hierarchy, which is connected with the state government. However, there is less social stratification in local monastic communities than in Thai society at large.

24. For parallels in other approaches to environmental ethics, see Duane Elgin (*Voluntary Simplicity* [New York: William Morrow, 1981]) on simplicity; Warwick Fox (*Toward a Transpersonal Ecology: Developing New Foundations for Environmentalism* [Boston: Shambhala Publications, 1990]) on self-realization; Arne Naess (*Ecology, Community, and Lifestyle: Outline of an Ecosophy*, trans. David Rothenberg [Cambridge: Cambridge University Press, 1989]) on community; and E. O. Wilson and Stephen Kellert, eds., *The Biophilia Hypothesis* (Washington, D.C.: Island Press, 1993), on biophilia.

25. A very useful brief description of the temple and monkhood in Thailand is provided by Gerald Roscoe, *The Monastic Life: Pathway of the Buddhist Monk* (Bangkok: Asia Books, 1992). For more detailed studies of various aspects of these and related themes, see Martin Boord, "Buddhism," in *Sacred Place*, ed. Jean Holm and John Bowker (London: Pinter Publishers, 1994), 8–32; Susan Marie Darlington, *Buddhism, Morality, and Change: The Local Response to Development in Northern Thailand* (Ann Arbor, Mich.: University Microfilms International, 1990); Carla Deicke Grady, "A Buddhist Response to Modernization in Thailand: With Particular Reference to Conservationist Forest Monks" (Ph.D. diss., University of Hawaii, 1995); P. Jackson, *Buddhadasa: A Buddhist Thinker for the Modern World* (Bangkok: Siam Society, 1988); *Radical Conservatism: Buddhism*

in the Contemporary World, ed. Sulak Sivaraksa et al. (Bangkok: Thai Inter-Religious Commission for Development and International Network of Engaged Buddhists, 1990); Donald K. Swearer, *Wat Haripunjaya: A Study of the Royal Temple of the Buddha's Relic, Lamphun, Thailand* (Missoula, Mont.: Scholars Press, 1976); Stanley Jeyaraha Tambiah, *World Conqueror and World Renouncer: A Study of Buddhism and Polity in Thailand against a Historical Background* (New York: Cambridge University Press, 1976), and *The Buddhist Saints of the Forest and the Cult of Amulets* (New York: Cambridge University Press, 1984); J. L. Taylor, *Forest Monks and the Nation-State: An Anthropological and Historical Study in Northeastern Thailand* (Singapore: Institute of Southeast Asian Studies, 1993); Kamala Tiyavanich, *Forest Recollections: Wandering Monks in Twentieth-Century Thailand* (Honolulu: University of Hawaii Press, 1997); and K. E. Wells, *Thai Buddhism: Its Rites and Activities* (Bangkok: Suriyabun Publishers, 1975).

26. Elsewhere we argue that collectively as well as accumulatively over time the temples may help promote both the *ex situ* and *in situ* conservation of biodiversity in Thailand. There are a large number of temples (around 29,002) distributed throughout Thailand, and many temples are ancient (some even centuries old). The area surrounding temples is considered sacred space; temples are often associated with trees, groves, and/or forests which also provide habitat for animal species, and these ecological phenomena are considered sacred. Temples often provide *habitat islands* in a sea of wet rice paddies and other agroecosystems. During our fieldwork in southern Thailand during the summers of 1994 and 1995, we found small but healthy forests associated with temples; entire mountain forests protected by temples or shrines; sections of community forests donated by villagers to the local temple to conserve them intentionally for future generations; and areas of forest restoration associated with the initiatives of Buddhist monks. Thus we hypothesize that a temple may serve one or more of these conservation functions: forest reserves, botanical gardens, germ plasm banks, medicinal plant collections, zoological gardens, wildlife sanctuaries, restoration ecology, model of green society, environmental education, and environmental action (Leslie E. Sponsel and Poranee Natadecha-Sponsel, "The Role of Sacred Places in the Conservation of Biodiversity," in *Ecology, Ethnicity, and Religion in Thailand*, ed. Sponsel and Natadecha-Sponsel, forthcoming; and Sponsel and Natadecha-Sponsel, "The Role of Buddhism for Creating a More Sustainable Society in Thailand"). For a pioneering ethnobotanical study of temple yards, see Shengji Pei, "Managing for Biological Diversity in Temple Yards and Holy Hills: The Traditional Practices of the Xishuangbanna Dai Community, Southwestern China," in *Ethics, Religion, and Biodiversity: Relations Between Conservation and Cultural Values*, ed. Lawrence S. Hamilton (Cambridge: White Horse Press, 1993), 118–32.

27. See Mehden, "The Impact of Modernization on Religion," 81–82; and Swearer, *The Buddhist World of Southeast Asia*, 116–18.

28. S. Suksamran, *Political Buddhism in Southeast Asia* (London: Hurst, 1977).

29. Roscoe, *The Monastic Life*, 13.

30. Department of Religious Affairs, *Annual Report of Religious Activities for 1967–91* (Bangkok: Ministry of Education, Department of Religious Affairs, 1992).

31. Roscoe, *The Monastic Life*, 12.

32. Ibid., 14.

33. Visalo, "The Forest Monastery," 297.

34. P. A. Payutto, *A Buddhist Solution for the Twenty-First Century*, 2nd ed. (Bangkok: Sahathammik, 1993), 7; and Payutto, *Buddhist Economics*.

35. This account is based on Sanitsuda Ekachai, *Seeds of Hope: Local Initiatives in Thailand* (Bangkok: Thai Development Support Committee, 1994), 124–29. See also 72–83 in her book.

36. Mark Graham and Philip Round, *Thailand's Vanishing Flora and Fauna* (Bangkok: Finance Once Public, 1994), 71. See also Darlington, *Buddhism, Morality, and Change*; Susan Marie Darlington, "Monks and Environmental Conservation: A Case Study in Nan Province," *Seeds of Peace* 9, no. 1 (1993):7–10; and Darlington, "The Ordination of a Tree: The Buddhist Ecology Movement in Thailand," in *Ecology, Ethnicity, and Religion in Thailand*, ed. Leslie E. Sponsel and Poranee Natadecha-Sponsel, forthcoming; Joe Franke, "A Walk with the Monk Who Ordained Trees," *Shambhala Sun*, November 1995, 48–53; Grady, "A Buddhist Response to Modernization in Thailand"; Steve Magagnini, "If a Tree Falls. . .A Monk's Blessing for Thailand's Forest," *Amicus Journal* 16, no. 2 (summer 1994:12–14; and J. L. Taylor, "Social Activism and Resistance on the Thai Frontier: The Case of Phra Prajak Khuttajitto," *Bulletin of Concerned Asian Scholars* 25, no. 2 (1993):3–16.

37. See, for example, Harris, "How Environmentalist Is Buddhism?"

38. See also Mehden, "The Impact of Modernization on Religion," 99–104.

39. See Yoneo Ishii, *Sangha, State, and Society: Thai Buddhism in History* (Honolulu: University of Hawaii Press, 1986); Mehden, "The Impact of Modernization on Religion," 76–79; and Swearer, *The Buddhist World of Southeast Asia*, 102–4.

40. Santikaro, "Planting Rice Together," 17.

41. Franke, "A Walk with the Monk Who Ordained Trees," 50.

42. Santikaro, "Planting Rice Together," 20. See also Tim Ward, *What the Buddha Never Taught* (Berkeley: Celestial Arts Publishing, 1993), 225.

43. Mehden, "The Impact of Modernization on Religion," 83.

44. Roscoe, *The Monastic Life*, 27–29.

45. Irene Diamond and Gloria Feman Orenstein, eds., *Reweaving the World: The Emergence of Ecofeminism* (San Francisco: Sierra Club Books, 1990).

46. Sponsel and Natadecha-Sponsel, "The Relevance of Buddhism for the Development of an Environmental Ethic," 87; cf. Mehden, "The Impact of Modernization on Religion," 180–82, 186, 197–98, 205.

47. See J. Baird Callicott and Roger T. Ames, "Epilogue: On the Relation of Idea and Action," in *Nature in Asian Traditions of Thought: Essays in Environmental Philosophy*, ed. Callicott and Ames (Albany: State University of New York Press, 1989), 279–89; Stephen R. Kellert, "Culture," in his *The Value of Life: Biological Diversity and Human Society* (Washington, D.C.: Island Press, 1996), 131–52; Mehden, "The Impact of Modernization on Religion"; Swearer, *The Buddhist World of Southeast Asia*, 5–7; and Tuan Yi-Fu, "Discrepancies between Environmental Attitude and Behaviour: Examples from Europe and China," *Canadian Geographer* 12, no. 3 (1968):176–91.

48. Keyes, "Thai Religion"; Visalo, "The Forest Monastery." See also Jack Kornfield and Paul Breiter, eds., *A Still Forest Pool: The Insight Meditation of Achaan Chah* (Wheaton, Ill.: Theosophical Publishing House, 1985).

49. See the critical accounts of the monkhood in Thailand by Santikaro, "Planting Rice Together," and Ward, *What the Buddha Never Taught.*

50. United Nations University, *Asia's New Initiatives in the 1990s: The Peace Process, Economic Cooperation, Management of the Environment* (Tokyo: United Nations University Japan-ASEAN Forum, 1994).

51. For example, Visalo, "The Forest Monastery," 295–96.

52. See Mehden, "The Impact of Modernization on Religion," 106, 112–14, 154, 179–82, 195. See also Theodore Mayer, "Thailand's New Buddhist Movements in Historical and Political Context," in Bryan Hunsaker, Theodore Mayer, Barbara Griffiths, and Robert Dayley, *Loggers, Monks, Students, and Entrepreneurs: Four Essays on Thailand* (DeKalb, Ill.: Center for Southeast Asian Studies, Northern Illinois University, 1996), 33–66.

53. A. F. C. Wallace, "Revitalization Movements," *American Anthropologist* 58 (1956):264–81.

54. Mehden, "The Impact of Modernization on Religion," 184.

55. Leslie E. Sponsel, "Cultural Ecology and Environmental Education," *Journal of Environmental Education* 19, no. 1 (1987):31–42; Poranee Natadecha, "Nature and Culture in Thailand: The Implementation of Cultural Ecology and Environmental Education through the Application of Behavioral Sociology" (Ph.D. diss., University of Hawaii, 1991).

56. Donald K. Swearer, "Francis of Assisi, Moral Exemplars, and Environmental Ethics," in *Man and Culture: A Cross-Cultural Perspective* (Bangkok: Chulalongkorn University Press, 1993), 196.

57. Sanitsuda Ekachai, *Behind the Smile: Voices of Thailand* (Bangkok: Thai Development Support Committee, 1990), and *Seeds of Hope.*

58. See Christopher Key Chapple, *Nonviolence to Animals, Earth, and Self in Asian Traditions* (Albany: State University of New York Press, 1996); Kenneth Kraft, *Inner Peace, World Peace: Essays on Buddhism and Nonviolence* (Albany: State University of New York Press, 1992); Glenn D. Paige and Sarah Gilliatt, eds., *Buddhism and Nonviolent Global Problem-Solving* (Honolulu: Center for

Global Nonviolence and Spark M. Matsunaga Institute for Peace, 1991); and Santikaro, "Planting Rice Together," 20.

59. Darlington, *Buddhism, Morality, and Change*, "Monks and Environmental Conservation," and "The Ordination of a Tree."

60. See note 7 above.

61. Santikaro Bhikkhu, "Buddhadasa Bhikkhu: Life and Society through the Natural Eyes of Voidness," in *Engaged Buddhism: Buddhist Liberation Movements in Asia*, ed. Christopher S. Queen and Sallie B. King (Albany: State University of New York Press, 1996), 147–94; and Sivaraksa et al., *Radical Conservatism: Buddhism in the Contemporary World.*

Mahāyāna Buddhism and Ecology:
The Case of Japan

The Jeweled Net of Nature

Paul O. Ingram

> Most significant and profound is the teaching of the
> ultimate path of Mahayana. It teaches salvation of
> oneself and others. It does not exclude even animals
> or birds. The flowers in the spring fall beneath its
> branches; Dew in autumn vanishes before the
> withered grass.
>
> —*Sango shiki* (Indications of the goals of the
> three teachings)[1]

During my last visit to Japan I was invited by three Shingon
Buddhist lay scholars to a restaurant outside Osaka specializing in
the preparation and serving of a deadly toxic fish known as *fugu*.
Though it has a certain Russian-roulette quality, eating *fugu* is
considered by many Japanese to be a highly aesthetic experience.

Of course, I declined; my aesthetic tastes run in different
directions. Still, the experience of watching my friends eat *fugu*
made me wonder about the condition that we, in chauvinistic
shorthand, refer to as "human." Beings who will one day vanish
from the earth in that ultimate subtraction of sensuality called death,
we spend so much of our lives courting it: fomenting wars,
watching, with sickening horror, movies in which maniacs slice and
dice their victims, or hurrying to our own deaths in fast cars, through
cigarette smoking, or by committing suicide. Death obsesses us, as
well it might, but our responses are so strange.

This is particularly true of our response to nature. All we have
to do is look in a mirror. The face that pins us with its double gaze
reveals a frightening secret: we look into a predator's eyes. It's rough
out there in nature, whether in the wilds of a rain forest or an urban

jungle, partly because the earth is jammed with devout human predators unlike all others: we not only kill for food, we kill each other along with the natural forces that nourish life on this planet.

We stalk and kill nature even as we know what contemporary ecological research makes plain: that we are enfolded in a living, terrestrial environment in which all living and nonliving things are so mutually implicated and interrelated that no distinct line separates life from nonlife.[2] This conclusion is not only a biological claim; it is also a claim about the nature of reality. Of necessity, ecological research alters our understanding of ourselves, individually, and of human nature, generally. Or at least it ought to. For not only do "ecology and contemporary physics complement one another conceptually and converge toward the same metaphysical notions,"[3] so also do contemporary process theology and Buddhist teachings and practices. The question is, how can we, the most efficiently aggressive predators in nature, train ourselves to act according to what this research shows?

It is least of all a matter of technology and mostly a matter of vision, that sense of reality—"the way things really are"—according to which we most appropriately structure our relation to nature. For, as Proverbs 29:18 warns, "Where there is no vision, the people perish." My thesis is this: Dialogical encounter with Buddhist traditions—in this case illustrated by the esoteric teachings of the Japanese Buddhist monk Kūkai (774–835)—and Western ecological models of reality, as seen emerging in the natural sciences and Christian process theology, may energize an already evolving global vision through which to refigure and resolve the current ecological crisis. What is at stake is nothing less than the "liberation of life."[4]

But first, some remarks on mainstream Christian teaching about nature. In 1967, Lynn White, Jr.'s, controversial essay, "The Historical Roots of Our Ecologic Crisis,"[5] started a debate that raged through the 1970s among theologians, philosophers, and scientists. One focal point of this debate was White's recommendation for reforming the Christian Way in order to lead humanity out of the ecological shadow of death he thought "mainstream Christianity" originally created. Specifically, he recommended that mainstream Christianity endorse a "Franciscan worldview" and "panpsychism" in order to reconstruct, deliberately, a contemporary Western environmental ethic.[6]

Initial reaction to White's essay focused on identifying the Christian worldview. Surprisingly, there was little Christian bashing; more surprising, most Christian discussion agreed with White's characterization of Christian tradition. But there was little agreement about how to reconstruct a distinctively Christian view of nature, or indeed, whether it could or should be reconstructed.

Recently, the structure of "mainstream" Christian tradition roughly caricatured by White was formulated into a typology by J. Baird Callicott and Roger T. Ames:[7] 1) God transcends nature; 2) nature is a creation, an artifact, of a divine craftsman-like male creator; 3) human beings are exclusively created in God's image and therefore are essentially segregated from the rest of nature; 4) human beings are given dominion by God over nature; 5) God commands humanity to subdue nature and multiply the human species; 6) nature is viewed politically and hierarchically—God over humanity, humanity over nature, male over female—which establishes an exploitive ethical-political pecking order and power structure; 7) the image of God-in-humanity is the ground of humanity's *intrinsic* value, but nonhuman entities lack the divine image and are religiously and ethically disenfranchised and possess merely *instrumental* value for God and human beings; 8) the biblical view of nature's instrumental value is compounded in mainline Christian theology by an Aristotelian-Thomistic teleology that represents nature as a support system for rational human beings.

The upshot of this seems clear. The great monotheistic traditions of the West are the major sources of Western moral and political attitudes. Christianity doctrinally focuses on humanity's uniqueness as a species. Thus, if one wants theological license to increase radioactivity without constraint, to consent to the bulldozer mentality of developers, or to encourage unbridled harvest of old-growth forests, historically there has been no better scriptural source than Genesis, chapters 1 and 2. The mythological injunctions to conquer nature, the enemy of God and humanity, are here.

However, placing the full blame for the environmental crisis on the altar of the Christian Way is far too simplistic. Historically, the biblical creation story was read through the sensitivities of Greco-Roman philosophy; in fact, the legacy of Greco-Roman contributions to the ecological crisis may be more powerfully influential than distinctively biblical contributions.

Furthermore, Greek philosophical anthropology assumed an atomistic worldview, paradigmatically expressed in Plato and given its modern version by Descartes. Human nature is dualistic, composed of body and soul. The body, especially in Descartes's version, is like any other natural entity, exhaustively describable in atomistic-mechanistic language. But the human soul resides temporarily in the body—the ghost in the machine—and is otherworldly in nature and destiny. Thus, human beings are both essentially and morally segregated from God, nature, and each other. Accordingly, the natural environment can and should be engineered to human specifications, no matter what the environmental consequences, without either human responsibility or penalty.

Here we have it in a nutshell. The contemporary ecological crisis represents a failure of prevailing Western ideas and attitudes: a male-oriented culture in which it is believed that reality exists only as human beings perceive it (Berkeley); whose structure is a hierarchy erected to support humanity at its apex (Aristotle, Aquinas, Descartes); to whom God has given exclusive dominance over all life-forms and inorganic entities (Genesis 1–2); in which God has been transformed into humanity's image by modern secularism (Genesis inverted). It seems unlikely that mainstream Christian tradition, married as it is in the West to the traditions of Greco-Roman philosophy, is capable of resolving the ecological crisis that Christian reading of Genesis 1–2 through Greco-Roman philosophy created.

However, the traditional Western-Christian paradigm of nature is being challenged by new ecological models and theoretical explanations of the interconnectedness of humanity and nature that are developing within the natural sciences.[8] Recent Christian theological discussion, most notably process theology, also focuses on these same scientific models, recognizing the inadequacies of traditional Christian and secular views of nature.[9] Of course, there are a number of Western versions of this emerging ecological paradigm; no two are exactly alike in their technical details or explanatory categories. Even so, it is possible to abstract three principles these paradigms share.[10]

The first principle is holistic unity—nature is an "ecosystem" whose constituent elements exist in constantly changing, interdependent causal relationships. What an entity is, or becomes, is a direct function of how it relates with every other entity in the universe at every moment of space-time. Second is the principle of interior life movements—all living entities possess a life force intrinsic to their own natures that is not imposed from other things or from God but is derived from life itself. That is, life is an emerging field of force supporting networks of interrelationship and interdependency ceaselessly occurring in all entities in the universe. Or, to invert traditional Christian images of God, God does not impose or give life; God is the chief exemplar of life. The third principle—that of organic balance—means that all things and events at every moment of space-time are interrelated bipolar processes that proceed toward balance and harmony between opposites.

Similar organic principles have always been structural elements of the Buddhist worldview. The Shingon (Chinese, *chen-yen*, or "truth word") "esoteric" (Japanese, *mikkyō*, or "secret teaching") transmission established in Japan by Kūkai in the ninth century particularly embraces these elements.[11] Kūkai's Buddhist environmental paradigm is summarized in the first stanza of a two-stanza poem he wrote in Chinese in *Attaining Enlightenment in This Very Existence (Sokushin jōbutsu gi)*:

> The Six Great Elements are interfused and are in a state of eternal harmony;
> The Four Mandalas are inseparably related to one another.
> When the grace of the Three Mysteries is retained, (our inborn three mysteries will) quickly be manifested.
> Infinitely interrelated like the meshes of Indra's net are those we call existences.[12]

The first line, "The Six Great Elements are interfused and are in a state of eternal harmony," presupposes two propositions upon which Kūkai's Buddhist understanding of nature rests: 1) the Buddha, Dainichi Nyorai, or "Great Sun" (Sanskrit, Mahāvairocana Tathāgata), and the Six Great Elements are interfused; and 2) Dainichi and the universe coexist in a state of timeless nondual harmony.

Kūkai's buddhology and subsequent Shingon doctrinal formulation assumed standard Mahāyāna "three-body" theory (Sanskrit, *trikāya*; Japanese, *sanshin*), but with a difference. Prior to Kūkai's teacher, Hui-kuo, Dainichi was symbolized as one of a number of *saṃbhogakāya* ("body of bliss") forms of absolute reality called

dharmakāya (*Dharma* or teaching body) that all Buddhas comprehend and manifest when they become "enlightened ones." But
in Exoteric Buddhist teaching and Esoteric Buddhist tantra prior to
Hui-kuo and Kūkai, the *dharmakāya* is ultimate reality, beyond
names and forms, utterly beyond verbal capture by doctrines, while
yet the foundational source of all Buddhist thought and practice.
Thus, *saṃbhogakāya* forms of Buddhas are not "historical Buddhas"
(*nirmāṇakāya*), of whom the historical Śākyamuni is an example:
they exist in nonhistorical realms of being, forever enjoying their
enlightened bliss, as objects of human veneration and devotion.
Normally, *bodhisattva*s and nonhistorical Buddhas, including
Dainichi, were represented as *saṃbhogakāya* forms of the eternal
dharmakāya.

It was probably Hui-kuo who first identified Dainichi as the
dharmakāya Buddha and who taught that the *Dainichi-kyō* and the
Kongocho-kyō, which according to Shingon teaching embody the
fullest expression of truth, were as Dainichi preached, not the
historical Śākyamuni.[13] Kūkai, following Hui-kuo, transformed
Dainichi into a personified, uncreated, imperishable, beginningless
and endless Ultimate Reality. He reasoned that as the sun is the
source of light and warmth, Dainichi is the "Great Luminous One"
at the source of enlightenment and unity underlying the diversity
of the phenomenal world. And, since Buddha-nature is within all
things and events in space-time—an idea Kūkai also accepted—the
implication is that Dainichi is the ultimate reality "originally" within
all sentient beings and nonsentient natural phenomena. As Kūkai
explained it:

> Where is the Dharmakaya? It is not far away; it is in our own body.
> The source of wisdom? In our mind; indeed, it is close to us![14]

As a Buddhist, Kūkai also accepted the doctrine of dependent
co-origination (Sanskrit, *pratītya-samutpāda*), but he interpreted this
teaching according to his notion that reality is constituted by the
Six Great Elements in ceaselessly interdependent and interpenetrating interaction: earth, water, fire, wind, space, and consciousness
or "mind" (Sanskrit, *citta*; Japanese, *shin*). The adjective "great"
signifies the universality of each element. The first five elements
stand for all material realities, and the last, "consciousness," for the
Body and Mind of Dainichi.

All Buddhas and unenlightened beings, all sentient and non-sentient beings, all material "worlds" are "created" by the ceaseless interaction of the Six Great Elements. This means that all phenomena are identical in their constituent self-identity; all are in a state of constant transformation; and there are no absolute differences between human nature and the natural order, body and mind, male and female, enlightenment and ignorance. In short, reality—the way things really are—is nondual. In Kūkai's words:

> Differences exist between matter and mind, but in their essential nature they remain the same. Matter is no other than mind; mind, no other than matter. Without any obstruction, they are interrelated. The subject is the object; the object, the subject. The seeing is the seen, and the seen is the seeing. Nothing differentiates them. Although we speak of the creating and the created, there is in reality neither the creating nor the created.[15]

The problem raised, then, is: how do we train ourselves to experience this eternal cosmic harmony and attune ourselves to it as it occurs? This "how" is expressed in the second line of the stanza: "The Four Mandalas are inseparably related to one another." Involved here is the practice of meditation, which in Shingon tradition is a skillful method (*upāya*) of integrating our body, speech, and mind (the "three mysteries," or *sanmitsu*) with the eternal harmony of Dainichi's Body, Speech, and Mind. In this sense, Shingon meditation is a process of imitation of Dainichi's enlightened harmony with nature through ritual performance of *mudrā*s (Body), *mantra*s (Speech), and *maṇḍala*s (Mind).

Shingon training involves a number of *maṇḍala*s, but Kūkai's poem refers to four. The "Mahā-maṇḍalas" or "Great Maṇḍalas" (Japanese, *daimandara*) are circular portrayals of Buddhas, *bodhisattva*s, and deities in anthropomorphic form painted in the five Buddhist colors—yellow, white, red, black, and blue or blue green. These colors correspond to five of the Six Great Elements: earth is yellow, water is white, fire is red, wind is black, space is blue. Since consciousness is nonmaterial, it is colorless and cannot be depicted in the *maṇḍala*. But Kūkai also taught that there is perfect interpenetration of the Six Great Elements, so that consciousness is present in the five colors and pervades the painting. Thus,

Mahā-maṇḍalas symbolize the universe as the physical extension of Dainichi.

The second *maṇḍala* is the Samaya-maṇḍala. *Samaya* is a Sanskrit word meaning "a coming together and agreement." So Samaya-maṇḍalas express the ontological unity underlying the diversity of all things in space-time as forms of Dainichi's *Dharma* body. Accordingly, every thing and event in the universe is a *samaya* or "a coming together and agreement" of this ontological unity—all things and events are forms of Dainichi—experienced from the perspective of Dainichi as well as of all Buddhas.[16]

The third *maṇḍala*, the Dharma-maṇḍala, is the same circle as the Mahā-maṇḍala and the Samaya-maṇḍala, but "viewed" as the sphere where "revelation" of absolute truth—the *Dharma*—takes place. Thus Dharma-maṇḍalas portray Dainichi Nyorai's continual communication of the *Dharma* throughout all moments of space-time to all sentient and nonsentient beings. The universe is Dainichi's "sound-body." Dharma-maṇḍalas represent the totality of the sound of the *Dharma* as Dainichi continually discloses or "preaches" it throughout the universe as depicted in "seed syllables" (Sanskrit, *bija*; Japanese, *shuji*) written in Sanskrit letters.

Finally, Karma-maṇḍalas are the same circle viewed from the perspective of Dainichi's action in the realm of *saṃsāra*. Since, as Kūkai taught, all things and events, all transformations in the flux of nature, interpenetrate the actions of Dainichi's *Dharma* body, every change in any form or entity is simultaneously an action of Dainichi. Conversely, every action of Dainichi is simultaneously the action of all things and events in the universe.[17]

In summary, the Four Maṇḍalas symbolize Dainichi Nyorai's "extension, intention, communication, and action."[18] "Extension" is Dainichi's compassionate wisdom; "communication" is his intended "self-revelation" as the "preaching of the *dharmakāya*" in all things and events in space-time; and his "action" is all movement in the universe.

The third line of the stanza, "When the grace of the Three Mysteries is retained, (our inborn three mysteries will) quickly be manifested," summarizes Kūkai's conception of Esoteric Buddhist practice. In relation to Dainichi Nyorai, the Three Mysteries stand for suprarational activities or macrocosmic functions of Dainichi's Body, Speech, and Mind at work in all things and events. Through the Mystery of Body, Dainichi's "suchness" is incarnate within the patterns and forms of all natural phenomena; the Mystery of Speech refers to Dainichi's continual "preaching" or "revelation" of the *Dharma* through every thing and event in space time; the Mystery of Mind refers to Dainichi's own enlightened experience of the suchness of all natural phenomena as interdependent forms of the *dharmakāya*.[19] In this way, Kūkai personified the Three Mysteries as interrelated forms of Dainichi's enlightened compassion toward all sentient and nonsentient beings.

Finally, in the stanza's fourth line, "Infinitely interrelated like the meshes of Indra's net are those we call existences," Kūkai employed the well-known Buddhist metaphor of Indra's net. As every jewel of Indra's net reflects all others, and as all jewels are reflected in a single jewel, so existence *is* Dainichi Nyorai: seemingly discrete entities are interdependent forms of Dainichi, the one ultimate reality underlying the diversity of all natural phenomena. Or, in Kūkai's words:

> Existence is my existence, the existences of the Buddhas, and the existences of all sentient beings. . . . The Existence of the Buddha [Mahāvairocana] is the existences of the sentient beings and vice versa. They are not identical but are nevertheless identical; they are not different but are nevertheless different.[20]

That Kūkai's Esoteric Buddhist teachings assert an ecological conception of nature quite different from mainstream Christian tradition is quite evident. First, Christian tradition understands and explains the universe in terms of a divine plan with respect to its creation and final end. Kūkai's universe is completely nonteleological. For him, the universe has neither beginning nor end, no creator, and no purpose. The universe just is, to be taken as given, a marvelous fact which can be understood only in terms of its own inner dynamism.

Second, mainstream Christian teaching and our Greek philosophical heritage have taught the West that nature is a world of limited, external, and special relationships. We have family relationships, marital relationships, relationships with a limited number of animal species, and occasional relationships with inanimate objects, most of which are external. But it is hard for us to imagine how any one thing is internally related to everything. How, for example, are we related to a star in Orion? How are Euro-Americans related to Lakota Native Americans or Alaskan Inuit? How are plants and animals related to us, other than externally as objects for exploitation? In short, those trained in Western philosophical traditions generally find it easier to think of isolated beings and insulated minds, rather than of one reality ontologically interconnecting all things and events. In contrast, Kūkai's universe is a universe of nondual-identity-in-difference, in which there is total interdependence: what affects and effects one item in the cosmos affects and effects every item, whether it is death, ignorance, enlightenment, or sin.

Finally, the mainstream Christian view of existence is one of rigid hierarchy, in which a male creator-god occupies the top link in the chain of being, human beings follow, and nature—animals, plants, rocks—are at the bottom. Kūkai's universe, however, posits no hierarchy. Nor does it have a center—or if it does, it is everywhere. In sum, Kūkai's universe leaves no room for anthropocentric biases endemic to Hebraic and Christian tradition, as well as those modern movements of philosophy having roots in Cartesian affirmation of human consciousness divorced from dead nature.

It is at this point that Kūkai's Esoteric Buddhist worldview makes contact with the vision and work of earlier Western physicists such as Michael Faraday and James Clerk Maxwell, later physicists such as Albert Einstein and Niels Bohr, and process philosophy and Christian process theology. Like Western new physics and process thought, Kūkai's worldview also characterizes nature as an "aesthetic order" that cognitively resonates with contemporary Western ecological ideas.

According to Roger Ames,[21] an "aesthetic order" is a paradigm that 1) proposes plurality as prior to unity and disjunction to conjunction, so that all particulars possess real and unique individuality; 2) focuses on the unique perspective of concrete particulars as the source of emergent harmony and unity in all interrelationships; 3) entails movement away from any universal characteristic to concrete particular detail; 4) apprehends movement and change in the natural order as a processive act of "disclosure," and hence describable in qualitative language; 5) perceives that nothing is predetermined by preassigned principles, so that creativity is apprehended in the natural order in contrast to being determined by God or chance; and 6) understands "rightness" to mean the degree to which a thing or event expresses, in its emergence toward novelty as this exists in tension with the unity of nature, an aesthetically pleasing order.

In contrast to the aesthetic order implicit in Kūkai's view of nature and contemporary science and process thought, the "logical order" of mainline Christianity characterized by Ames 1) assumes preassigned patterns of relatedness, a "blueprint" wherein unity is prior to plurality and plurality is a "fall" from unity; 2) values concrete particularity only to the degree it mirrors this preassigned pattern of relatedness; 3) reduces particulars to only those aspects needed to illustrate the given pattern, which necessarily entails moving away from concrete particulars toward the universal; 4) interprets nature as a closed system of predetermined specifications and therefore reducible to quantitative description; 5) characterizes being as necessity, creativity as conformity, and novelty as defect; and 6) views "rightness" as the degree of conformity to preassigned patterns.[22]

A number of examples of logical order come to mind: Plato's realm of Ideas, for instance, constitutes a preassigned pattern that charts particular things and events as real or good only to the degree they conform to these preexistent ideas. But aesthetic orders such as Kūkai's or process philosophy's are easily distinguishable from a logical order. In both, there are no preassigned patterns in things and events in nature. Creativity and order work themselves out through the arrangements and relationships of the particular constituents in the natural order. Nature is a "work of art" in which "rightness" is defined by the comprehension of particular details that constitute it as a work of art.

Of course, the technical details of the "aesthetic order" portrayed by Kūkai's ecological paradigm, and, for example, those of Christian process theology, are not identical. This much, however, should be noted: in spite of important technical differences, two common conceptualities are foundational in Kūkai's worldview and Whiteheadian process theology. The first is that there is continuity within nature. Kūkai portrayed this continuity in his doctrines of the Three Mysteries and the Six Great Elements. For both Kūkai and Whiteheadian thought, nature's continuity extends internal relatedness—a metaphysical relatedness in which individuals and societies are constituted by relationships of interdependence—to organic and inorganic nature. The second shared teaching is that human beings have a vital connection with nature, since all of nature is interconnected. This corresponds to Kūkai's image of Indra's jeweled net, as well as to his doctrine of the Six Great Elements.

Alfred North Whitehead's definition of "living body" gives some precision to these similarities. The living body, he writes, is "a region of nature which is itself the primary field of expression issuing from each of its parts."[23] Those entities that are centers of expression and feeling are alive, and Whitehead clearly applies this description to both animal and vegetable bodies. Also, this same definition of living body is an expansion of his definition of the human and animal body; the distinction between animals and vegetables is not a sharp one.[24]

Whitehead also contends that precise classification of the differences between organic and inorganic nature is not possible; although such classification might be pragmatically useful for scientific investigation, it is dangerous for nature. Scientific classifications often obscure the fact that "different modes of natural existence often shade off into each other."[25] The same point was made in *Process and Reality*, where Whitehead noted that there are no distinct boundaries in the continuum of nature, and thus no distinct boundaries between living organisms and inorganic entities; whatever differences there are is a matter of degree. This does not mean that differences are unimportant; even degrees of difference affirm the continuity of all nature.[26]

This point is central to Whiteheadian biologist Charles Birch and process theologian John Cobb's definition of "life." They raise the issue of the boundaries between animate and inanimate in light of the ambiguity of "life" on hypothetical boundaries.[27] Viruses are particularly good examples of entities possessing the properties of life and nonlife. Another example is cellular organelles, which reproduce but are incapable of life independent of the cell that is their environment.

The significance of these examples for the ecological model of life Birch and Cobb propose is that every entity is internally related to its environment. Human beings are not exceptions to the model, nor, in Cobb's opinion, is God, who is the chief example of what constitutes life.[28] Kūkai's views are similar: every entity in nature is internally related to its environment and to Dainichi. Although Dainichi is not a reality Christians or Shingon Buddhists name as God, like God, Dainichi is the chief example of what constitutes life.

As there is continuity between organic and inorganic in White-headian process thought, so too there is continuity between human and nonhuman. Whitehead underscored this continuity by including "higher animals" in his definition of "living persons." Both human beings and animals are living persons characterized by a dominant occasion of experience which coordinates and unifies the activities of the plurality of occasions and enduring objects which ceaselessly form persons. Personal order is linear, serial, object-to-subject inheritance of the past in the present. Personal order in human beings and in nature is one component of what Whitehead called "the doctrine of the immanence of the past energizing the present."[29] This linear, one-dimensional character of personal inheritance from the past is called the "vector-structure" of nature. A similar picture of nature evolves in Kūkai's notions of the Six Great Elements and the Three Mysteries.

The question could be asked: Why is it important for Western organic environmental paradigms to encounter Asian versions of organic views of nature such as Kūkai's? The answer is: Because what people *do* to the natural environment corresponds to what they *think* and *experience* about themselves in relation to the things around them.

Even at the level of empirical confirmation of scientific theory, it seems evident that "the ruination of the natural world is directly related to the psychological and spiritual health of the human race since our practices follow our perceptions."[30] Culture and world-view, faith and practice merge in language and indicate perceptions in persons and in societies. By relating to nature as a "thing" separate from ourselves or as separate from God, we not only have engendered but also perpetuate the environmental nightmare through which we are now living. The Christian term for this separation of ourselves from nature is "original sin"; the Buddhist word is "desire" (*taṇhā*).

The environmental destructiveness of Western rationalism's hyper-*yang* view of its own culture and of nature has been to a large extent delayed. But the ecological limits of the earth are now stretched and, in some cases, broken. Dialogue with Asian views of nature such as Kūkai's can foster the process of Western self-critical "consciousness raising" by providing alternative places from which to imagine new possibilities. In so doing, we might discern deeper organic strata within our own inherited cultural biases and assumptions and apprehend that we neither stand against nor dominate nature.

Like any particular dialogue, dialogue between Buddhists and Christians about nature has an inner and an outer dimension. Discussion of organic paradigms must not remain at the level of verbal abstraction. Buddhists can understand and appreciate the conceptions and technical language of Christian process views; process theologians can understand and appreciate Buddhist conceptions of nature. Both may be conceptually transformed. But this is an outer dialogue. Important as such dialogue is, it is incomplete if divorced from an inner dialogue about how Buddhists and Christians can personally experience nonduality between themselves and nature. For to the degree we experience the realities to which Buddhist and process Christian concepts of nature point, we are energized to live according to the organic structures of nature that outer dialogue conceptually reveals.

It's like the union of lyrics with music in a great chorale: the "music" of inner dialogue "enfleshes" the abstract lyrics of outer dialogue. What inner dialogue teaches is that we can live in whatever way we choose. In "Living Like Weasels" the poet and essayist Annie Dillard says just that: "We can live any way we want. People take vows of poverty, chastity, and obedience—even of silence—by choice." People destroy the environment—by choice—because they experience it only as a machine. Choosing to experience nature organically "is to stalk your calling in a certain skilled and supple way, to locate the most tender and live spot and plug into that pulse. This is yielding, not fighting"[31]—yielding to nature, not dominating nature.

From Kūkai's perspective transformed by encounter with Christian process thought, outer and inner dialogue means to follow our collective path with embodied detachment. As Annie Dillard has written:

> I think it would be well, and proper, and obedient, and pure, to grasp
> your one necessity and not let it go, to dangle from it limp wherever

it takes you. Then even death, where you're going no matter how you live, cannot you part.[32]

In so doing, we discover there was nothing to grasp all along, because we *are* nature, looking at ourselves.

Or, from a Christian process theological perspective transformed by inner and outer dialogue with Kūkai: God does not demand that we give up our personal dignity, that we throw in our lot with random people, that we lose ourselves and turn from all that is not God. For God is the "life" of nature, *intimior intimo meo*, as Augustine put it—"more intimate than I am to myself." God, like the stars, needs nothing, demands nothing. It is life with God that demands these things. Of course, we do not have to stop abusing the environment; not at all. We do not have to stop abusing nature—unless we want to know God. It is like sitting outside on a cold, clear winter's night. We are not obligated to do so; it may be too cold. If, however, we want to look at the stars, we will find that darkness is necessary. But the stars neither require nor demand it.

Notes

1. *Kūkai: Major Works*, translated, with an account of his life and a study of his thought, by Yoshito S. Hakeda (New York: Columbia University Press, 1972), 139. All citations from Kūkai's works in this essay are from Hakeda's translation, although I have checked them against the Chinese text in Yoshitake Inage, ed., *Kōbō daishi zenshū* (The complete works of Kōbō Daishi), 3rd ed. rev. (Tokyo: Mikkyō Bunka Kenkyūsha, 1965), Although Hakeda's volume does not translate all of Kūkai's works, it remains the best English translation of Kūkai's most influential writings in print. Since I cannot improve on Hakeda's translations, I have cited his with gratitude.

2. Charles Birch and John B. Cobb, Jr., *The Liberation of Life* (Denton, Tex.: Environmental Ethics Books, 1990), chap. 3.

3. J. Baird Callicott, "The Metaphysical Implications of Ecology," in *Nature in Asian Traditions of Thought: Essays in Environmental Philosophy*, ed. J. Baird Callicott and Roger T. Ames (Albany: State University of New York Press, 1989), 51.

4. See Birch and Cobb, *The Liberation of Life*, chaps. 1–2.

5. Lynn White, Jr., "The Historical Roots of Our Ecologic Crisis," *Science* 155 (1967):1203–7.

6. Ibid., 1206–7.

7. J. Baird Callicott and Roger T. Ames, "Introduction: The Asian Traditions as a Conceptual Resource for Environmental Philosophy," in *Nature in Asian Traditions of Thought: Essays in Environmental Philosophy*, ed. J. Baird Callicott and Roger T. Ames (Albany: State University of New York Press, 1989), 3–4.

8. See E. A. Burtt, *The Metaphysical Foundations of Modern Science* (Garden City, N.Y.: Anchor Books, 1954). Also see Alfred North Whitehead, *The Concept of Nature* (Cambridge: Cambridge University Press, 1971); two recent studies by Kenneth Boulding, *The World as a Total System* (Beverly Hills, Calif.: Sage Publications, 1985) and *Ecodynamics* (Beverly Hills, Calif.: Sage Publications, 1981); and two works by Fritjof Capra, *The Tao of Physics* (Boulder, Colo.: Shambhala Publications, 1975) and *The Turning Point* (New York: Bantam Books, 1982).

9. See Birch and Cobb, *The Liberation of Life*; Richard H. Oberman, *Evolution and the Christian Doctrine of Creation* (Philadelphia: Westminster Press, 1967); and a series of wonderful essays edited by Ian Barbour, *Earth Might Be Fair: Reflections on Ethics, Religion, and Ecology* (Englewood Cliffs, N.J.: Prentice Hall, 1972), especially Huston Smith's essay, "Tao Now: An Ecological Statement," 66–69.

10. Much of what follows is based on previous research published in my essay "Nature's Jeweled Net: Kukai's Ecological Buddhism," *Pacific World* 6 (1990): 50–64.

11. Kūkai (774–835), "Empty Sea," is commonly known as Kōbō daishi, an honorific title posthumously awarded to him by the Heian Court. "Kōbō" means "to widely transmit the Buddha's teachings," and "daishi" means "great teacher." Widely revered in his own time, Kūkai remains a figure of profound reverence in Japan today, both as a Buddhist master and a cultural hero. In 804, Kūkai traveled to China to study Buddhism, and while there he visited many eminent teachers, among whom was the esoteric master Hui-kuo (746–805). He became Hui-kuo's favorite disciple. Presumably, Kūkai's understanding of Hui-kuo's teachings was so impressive that Hui-kuo declared Kūkai his *dharma* heir before he died. Kūkai's study in China lasted only thirty months, and he returned to Japan at the age of thirty-three as the eighth patriarch of the Shingon School. See Hakeda, *Kūkai*, 10–15.

12. Ibid., 227.

13. Ibid., 81–82. Also see Taiko Yamasaki, *Shingon: Japanese Esoteric Buddhism*, trans. Richard Peterson and Cynthia Peterson, ed. Yasuyoshi Morimoto and David Kidd (Boston: Shambhala Publications, 1988), 62–64.

14. Hakeda, *Kūkai*, 82.

15. *Sokushin jōbutsu gi* (Attaining enlightenment in this very existence), in ibid., 229–30.

16. Samaya-maṇḍalas portray each of the Buddhas, *bodhisattvas*, and deities in some *samaya* or "symbolic" form, such as a jewel, sword, or lotus, that embodies the special quality of the individual Buddha, *bodhisattva*, or deity portrayed.

17. Karma-maṇḍalas portray the "actions of awe-inspiring deportment" (*rijigyo*) of all Buddhas and *bodhisattvas* in three-dimensional figures representing each particular Buddha and *bodhisattva* painted in the five colors of the Mahā-maṇḍala.

18. Hakeda, *Kūkai*, 91.

19. Adrian Snodgrass, "The Shingon Buddhist Doctrine of Interpenetration," *Religious Traditions* 9 (1986):66–68; and Yamasaki, *Shingon*, 106.

20. *Sokushin jōbutsu gi* (Attaining enlightenment in this very existence), in Hakeda, *Kūkai*, 232.

21. Roger T. Ames, "Putting the *Te* Back into Taoism," in *Nature in Asian Traditions of Thought: Essays in Environmental Philosophy*, ed. J. Baird Callicott and Roger T. Ames (Albany: State University of New York Press, 1989), 117.

22. Ibid., 116.

23. Alfred North Whitehead, *Modes of Thought* (New York: Macmillan, 1938), 31.

24. Ibid., 31–34.

25. Ibid., 25.

26. Alfred North Whitehead, *Process and Reality* (New York: Free Press, 1978), 109 and 179.

27. Birch and Cobb, *The Liberation of Life*, 92.

28. Ibid., 176–78, 195–200.

29. Alfred North Whitehead, *Adventures of Ideas* (New York: Macmillan, 1933), 188.

30. Jay C. Rochelle, "Letting Go: Buddhism and Christian Models," *Eastern Buddhist* 9 (autumn 1989):45.

31. Annie Dillard, *Teaching a Stone to Talk: Expeditions and Encounters* (New York: Harper and Row, 1982), 16.

32. Ibid.

The Japanese Concept of Nature in Relation to the Environmental Ethics and Conservation Aesthetics of Aldo Leopold

Steve Odin

Introduction

Taoism, with its metaphysics of nature as creative and aesthetic transformation, and East Asian Buddhism, with its view of nature as an aesthetic continuum of organismic interrelationships, have been sources of inspiration for environmental philosophy, recently consolidated in an anthology edited by J. Baird Callicott and Roger T. Ames, entitled *Nature in Asian Traditions of Thought: Essays in Environmental Philosophy*.[1] Here I focus especially on the concept of nature in Japanese Buddhism as a valuable complement to the environmental philosophy of Aldo Leopold. In this context I clarify the hierarchy of normative values whereby a land ethic is itself grounded in a land aesthetic in the ecological worldviews of both Japanese Buddhism in the East and Aldo Leopold in the West.

The Environmental Philosophy of Aldo Leopold

The Land Ethic

One can point to various sources for the newly emerging field of "environmental ethics," for instance, the romantic movement, beginning with Rousseau and running through Goethe and the romantic poets (Blake, Wordsworth, Coleridge, Shelley), continuing in America through the transcendentalism of Whitman, Emerson, and Thoreau, as well as later conservationists such as John Muir and Gary Snyder. However, the *locus classicus* for environmental ethics as a distinctive branch of philosophy is widely regarded by those in the discipline as a volume by Aldo Leopold entitled *A Sand County Almanac*, first published in 1949, and in particular the capstone essay of this work, called "The Land Ethic."[2] According to Leopold's threefold division of ethics, "The first ethics dealt with the relation between individuals. . . . Later accretions dealt with the relation between the individual and society."[3] It is here that he makes a significant leap by enlarging the field of ethics to include a third element: namely, the relation of humans to the land. In Leopold's words:

> There is yet no ethic dealing with man's relation to land and to the animals and plants which grow upon it. . . . The land-relation is still strictly economic, entailing privileges but not obligations. The extension of ethics to this third element in human environment is, if I read the evidence correctly, an evolutionary possibility and an ecological necessity.[4]

Leopold defines ethics in terms of his key notion of "community." An individual is always contextually located in a social environment, or as Leopold puts it, in communities of interdependent parts that evolve "modes of cooperation," called *symbioses* by ecologists. However, while in the past ethical discourse has been confined to the human community so as to pertain solely to the relation between individuals and society, environmental ethics extends this into the realm of the "biotic community" of soil, plants, and animals so as to include the symbiotic relation between humans and the land. He writes:

> All ethics so far evolved rest upon a single premise: that the individual is a member of a community of interdependent parts. . . . The land ethic simply enlarges the boundaries of the community to include soils, waters, plants, and animals, or collectively: the land.[5]

Leopold goes on to argue that "a land ethic changes the role of *Homo sapiens* from conqueror of the land-community to plain member and citizen of it."[6] Further, his land ethic redefines conservation from maximizing the utility of natural resources to "a state of harmony between men and land."[7] For Leopold, the principles of a land ethic not only impose obligations in the legalistic sense, but also entail the evolution of what he calls an "ecological conscience,"[8] understood as an "extension of the social conscience from people to land."[9] According to Leopold, then, a land ethic reflects the existence of an ecological conscience, and this in turn reflects an inner conviction of individual responsibility for the health of the land.[10]

The Conservation Aesthetic

In Aldo Leopold's ecological worldview his "land ethic" is inseparable from what he calls a "land aesthetic."[11] As Leopold writes in the original 1947 foreword to his work, "These essays deal with the ethics and esthetics of land."[12] It is significant that Leopold's *A Sand County Almanac* ends with an essay entitled "Conservation Esthetic."[13] For Leopold, it is the beauty or aesthetic value intrinsic to nature that places a requirement upon us to enlarge ethics to include the symbiotic relation between humans and land, to extend the social conscience from the human community to the biotic community, and thereby to establish an ecological harmony between people and their natural environment of soil, plants, and animals. The importance of this land aesthetic as the ground for a land ethic is further indicated by Leopold in his 1948 foreword to *A Sand County Almanac*, where he asserts that the essays contained in his work "attempt to weld three concepts": 1) "That land is a community. . .the basic concept of ecology"; 2) "that land is to be loved and respected. . .an extension of ethics"; and 3) "that land yields a cultural harvest" or, as he alternatively puts it, an "esthetic harvest."[14] According to Leopold, the norm for behavior in relation to land use is whether or not our conduct is aesthetically as well as ethically right. The beauty of the land is, therefore, one of the fundamental criteria for determining the rightness of our relationship to it: "A thing is right when it tends to preserve the integrity, stability, and beauty of the biotic community."[15] Hence, the architectonic structure of *A Sand County Almanac* suggests a kind of Peircean hierarchy of normative values whereby environmental ethics is itself grounded in the axiology of a conservation aesthetics.[16] In other words, our moral love and respect for nature is based on an aesthetic appreciation of the beauty and value of the land. It should be noted that Eugene C. Hargrove has pursued a similar line of reasoning, arguing that not only the land ethic but the historical foundation of all broad Western environmental sentiments is ultimately aesthetic.[17] Indeed, this aesthetic foundation for a land ethic is one of the deepest insights into the human/nature relation developed in the ecological worldviews both East and West.

Japanese Buddhism: An Asian Resource for Environmental Ethics

The principles of environmental ethics articulated by Aldo Leopold find a powerful source of support in the concept of living nature formulated by traditional Japanese Buddhism. A profound current of ecological thought runs throughout the Kegon, Tendai, Shingon, Zen, Pure Land, and Nichiren Buddhist traditions as well as modern Japanese philosophy. In what follows I briefly present the Japanese concept of nature as an aesthetic continuum of interdependent events based on a field paradigm of reality. In this context I show how the Japanese concept of nature entails an extension of ethics to include the relation between humans and the land. Moreover, I argue that the land ethic is itself grounded in a land aesthetic in the Japanese Buddhist concept of nature as well as for Aldo Leopold. I further seek to clarify the soteric concept of nature in Japanese Buddhism wherein the natural environment becomes the ultimate locus of salvation for all sentient beings. Finally, I argue that the Japanese Buddhist concept of nature represents a fundamental shift from the egocentric to an ecocentric position, that is, a non-anthropocentric standpoint which is nature centered as opposed to human centered.

The Field Model of Nature in Ecology and Japanese Buddhism

The environmental ethics of Aldo Leopold arises from a metaphysical presupposition that things in nature are not separate, independent, or substantial objects, but relational fields existing in mutual dependence upon each other, thus constituting a synergistic ecosystem of organisms interacting with their environment. According to Leopold's field concept of nature, the land is a single living organism wherein each part affects every other part, and it is this simple fact which imposes certain moral obligations upon us in relation to our environment. As J. Baird Callicott argues in "The Metaphysical Implications of Ecology," at the metaphysical level of discourse, ecology implies a paradigm shift from atomism to field theory.[18] In this context he underscores various metaphysical overtones in the "field theory of living nature adumbrated by Leopold."[19] Callicott, following the insights of Leopold, argues that "object-ontology is inappropriate to an ecological description of the natural environment," and adds, "Living natural objects should be regarded as ontologically subordinate to 'events'. . .or 'field patterns.' "[20] According to Callicott, in the worldview of ecology, as in the new physics, organisms in nature are a "local perturbation, in an energy flux or 'field' " so that the "subatomic microcosm" is analogous to the "ecosystemic macrocosm," "moments in [a] network" or "knots in [a] web of life."[21] He further points out that for the Norwegian environmental philosopher Arne Naess, ecology suggests "a relational total field image [in which] organisms [are] knots in the biospherical net of intrinsic relations."[22] It should be noted that in the Western philosophical tradition, the field concept of nature implied by ecology has received its fullest systematic expression in the process metaphysics and philosophy of organism developed by Alfred North Whitehead, which elaborates a pan-psychic vision of nature as a creative and aesthetic continuum of living field events arising through their causal relations to every other event in the continuum.[23]

The primacy accorded to "relational fields" over that of the "substantial objects" implicit in the ecological worldview is also at the very heart of the organismic paradigm of nature in East Asian philosophy, especially Taoism and Buddhism. In his article "Putting the *Te* Back into Taoism," Roger T. Ames interprets the key ideas of *te* and *tao* in the Taoist aesthetic view of nature as representing a "focus/field" model of reality with clear implications for an environmental ethic.[24] Likewise, Izutsu Toshihiko in *Toward a Philosophy of Zen Buddhism* has clearly explicated what he refers to as "the field structure of Ultimate Reality" in traditional Japanese

Zen as well as Kegon (Chinese, Hua-yen) Buddhism, in which each event in nature is understood as a concentrated focus point for the whole field of emptiness (*kū*) or nothingness (*mu*), comprehended in Buddhist philosophy as a dynamic network of causal relationships, in other words, the process termed "dependent co-origination" (*engi*).[25] Moreover, this traditional Zen and Kegon Buddhist field model of reality has been reformulated in terms of the concept of *basho* or "field" (locus, matrix, place) in the modern Japanese syncretic philosophy of Nishida Kitarō (1870–1945) and the Kyoto School: namely, what Nishida calls *mu no basho*, the field of nothingness.[26] Nishida's concept of *basho* or field was itself profoundly influenced by Lask's scientific *Feldtheorie* (field theory). As Matao Noda has observed, "In this connection the modern physical concept of field of force, taken by Einstein as a cosmic field, seems to have suggested much to Nishida."[27]

The primacy of *basho* or relational fields in modern Japanese philosophy has been developed specifically with regard to the human/nature relationship in the ethics of Watsuji Tetsurō (1889–1960), Nishida's younger colleague in the philosophy department at Kyoto University. In his work *Ethics as Anthropology* (*Ningen no gaku toshite no rinrigaku*), Watsuji calls his "ethics" (*rinrigaku*) the science of the person, based upon the Japanese concept of human nature as *ningen*, whose two *kanji* characters express the double structure of selfhood as being both "individual" and "social."[28] Accordingly, the "person" as *ningen* means not simply the individual (*hito*) but also the "relatedness" or "betweenness" (*aidagara*) in which people are located. In his book entitled *The Body*, the Japanese comparative philosopher Yuasa Yasuo clearly expresses the relation of Watsuji's concept of person (*ningen*) as the life-space of "betweenness" in which people are situated to the general idea of *basho* as a relational field or spatial locus. He writes: "But what does it mean to exist in betweenness (*aidagara*)? . . . Our betweenness implies that we exist in a definite, spatial *basho* (place, topos, field)."[29] However, Watsuji's ethics based on the double structure of personhood as *ningen* does not emphasize the spatial locus of relationships between individual and individual or between the individual and the social only; rather, he further extends his moral considerations to the relationship between the individual and nature. In *Climate, an Anthropological Consideration* (*Fūdo ningengakuteki kosatsu*), Watsuji develops as his main philosophical theme the embodied spatiality of human existence in various social environments, so that the individual both influences and is influenced by the family, the community, and ultimately the natural environment of a *fūdo* or "climate."[30] As Yuasa puts it, "Watsuji wrote a book called *Climate* in which he said that to live in nature as the space of the life-world—in other words, to live in a 'climate'—is the most fundamental mode of being human."[31] Hence, Watsuji clearly formulates an ethics in which the individual must be conceived as being situated in a spatial field of relatedness or betweenness not only to human society but also to a surrounding climate (*fūdo*) of living nature as the ultimate extension of embodied subjective space in which man dwells. Watsuji's ethical philosophy is, therefore, one of the most suggestive Asian resources for environmental ethics as outlined by Aldo Leopold, in which morality is enlarged so as to include not simply individual/individual and individual/social relations, but also the encompassing human/nature relation as a major extension of practical ethics.

The Japanese Concept of Nature: A Unity of Onozukara/Mizukara

The extension of ethics to include the human/nature relationship in the philosophy of Watsuji Tetsurō itself reflects a traditional Japanese concept of living nature as a unity of *onozukara* (nature) and *mizukara* (self). The Japanese term for nature, *shi-zen* (also pronounced *ji-nen*), originally derived from the Chinese word *tsu-jan*, corresponds to the English word "nature," which comes from the Latin *natura,* which was used by the Romans to translate the Greek term *physis*. As various scholars have pointed out, the Japanese concept of *shizen/jinen* can be compared to the ancient Greek concept of nature through Heidegger's uncovering of the original Greek understanding of *physis* as that which presences or unfolds of itself into primordial appearance as openness, unhiddenness, and nonconcealment. In ancient Japanese, a common expression for *shizen/jinen* was *onozukara*, which like Greek *physis* indicates "what-is-so-of-itself." *Onozukara*, written with the first of the two characters for *shi-zen*, also stands for another original

Japanese term, *mizukara* or "self." The implications of this connection have been clarified by Hubertus Tellenbach and Kimura Bin in their article "The Japanese Concept of Nature":

> As of itself *Onozukara* expresses an objective state. . . . *Mizukara* as self expresses, on the other hand, a subjective state. . . . That the Japanese believe they can express these seemingly autonomous terms by means of a single character points towards a deeper insight by which they apprehend *Onozukara* and *Mizukara*, nature and self, as originating from the same common ground.[32]

Consequently, in the Japanese concept of nature as a unity of *onozukara/mizukara*, both self and nature are grounded in a common field of reality as the subjective and objective aspects of a single continuum or relational matrix.

One of the most interesting expressions of this traditional Japanese view of nature as a unity of *onozukara/mizukara* is to be found in the concept of *eshō funi* or "oneness of life and its environment" formulated by Nichiren Daishonin (1222–1282) and his followers in the Nichiren Shōshū sect of Buddhism. Nichiren is most famous for his apocalyptic teaching that enlightenment can be attained in the Latter Day of the Law (*mappō*) only by reciting the *daimoku* or title of the Lotus Sutra, *Myōhō renge kyō*, which he inscribed in a *maṇḍala* called the *Daigohonzon* for the purpose of awakening Buddhahood in all sentient beings. In his eschatological and apocalyptic teaching about *mappō*, Nichiren prophesied that not only human social disasters, like civil wars and foreign invasions, but also such natural catastrophes as floods, fires, earthquakes, droughts, plagues, and other calamities would all result from a failure of people to follow the Mystic Law of cause/effect, which he called *Myōhō renge kyō*. For Nichiren, *Myōhō renge kyō* is the Mystic Law of life itself which embodies the supreme principle of Tendai (Chinese, T'ien-t'ai) Buddhism known as *ichinen sanzen*, "three thousand worlds in one life-moment." Moreover, as the embodiment of *ichinen sanzen*, the Mystic Law of *Myōhō renge kyō* contains the principle of *eshō funi*, the "oneness of life and its environment." In his text, "The True Entity of Life," Nichiren writes: "Where there is an environment, there is life within it. Miao-lo states, 'Both life (*shōhō*) and its environment (*ehō*) always manifest

Myōhō renge kyō.'"[33] By this view, both the subjective human being and its objective environment are two aspects of a single reality, the true entity of life, in other words, the Mystic Law of *Myōhō renge kyō*. In his exegesis of the above passage by Nichiren, Ikeda Daisaku concludes: "People (*shōhō*) and their environments (*ehō*) are inseparable. . . . Both are aspects of the Law of *Myōhō renge kyō*. . . . Thus we can see the powerful principle in Buddhism that a revolution within life (*shōhō*) always leads to one in the environment (*ehō*)."[34] From this insight it follows that at the level of practice, the inseparability of life and its environment is discovered by fusing with the Mystic Law, which in Nichiren Buddhism is caused by reciting the mantric formula "*Namu myōhō renge kyō*." Furthermore, chanting "*Namu myōhō renge kyō*" is thought to produce a "human revolution," that is to say, a transformation of subjective selfhood which in turn effects a corresponding change in the objective environment, thereby resulting in the metamorphosis of nature into a Buddha land of peace and harmony. Hence, according to Nichiren Buddhism, the principle of *eshō funi* constitutes the doctrinal foundation for an ecological worldview based on the inseparability of life and its environment.

The Kegon Infrastructure of Nature in Zen Buddhism

In the case of Nichiren Buddhism, the concept of nature as a cosmic field in which life and its environment are integrated is explained by invoking the master concept of Tendai Buddhism, namely, *ichinen sanzen*, "three thousand worlds in one life-moment." However, in Zen Buddhism, this kind of field theory of nature is elaborated in terms of an analogous Kegon (Hua-yen) Buddhist doctrinal formula known as *riji muge* (Chinese, *li-shih wu-ai*), the "interpenetration of part and whole." Like the *ichinen sanzen* principle of Tendai, the *riji muge* principle of Kegon articulates a microcosmic/macrocosmic paradigm of reality which depicts nature as a sacred matrix of interrelationships. This Kegon infrastructure underlies not only traditional Zen Buddhist teachings but also the modern Japanese philosophy of Nishida and the Kyoto School.[35] The profound ecological worldview implicit in the Kegon or Hua-yen vision of organismic interrelatedness is discussed by Francis H.

Cook in his essay "The Jewel Net of Indra." At the outset he writes:

> Only very recently has the word "ecology" begun to appear in our
> discussion, reflecting the arising of a remarkable new consciousness
> of how all things live in interdependence. . . . The ecological
> approach. . . views existence as a vast web of interdependencies in
> which if one strand is disturbed, the whole web is shaken.[36]

Cook goes on to situate the ecological model of organismic
interdependence in a wider context by discussing the relationship
between humans and nature in the "cosmic ecology" of Hua-yen
Buddhism.[37] He presents the Chinese Hua-yen vision of nature in
terms of the microcosmic/macrocosmic paradigm expressed by the
famous metaphor of Indra's net, which depicts a cosmic web of
dynamic causal interrelationships wherein at every intersection in
the latticework there is a glittering jewel reflecting all the other
jewels in the net, infinite in number.[38] In the pattern of inter-
connectivity depicted by Indra's net, each and every event in nature
arises through an interfusion of the many and the one, thus being
likened to a shining jewel which both contains and pervades the
whole universe as a microcosm of the macrocosm. By this view,
all events arise through their functional relationships to all the other
events and to the whole so that each thing is interconnected to
everything else in the aesthetic continuum of nature. This relational
cosmology is codified by the famous doctrinal formulas of Kegon
Buddhism, named *riji muge* (Chinese, *li-shih wu-ai*) or "interpenetration of part and whole" and *jiji muge* (Chinese, *shih-shih wu-ai*)
or "interpenetration of part and part." In such a manner Hua-yen
Buddhism has established a compelling axiological cosmology,
according to which, given that everything functions as a causal
condition for everything else, there is nothing which is not of value
in the great harmony of nature. This view further entails a morality
of unconditional compassion and loving kindness for all sentient
beings in nature. Hence, it can be argued that Hua-yen Buddhism
has provided an explicit, comprehensive, and systematic relational
cosmology which fully supports the fundamental principles of
ecological ethics propounded by Aldo Leopold and other environmental philosophers, whereby the atomistic paradigm of nature is
wholly abandoned in favor of a model of organismic interdependence.

The Aesthetic Concept of Nature in
Japanese Buddhism

Scholars of Asian civilization have often pointed to the primacy of aesthetic value as the distinguishing feature of traditional Japanese Buddhist culture. During the medieval period of Japanese history (ca. 950–1400), art and religion were fused to the extent that spiritual and aesthetic values became virtually identified in what was called *geidō*—the "*tao* (or Way) of art." In the concept of nature developing out of Japanese *geidō*, the natural environment is seen as laden not only with aesthetic but also religious values so that it becomes the ultimate ground and source of salvation itself. This Japanese aesthetic concept of nature has long been articulated by a lexicon of technical terms based on the canons of art and literature, including *aware*, *yūgen*, *wabi*, *sabi*, and *yojō*. In Japanese Buddhism nature is conceived not in eternalist or substantialist terms as static being, but through process categories as a dynamic becoming, that is, *mujō* or "impermanence." Yet, as opposed to a nihilistic view of becoming, Japanese Heian poetics affirms the positivity of nature as a flux of impermanence with the aesthetic value notion of *aware*, the sorrow-tinged appreciation of transitory beauty. In this way the Japanese value-centric concept of nature as creative and aesthetic process is a worldview based on the Middle Way between eternalism on the one side and nihilism on the other. Moreover, in the *waka* poetry of Fujiwara Teika, the *sumie* monochrome inkwash paintings of Sesshū, and the *Noh* drama of Zeami, the beauty of *yūgen* or "mysterious depths" was evoked by visions of nonsubstantial phenomena in nature fading into the background field of *mu* or nothingness. In *chanoyu*, or the tea ceremony of Sen no Rikyū, nature is described in terms of *wabi*, the beauty of simplicity and poverty, while the *haiku* poetry of Bashō conjures the feeling of *sabi*, the beauty of the solitude and tranquility of events in nature. All of these aesthetic value categories are regarded as aspects of *yojō* or "overtones of feeling," reflecting a deeply emotional and artistic sensitivity to the sublime beauty of nature as a continuum of organismic relationships and dynamic processes.

In "Love of Nature," the final chapter of his book *Zen and Japanese Culture*, D. T. Suzuki underscores the Kegon or Hua-yen

(Sanskrit, Avataṃsaka) infrastructure underlying the traditional aesthetic concept of nature in Japanese Zen Buddhism. Suzuki writes, "The balancing of unity and multiplicity or, better, the merging of self with others in the philosophy of Avatamsaka (Kegon) is absolutely necessary to the aesthetic understanding of Nature."[39] According to the organismic paradigm of Zen and Kegon Buddhism, nature is to be comprehended as an undivided aesthetic continuum wherein each momentary and unsubstantial event arises through the harmonic interfusion of oneness and multiplicity, unity and plurality, or subjectivity and objectivity, thus emerging as a cosmic field of relationships which both contains and pervades the universe as a microcosm-qua-macrocosm. Because for Zen there is a mutual containment or reciprocal penetration of subject and object, there is said to be a continuity or interfusion between humans and nature. In Suzuki's words:

> Zen proposes to respect Nature, to love Nature, to live its own life; Zen recognizes that our Nature is one with objective Nature...in the sense that Nature lives in us and we in Nature. For this reason, Zen asceticism advocates simplicity, frugality, straightforwardness, virility, making no attempt to utilize Nature for selfish purposes.[40]

I would like to make two observations about this passage concerning the relation of Zen to Aldo Leopold's environmental ethics. First, as Suzuki points out, the insight that humans and nature are interdependent has led to Zen ideals of simplicity, frugality, and poverty in relation to land use so that nature is not exploited out of selfish motivations. Hence, in his famous work *Small Is Beautiful: Economics As If People Mattered*, E. F. Schumacher synthesizes the environmental ethics of Leopold with the Zen ecology of nature to develop what he calls a "Buddhist economics" oriented toward attaining given ends with minimal consumption.[41] Second, the Zen Buddhist love and respect for nature described by Suzuki in this passage directly accords with a major theme in the environmental philosophy of Leopold, namely, "that land is to be loved and respected [a]s an extension of ethics."[42] This love and respect for the natural world, viewed as an extension of ethics, is itself directly related to the aesthetic and religious concept of nature. From a comparative standpoint, these connections can be helpful in illuminating the axiological foundations underlying the ecological worldview of Aldo Leopold, in which the land ethic is grounded in a conservation aesthetic.

The Salvific Function of Nature in Japanese Buddhism

The religio-aesthetic concept of nature as a continuum funded with value and beauty is a correlate to what can be referred to as the "salvific function" of nature in traditional Japanese Buddhism. A paradigm of one who endeavors to find salvation through nature is provided by a novel entitled *Kusamakura* (Grass pillow) by Sōseki Natsume.[43] This novel describes the *haiku* journey of a twentieth-century artist-poet from Tokyo who ventures into the solitude of a mountain wilderness for the sole purpose of attaining Zen *satori* or enlightenment through the tranquil beauty of nature. By exercising aesthetic detachment the poet hero of *Kusamakura* attempts to envision all things in the landscape as displaying the religio-aesthetic value of *yūgen*, "mystery and depth," such that everything in nature is transformed into a scene from a monochrome *sumie* inkwash painting, a *Noh* drama, or a *haiku* poem. In this way, living nature is prized not only for its beauty but also for its salvific function as the ultimate locus for spiritual awakening.

Sōseki's artist hero is a modern literary prototype for a long and profound tradition of Japanese figures seeking salvation through nature by means of the religio-aesthetic path of *geidō* or the "*tao* of art," including Teika, Saigyō, Bashō, Sesshu, and Sen no Rikyū. In his article "Probing the Japanese Experience of Nature," Omine Akira traces this soteric concept of nature in the Japanese literary tradition beginning with the earliest eighth-century anthology, called the *Man'yo-shu* (Collection of myriad leaves), and running through Saigyō (1118–1190), Ippen (1239–1289), and Bashō (1644–1694) as set in the context of the Japanese Buddhist worldview formulated by Zen master Dōgen (1200–1253) as well as the founder of True Pure Land Buddhism (Jōdō Shinshū), Shinran (1173–1263). Omine emphasizes the religio-aesthetic concept of nature in this tradition as having two aspects: "nature as companion and nature as Buddha."[44] When viewed as friend or companion, nature holds the significance of the Buddhist terms "sentient being" or "living things" (*shujo*), such that mountains and rivers, stones and trees, flowers and birds all have the potential for enlightenment and tread the path to Buddhahood together. The other aspect is nature, just as it is, as sacred Buddha.[45] In this context, he quotes directly from Dōgen's "Sutra of Mountains and Waters" (*Sansui-kyō*), the twenty-ninth chapter of *Shōbōgenzō*: "Mountains and rivers right now are the emerging presence of the ancient Buddhas."[46] As implied by Dōgen's theories of *hosshin seppō*, "the Dharmakaya expounds the dharma," and *genjōkōan*, "presencing things as they are," mountains, rivers, and all phenomena in nature are presencing forth in their suchness so as to disclose the Buddha-nature inherent in all things, understood in Dōgen's Buddhist philosophy of *uji* or "being-time" as *mujō-busshō*, "impermanence-Buddha-nature." Omine further makes reference to Shinran's Pure Land theory of salvation by the grace of "Other-power" (*tariki*), reformulated in later writings through his famous doctrine *jinen honi*, "naturalness." To be saved by Buddha, to be born in the Pure Land, is simply a function of *jinen* (*shizen*), "nature," defined by Shinran as "from the very beginning made to become so."[47] Omine concludes with his assessment that Shinran's Pure Land Buddhist notion of *jinen honi* reflects an ancient Japanese concept of living nature as the ground and source of human salvation.

The soteriological function of nature in the poetics of Saigyō and the Japanese literary heritage as understood against the background of traditional Buddhist philosophy has also been developed in a fine scholarly essay by William R. LaFleur, "Saigyō and the Buddhist Value of Nature."[48] LaFleur demonstrates that Saigyō must be interpreted in the historical context of a Buddhist tradition including both Saichō (767–822) and Kūkai (774–835) which regards "nature as a locus of soteriological value."[49] This tradition emphasizes the capacity of nature to provide solace and some type of "salvation" for individuals looking for a locus of value other than that provided by city life.[50] Buddhist philosophers in this tradition underscore the potential Buddhahood of all things in nature so as to dissolve the older distinction between sentient (*yūjō*) and insentient (*mujō*) beings.[51] LaFleur argues that Buddhism in Japan developed arguments on behalf of the Buddhahood potentialities of the natural world, because it was compelled to accommodate itself to the longstanding and pre-Buddhist (Shintō) attribution of high religious value to nature as the locus of salvation.[52] He summarizes the soteric function of nature depicted in the poetry of Saigyō as follows:

> The natural "images" in Saigyō's poetry are not something which must themselves be transcended. . . . For Kūkai and for Saigyō, there is no beyond. The concrete phenomenon. . . is both the symbol and the symbolized. It is the absolute which theorists might call "Emptiness," but which is, in fact, nothing other than the phenomenon itself.[53]

Hence, as LaFleur emphasizes here, the understanding of the religio-aesthetic function of poetic symbols in Saigyō and the Japanese tradition of nature poetry is derived from the Mikkyō (Tantric) tradition of Saichō and Kūkai wherein Buddhahood can be revealed only through "expressive symbols" (*monji*). In accord with Japanese Mikkyō Buddhism, the aesthetic and spiritual symbols of Saigyō's nature poetry do not point beyond themselves to a transcendent or supra-sensible reality over and above the natural world, but fully contain the reality which they symbolize.

In the final analysis, this traditional soteric concept of nature in Japan is itself grounded in a Mahāyāna Buddhist metaphysic of Emptiness (Japanese, *kū*; Sanskrit, *śūnyatā*), wherein the mountains and rivers of the natural world, just as they are here and now, are the revelation of impermanence-Buddha-nature in the dynamic and nonsubstantial flux of being-time. According to the Japanese Buddhist doctrine of emptiness, there is nothing which is "more real" beyond the interdependence of everything in nature. The Buddhist metaphysics of emptiness, with its explicit identification of *saṃsāra* and *nirvāṇa*, therefore results in the complete dialectical interfusion of transcendence and immanence, absolute and relative, or sacred and profane. In this way, Japanese Buddhism overturns all models of transcendence and dualism so as to effect a radical paradigm shift from "otherworldliness" to "this-worldliness." For Japanese Buddhism, ultimate reality is to be found not in a transcendent beyond as in the conventional Judeo-Christian paradigm, but in fields of interrelationships which confer to each event a boundless depth of aesthetic and religious value. It is in this philosophical context that nature becomes the "locus of salvation" in traditional Japanese Buddhism as reflected by poet-seers following the religio-aesthetic path of *geidō* in Japan.

Conclusion: An East-West Gaia Theory of Nature

In East Asia the delicate harmony between humans and nature has
long been maintained through geomancy, what is known in China
as *feng shui*. In his book *Feng Shui: The Chinese Art of Designing
a Harmonious Environment*, Derek Walters defines *feng shui* as
follows: "A complex blend of sound commonsense, fine aesthetics,
and mystical philosophy, *Feng Shui* is a traditional Chinese technique which aims to ensure that all things are in harmony with their
environment."[54] Walters further explains that the geomantic philosophy of *feng shui* came to permeate every aspect of traditional
Japanese culture, including city planning, temple construction,
inkwash painting, flower arranging, and gardening. He adds:
"Indeed, there are few areas of Japanese thought which are not in
some way affected by the influence of *Feng Shui*."[55] Long before
the discovery of the earth's magnetic field and the modern physics
theory of lines of force, nature was conceived as an energy pattern
comprised of flowing *ch'i* (Japanese, *ki*) or vital-power, a grid
network of intersecting *yin/yang* forces, known as *lung-mei* or
"dragon and tiger" currents in the study of *feng shui*.[56] As Tu
Weiming puts it in "The Continuity of Being: Chinese Visions of
Nature," according to the Chinese "philosophy of *ch'i*," which later
spread to Japan, the earth forms one body as a single living organism
created out of the interfusion and convergence of numerous streams
of vital force which together establish the wholeness and continuity
of nature.[57]

Throughout *A Sand County Almanac* Aldo Leopold also describes
the land as "a single living organism," understood as an "energy
circuit," a "fountain of energy," a "flow of energy," and a "circuit
of life." He thus writes:

> Land, then, is not merely soil; it is a fountain of energy flowing
> through a circuit of soils, plants, and animals. . . . This inter-
> dependence between the complex structure of the land and its
> smooth functioning as an energy unit is one of its basic attributes.[58]

In this way, the ecological worldview of Aldo Leopold, along with
the geomantic philosophy of East Asia based on Taoism and
Buddhism, can be seen as providing theoretical support for what is

known in environmental philosophy as the Gaia theory. According to Gaia theory, the earth is a single living organism forming a vast biotic community in which a complex grid network of energy currents or lines of force constitutes nature as a synergistic ecosystem of symbiotic relationships in an interconnected web of life.[59] It is precisely such an East-West Gaia theory of living nature which might point a way toward healing our plundered planet, overcoming today's environmental crisis, and establishing a harmony between man and the land.

Notes

1. J. Baird Callicott and Roger T. Ames, eds., *Nature in Asian Traditions of Thought: Essays in Environmental Philosophy* (Albany: State University of New York Press, 1989).

2. Aldo Leopold, *A Sand County Almanac: With Essays on Conservation from Round River* (New York: Ballantine Books, 1966). See especially "The Land Ethic," 237–64. All citations refer to this edition unless otherwise noted.

3. Ibid., 238.

4. Ibid., 238–39.

5. Ibid., 239.

6. Ibid., 240.

7. Ibid., 243.

8. Ibid.

9. Ibid., 246.

10. Ibid., 258.

11. For a careful scholarly analysis of Aldo Leopold's land ethic in relation to his land aesthetic as formulated in *A Sand County Almanac*, see J. Baird Callicott, "The Land Aesthetic," in *Companion to* A Sand County Almanac, ed. Callicott (Madison: University of Wisconsin Press, 1987), chap. 7, 157–71.

12. Aldo Leopold, foreword to Callicott, *Companion to* A Sand County Almanac, 281. This statement is the opening sentence in Aldo Leopold's original foreword to "Great Possessions" (the author's own title of what later became *A Sand County Almanac*), dated 31 July 1947. However, the revised foreword dated 4 March 1948 is the one with which readers are now familiar.

13. "Conservation Esthetic" is the final essay in the enlarged edition of *A Sand County Almanac*, subtitled *With Essays on Conservation from Round River*, edited by Luna B. Leopold, the author's son, and first published by Ballantine Books in 1966. "The Land Ethic" is the final essay in the original edition, subtitled *And Sketches Here and There*, published posthumously by Oxford University Press in 1949. However, "Great Possessions," Leopold's manuscript of the book, conformed to the arrangement of the enlarged edition of 1966. According to Dennis Ribbens, "possibly the most important change to the manuscript after Leopold's death was the decision to shift 'The Land Ethic' from its original first position in Part III to its present final position" (Dennis Ribbens,"The Making of *A Sand County Almanac*," in *Companion to* A Sand County Almanac, 107). Confirming Ribbens's information, Curt Meine says that in the process of preparing Leopold's manuscript for publication, " 'The Land Ethic' was moved to the end of Part III, 'The Upshot' " (Curt Meine, *Aldo Leopold: His Life and Work* [Madison: University of Wisconsin Press, 1988], 524). Thus, it appears that the author himself intended for "Conservation Esthetic" to be the final essay of the book that became *A Sand County Almanac*.

14. Leopold, *A Sand County Almanac*, xix.

15. Ibid., 262.

16. In the context of developing his "architectonic" of theories, the American process philosopher Charles Sanders Peirce suggested a "hierarchy of the normative sciences" in which logic depends on ethics and ethics depends on aesthetics. See *Philosophical Writings of Peirce: Selected and Edited with an Introduction by Justus Buchler* (New York: Dover Publishing, 1955), 62.

17. Eugene C. Hargrove, *Foundations of Environmental Ethics* (Englewood Cliffs, N.J.: Prentice Hall, 1989). See especially chap. 3, "Aesthetic and Scientific Attitudes" (77–103), and the section entitled "The Aesthetics of Wildlife Preservation" (122–23) in chap. 4, "Wildlife Protection Attitudes."

18. J. Baird Callicott, "The Metaphysical Implications of Ecology," in Callicott and Ames, *Nature in Asian Traditions of Thought*, 51–64.

19. Ibid., 57.

20. Ibid., 58.

21. Ibid., 59.

22. Arne Naess, "The Shallow and the Deep, Long-Range Ecology Movement: A Summary," *Inquiry* 16 (1973):98.

23. For examples of Alfred North Whitehead's concept of nature as aesthetic and creative process in relation to this general philosophy of organism, see *The Concept of Nature* (Cambridge: Cambridge University Press, 1971); "The Romantic Reaction," in *Science and the Modern World* (New York: Collier Macmillan, 1925); "The Order of Nature" and "Organisms and Environment," in *Process and Reality* (New York: Free Press, 1978); and "Nature and Life," in *Modes of Thought* (New York: Macmillan, 1968).

24. Roger T. Ames, "Putting the *Te* Back into Taoism," in Callicott and Ames, *Nature in Asian Traditions of Thought*, 113–44.

25. Toshiko Izutsu, *Toward a Philosophy of Zen Buddhism* (Boulder, Colo.: Prajna Press, 1982), especially "The Field Structure of Ultimate Reality," 45–49.

26. Nishida first developed his notion of *basho*, or field, as the fundamental concept in his philosophy of nothingness in *Hataraku mono kara miru mono e* (From the acting to the seeing), published in 1927, included in *Nishida Kitarō zenshū* (Tokyo: Iwanami Shoten, 1947–53), 4:207–89.

27. Matao Noda, "East-West Synthesis in Kitarō Nishida," *Philosophy East and West* 4 (1955):350.

28. Watsuji Tetsurō, *Ningen no gaku toshite no rinrigaku* (Tokyo: Iwanami Shoten, 1936).

29. Yuasa Yasuo, *The Body: Toward an Eastern Mind-Body Theory*, ed. Thomas P. Kasulis, trans. Nagatomo Shigenori and Thomas P. Kasulis (Albany: State University of New York Press, 1987), 38.

30. Watsuji Tetsurō, *Fūdo ningengakuteki kosatsu*, 2d ed. (Tokyo: Iwanami Shoten, 1951). For an English translation, see *A Climate: A Philosophical Study*,

trans. Geoffrey Bownas (Tokyo: Japanese National Commission for UNESCO, 1961), now reprinted as a volume in the new series, Classics of Modern Japanese Thought and Culture (Westport, Conn.: Greenwood Press, 1989).

31. Yuasa Yasuo, *The Body: Toward an Eastern Mind-Body Theory*, 38.

32. Hubertus Tellenbach and Bin Kimura, "The Japanese Concept of Nature," in Callicott and Ames, *Nature in Asian Traditions of Thought*, 154–55.

33. Daisaku Ikeda, *Selected Lectures on the Gosho* (Tokyo: Nichiren Shōshū International Center, 1979), 22.

34. Ibid., 23.

35. For an account of Hua-yen (Japanese, Kegon; Sanskrit, Avataṃsaka) Buddhism developed from the standpoint of Alfred North Whitehead's philosophy of organism, see Steve Odin, *Process Metaphysics and Hua-yen Buddhism* (Albany: State University of New York Press, 1982).

36. Francis H. Cook, "The Jewel Net of Indra," in Callicott and Ames, *Nature in Asian Traditions of Thought*, 213.

37. Ibid., 214.

38. Ibid., 226.

39. D. T. Suzuki, *Zen and Japanese Culture* (Princeton: Princeton University Press, 1959), 354.

40. Ibid., 351–52.

41. E. F. Schumacher, *Small Is Beautiful: Economics As If People Mattered* (New York: Harper and Row, 1973). See especially pt. 1, chap. 4, "Buddhist Economics," 53–62.

42. Leopold, *A Sand County Almanac*, xix.

43. Sōseki Natsume, *Kusamakura* (Tokyo: Kodansha, 1972). For an English translation, see *The Three-Cornered World*, trans. Alan Turney (Tokyo: Charles E. Tuttle, 1968).

44. Omine Akira, "Probing the Japanese Experience of Nature," translated from *Nihon-teki Shizen no Keifu* by Dennis Hirota, *Chanoyu Quarterly: Tea and the Arts of Japan* 51 (1987).

45. Ibid., 7.

46. Ibid., 19.

47. Ibid., 28.

48. William R. LaFleur, "Saigyō and the Buddhist Value of Nature," in Callicott and Ames, *Nature in Asian Traditions of Thought*, 183–209.

49. Ibid., 196.

50. Ibid.

51. Ibid., 186–87.

52. Ibid., 195.

53. Ibid., 203.

54. Derek Walters, *Feng Shui: The Chinese Art of Designing a Harmonious Environment* (New York: Simon and Schuster, 1988), 8.

55. Ibid., 14.

56. Ibid., 10.

57. Tu Wei-ming, "The Continuity of Being: Chinese Visions of Nature," in Callicott and Ames, *Nature in Asian Traditions of Thought*, 67–78; also published in Tu Wei-ming, *Confucian Thought: Selfhood as Creative Transformation* (Albany: State University of New York Press, 1985).

58. Leopold, *A Sand County Almanac*, 253–54.

59. See J. E. Lovelock, *Gaia: A New Look at Life on Earth* (New York: Oxford University Press, 1979).

Voices of Mountains, Trees, and Rivers: Kūkai, Dōgen, and a Deeper Ecology*

Graham Parkes

Although environmental problems are now attaining global proportions, discussion of them tends to be conducted in quite parochial terms. Current debates for the most part presuppose a worldview with its roots in Europe—one informed by the Platonic/Judeo-Christian tradition as well as Cartesian philosophy and Newtonian science. Even though contemporary physics and biology are giving us a very different picture of the world from that envisaged by Newton and Descartes, the fact that these two figures enabled the development of modern technology has preserved the viability of their worldview and extended it over most of the globe. Belief in the natural superiority of human beings and justification for their domination of a supposedly soulless world stem from this religious and philosophical worldview, which continues to inform—even if in less arrogant forms—current debates in the ethics of environmental concern.

It may be a sign of progress when people begin to acknowledge the "rights" of beings other than humans, but the language is still too parochial. If the East Asian traditions, for example, contain nothing that corresponds to our conception of rights—and they do not—then talk of the rights of trees will have no more effect on Japanese timber interests than talk of human rights has on Chinese politicians. What is needed is a more radical revisioning of the human relation to the natural world, a shift toward a less hubristic attitude toward the environment upon which our existence depends.

It is fashionable in some ecologically correct circles to ascribe blame for the devastation of the earth to the combination of

Christianity and capitalism that made possible the enormous material achievements of the industrialized nations of the West. While such criticisms are often rather facile, it does seem reasonable to suppose that where people's lives are informed by ways of thinking that denigrate the physical world in favor of a purely spiritual realm (as with the Orphic strain in Platonism), or by cosmogonies according to which the natural world was created for the benefit of humans as the only beings made in the image of the creator (as in the Genesis story), or by soteriologies where the soul is alienated from the natural world and the crucial question concerns the individual's direct relation to God (as in Gnostic Christianity and "the American religion"), they are going to have relatively few qualms about exploiting the natural world for their own purposes.[1]

The corollary seems equally reasonable: that where worldviews prevail in which nature is regarded as the locus of ultimate reality or value, as a sacred source of wisdom, or as a direct manifestation of the divine, one can expect that, other things being equal, people will restrain themselves from inflicting gratuitous harm on the environment. The nature of the connection between a religious or philosophical worldview and actual behavior is difficult to determine since, for the most part, other things are precisely not equal. An individual's desire for material well-being may occlude his or her self-understanding vis-à-vis the cosmos, and the demands of culture—and of contemporary consumerist culture especially—may overwhelm one's reverence for the natural world. But rather than attempt to untangle that complex of difficult issues, let us simply suppose that someone concerned about the fate of the earth were to realize, experientially, the validity of a worldview in which nature is seen as sacred and a source of wisdom. That person would then naturally incline (by virtue of the meanings of such terms as "ultimate value," "wisdom," and "the divine") to care for the natural environment on an individual level; and the deeper the experiential realization, the more one could expect that care to expand into the collective sphere. And if one could then find a way of imparting such a realization to a wider audience, considerable progress could be made toward solving environmental problems.

A proposal for a revisioning of our relations to the natural world comes with the program of "deep ecology," but this movement, insofar as it has been acknowledged at all, is often rejected for being too radical or else simply incoherent.[2] While the hearts of the deep ecologists are surely in the right places, their minds are not always so clear—especially when they wander as far afield as East Asia. This is regrettable because the East Asian philosophical world is especially rich in resources for ecological thinking. In what follows, I shall outline some features of the philosophies of two of the foremost figures in Japanese Buddhism, Kūkai and Dōgen, which would appear to be eminently salutary for the natural environment. There will be a need to respond to some doubts that may arise in this context, and to protest briefly a tendency toward simpleminded appropriation by some deep ecologists of Dōgen's ideas. A final concern will be the extent to which these ideas might be practically applied in the task of mitigating the environmental crisis.

Kūkai

When Buddhism was transplanted from India to China during the first century of the common era, some thinkers there began to ask—perhaps under the influence of Taoist ideas—whether the Mahāyāna Buddhist extension of the promise of Buddhahood to "all sentient beings" did not go far enough. A long-running debate began in China during the eighth century, in which thinkers in the T'ien-t'ai school argued that the logic of Mahāyāna universalism required that the distinction between sentient and nonsentient be abandoned and that Buddha-nature be ascribed not only to plants, trees, and earth, but even to particles of dust.[3] (The contrast with the Christian tradition is striking, where Aristotle's musings on the vegetal soul were largely ignored and arguments over the reaches of salvation were restricted to the question of whether animals have souls.)

When Buddhist ideas from China began to arrive in Japan in the seventh century, they entered an ethos conditioned by the indigenous religion of Shintō, according to which the natural world and human beings are equally offspring of the divine. In Shintō the whole world is understood to be inhabited by *shin* (*kami*), or divine spirits. These are spirits not only of the ancestors but also of any phenomena that occasion awe or reverence: wind, thunder, lightning, rain, the sun, mountains, rivers, trees, and rocks. Such an atmosphere was naturally receptive to the idea that the earth and plants participate in Buddha-nature. Although the first Japanese thinker to use the phrase *mokuseki bussho* ("Buddha-nature of trees and rocks") was apparently Saichō (766–822), founder of the Tendai school, the first one in Japan to elaborate the idea of the Buddhahood of all phenomena and make it central to his thought was Kūkai (774–835).

In a passage of verse in his essay "On the Meanings of the Word *Hūṃ*" (*Unji gi*), Kūkai twice alludes to the awakened nature of vegetation (*sōmoku*):

> If trees and plants are to attain enlightenment,
> Why not those who are endowed with feelings? . . .
> If plants and trees were devoid of Buddhahood,
> Waves would then be without humidity.[4]

In a later work he argues for the Buddhahood of *sōmoku* on the grounds that it is included within the "Five Great Elements" (earth, water, fire, wind, space) that comprise the *dharmakāya* (*hosshin*), or "reality embodiment" of the cosmic Buddha Dainichi Nyorai (Mahāvairocana).[5] He qualifies this statement by adding that the Buddha-nature of plants and trees is not apparent to normal vision, but can be seen only by opening one's "Buddha eye."

In distinguishing his own Esoteric Buddhism from other schools, Kūkai makes a more comprehensive claim concerning natural phenomena:

> In Exoteric Buddhist teachings, the four great elements [earth, water, fire, and wind] are considered to be nonsentient beings, but in Esoteric Buddhist teaching they are regarded as the *samaya*-body of the Tathāgata.[6]

There seems to be an equivocation here, however, when Kūkai calls the natural elements the *samaya*-body of the Buddha, since this connotes not simple identity with the *dharmakāya* but a relation of symbolizing *and* participation at the same time. The ambiguity is brought out in another passage, where Kūkai writes:

> The existence of the Buddha [Mahāvairocana] is the existences of the sentient beings and vice versa. They are not identical but are nevertheless identical; they are not different but are nevertheless different.[7]

It is interesting to note a similar equivocation in the philosophy of a close contemporary of Kūkai's in the West, John Scotus Erigena. (Their lives overlap by twenty-five years.) Erigena's major treatise—the *Periphuseōn*, or *De Divisione Naturae*, from the year 865—is on nature, and he argues there that the natural world *is* God "as seen by Himself" (704c). His understanding of the relation between God and the natural world is informed throughout by a tension between his Catholic faith and his devotion to Greek philosophy, as exemplified in the tension in Neoplatonic theology generally between God's emanation throughout creation (*processio Dei per omnia*) and His remaining in Himself (*mansius in se ipso*). Insofar as Erigena regards natural creatures as "theophany," he believes that they will ultimately be restored to their source in God—even though this restoration takes place only via the resurrection of the human. Dainichi is, for Kūkai, an "emanation throughout creation"; but his non-identity with, or difference from, sentient beings would not consist in his "remaining in himself." To the extent that he is the *dharmakāya*, which is "beginningless and endless," he would transcend the totality of all things that are currently present—but he would not transcend the totality of all things that have been, will be, and could be.

The practical (or practice-oriented) aspect of Kūkai's Esoteric Buddhism involves entering into what he calls the "three mysteries," or "intimacies" (*sanmitsu*), of Dainichi Nyorai, which are body, speech, and mind. Thus, by adopting certain postures (*mudrā*s), by chanting certain syllables (*mantra*s), and by allowing the mind to abide in the state of *samādhi*, or concentration, the practitioner will come to experience direct participation in the *dharmakāya*. We can be sure that those who successfully practice such a philosophy, realizing their participation in the body of the cosmic Buddha simultaneously with the divinity of natural phenomena, will treat the natural world with the utmost reverence.

There is another feature of Kūkai's teaching which helps illuminate the idea that natural phenomena possess Buddha-nature, and that is his notion of *hosshin seppō*, the idea that "the *dharmakāya* expounds the *dharma*," or, "the Buddha's reality embodiment expounds the true teachings."[8] This idea emphasizes the radically *personal* nature of Dainichi Nyorai in drawing attention to the way he teaches the truth of Buddhism through all phenomena, and through *speech* as one of the three "intimacies." The element of intimacy, or mystery, comes in because Dainichi's teaching is strictly, as Kūkai often emphasizes, "for his own enjoyment." It is only in a loose sense that the cosmos "speaks" to us—for, properly speaking, Dainichi does not expound the teachings for our benefit. (The other embodiments of the Buddha—the *nirmāṇakāya* and the *saṃbhogakāya*—perform that function.)

Just as visualization plays an important role in the meditation practices of Kūkai's Shingon Buddhism, so the sacred nature of the world is also accessible to the sense of sight. As well as hearing the cosmos as a sermon, Kūkai sees, or reads, the natural world as scripture. As he writes in one of his poems:

> Being painted by brushes of mountains, by ink of oceans,
> Heaven and earth are the bindings of a sutra revealing the truth.[9]

In this respect there are remarkable parallels between Kūkai and the seventeenth-century German thinker Jakob Böhme. Not only is the natural world of paramount soteriological importance for them both, but their suggested ways of realizing this, by meditation on images and sounds, are interestingly comparable. In reverting to the root syllables of the Sanskrit in which the mystical aspects of early Buddhism were embodied, Kūkai employs them as sounds as well as visual images. Böhme is equally concerned with mystic syllables, in his native German as well as in the Latin and Hebrew of the alchemical and kabbalistic traditions. And just as for Kūkai nature is Dainichi Nyorai expounding the teachings for his own enjoyment, so for Böhme the natural world is the "corporeal being" of the Godhead in its joyous self-revelation.[10]

Dōgen

The philosophy of Dōgen (1200–1253) shares many roots with Kūkai's thought, and his understanding of the natural world is especially similar (no doubt owing to some influence). Parallel to Kūkai's identification of the *dharmakāya* with the phenomenal world is Dōgen's bold assertion of the nonduality of Buddha-nature and the world of impermanence generally. He rereads the line from the *Nirvāṇa Sūtra* "All sentient beings without exception have Buddha-nature" as "All is sentient being, all beings are Buddha-nature."[11] Dōgen thus argues that all beings are sentient being, and as such *are* Buddha-nature—rather than "possessing" or "manifesting" or "symbolizing" it. Again, however, the usual logical categories are inadequate for expressing this relationship. Just as Kūkai equivocates in identifying the *dharmakāya* with all things, so Dōgen says of all things and Buddha-nature: "Though not identical, they are not different; though not different, they are not one; though not one, they are not many."[12] Again as in Kūkai, while the natural world is ultimately the body of the Buddha, it takes considerable effort to be able to see this. Dōgen regrets that most people "do not realize that the universe is proclaiming the actual body of Buddha," since they can perceive only "the superficial aspects of sound and color" and are unable to experience "Buddha's shape, form, and voice in landscape."[13]

Perhaps in order to avoid the absolutist connotations of the traditional idea of the *dharmakāya*, Dōgen substitutes for Kūkai's *hosshin seppō* the notion of *mujō-seppō*, which emphasizes that even *nonsentient* beings expound the true teachings. They are capable of this sort of expression since they, too, are what the Buddhists call *shin* ("mind/heart"). And just as the speech of Dainichi Nyorai is not immediately intelligible to us humans, so, for Dōgen:

> The way insentient beings expound the true teachings should not be understood to be necessarily like the way sentient beings do. . . .
> It is contrary to the Buddha-way to usurp the voices of the living and conjecture about those of the non-living in terms of them.[14]

Only from the anthropocentric perspective would one expect natural phenomena to expound the true teachings in a human language.

While the practice followed in Dōgen's Sōtō Zen is less exotic than in Kūkai's Shingon, the aim of both is the integration of one's activity with the macrocosm. Whereas Kūkai's practice grants access to the intimacy of Dainichi's conversing with himself for his own enjoyment, Dōgen tells his students:

> When you endeavor in right practice, the voices and figures of streams and the sounds and shapes of mountains, together with you, bounteously deliver eighty-four-thousand gāthās. Just as you are unsparing in surrendering fame and wealth and the body-mind, so are the brooks and mountains.[15]

If we devote our full attention to them, streams and mountains can, simply by being themselves, teach us naturally about the nature of existence in general. And yet for Dōgen this process works only as a cooperation between the worlds of the human and the nonhuman and as "the twin activities of the Buddha-nature and emptiness."[16]

Kūkai's idea that heaven and earth are the bindings of a *sūtra* painted by brushes of mountains and ink of oceans is also echoed by Dōgen, who counters an overemphasis on study of literal scriptures in certain forms of Buddhism by maintaining that *sūtra*s are not just texts containing written words and letters.

> What we mean by the sutras is the entire cosmos itself...the words and letters of beasts...or those of hundreds of grasses and thousands of trees.... The sutras are the entire universe, mountains and rivers and the great earth, plants and trees; they are the self and others, taking meals and wearing clothes, confusion and dignity.[17]

As in Kūkai, natural phenomena are a source of wisdom and illumination, as long as we learn how to "read" them. But just as Kūkai claims that *all* phenomena, as the *dharmakāya*, expound the true teachings, so Dōgen says that it is not just natural phenomena that are *sūtra*s but also "taking meals and wearing clothes, confusion and dignity"—activities and attributes that distinguish humans from other beings. So, while Western thinkers like Erigena and Böhme talk of nature as "God's corporeal being" and of the language and voices of all created beings, both Dōgen and Kūkai would want to go further and ascribe Buddha-nature to *all* beings and not just to natural (as in God-created) beings.

I have been suggesting that where such a worldview as Kūkai's or Dōgen's—in which nature is regarded as sacred and a source of wisdom—prevails, people will tend to treat the environment with respect. But now the universalistic strain in their thinking might appear to detract from the ecologically beneficial features, since it would seem to entail that all human-made things—including such environmentally noxious substances as radioactive waste—are similarly sacred and worthy of reverence. This consideration leads into a complex of issues, the complexity of which should be acknowledged before a solution is suggested.

Problematic Issues

It is hard to retain one's composure in the face of talk about the "love of nature" that is often said to inform Japanese culture, in view of Japan's dismal environmental record in recent decades. In a short but pointed article Yuriko Saito examines three "conceptual bases for the alleged Japanese love of nature" and finds them wanting in their ability to "engender an ecologically desirable attitude" toward the natural world.[18] She argues that "the tradition of regarding nature as friend and companion, which serves the individual as refuge and restorative" is too anthropocentric to be able to value the natural world for its own sake rather than for the benefits it can afford human beings (3). Saito also shows how the *mono no aware* ("the pathos of evanescence") worldview that has conditioned so much of Japanese culture is too fatalistic to promote salutary ecological awareness, arguing that deforestation or pollution can, according to this view, be "accepted as yet another instance of transience" (5).

The third conceptual basis Saito considers is Zen Buddhism— with its idea of the harmony between human beings and nature— which, "as respectful of and sensitive to nature's aesthetic aspect as [it] might be," still "does not contain within it a force necessary to condemn and fight the human abuse of nature" (8). "If everything is Buddha nature because of impermanence," she argues, "strip-mined mountains and polluted rivers must be considered as manifesting Buddha nature as much as uncultivated mountains and unspoiled rivers." Similarly, the notion of "responsive rapport" between all things, which she associates with Dōgen, "makes it impossible for any intervention in nature to be disharmonious with it" (8).

These points about the anthropocentrism of nature-as-companionable-refuge philosophy and the fatalism of the *mono no aware* worldview are well taken, but not, I think, the criticism of Zen Buddhism. This last seems plausible initially, because when Mahāyāna distinguishes itself from early Buddhism in asserting that *nirvāṇa* is not different from *saṃsāra*, it appears to expose itself *eo ipso* to charges of quietism (or at least "anactivism"). For if this apparently imperfect world is actually *nirvāṇa*, then what is there to be done? In that case there would hardly be any need for activity, let alone activism. Let me begin to respond to such criticisms with reference to Kūkai; although Saito doesn't mention him, or Shingon Buddhism, her point about strip-mined mountains and polluted rivers "as manifesting Buddha nature" applies equally to such phenomena as part of the *dharmakāya*.

It is easy to see why for Kūkai certain kinds of things produced by humans would constitute the *dharmakāya*. Works of art, for example, are especially effective expositors of the *dharma*: "Since the esoteric Buddhist teachings are so profound as to defy expression in writing," he writes—a remark struggling readers will find consoling—"they are revealed through the medium of painting."[19] But while there is surely an important sense in which what we call "sick" buildings, for example, or toxic-waste dumps, are *speaking* to us, it may be hard to imagine them as the body of the Buddha or as expounding the true teachings. Since such insalubrious things are nevertheless part of the totality of beings, Kūkai would have to regard them as part of the *dharmakāya* and hence also as expositing the *dharma*. But the important question concerns his attitude toward such things: if he would advocate reverence toward sick buildings and toxic waste as part of the body of Dainichi, one might well doubt the wisdom of introducing his ideas into current debates about the environment.

Let us make the question more pointed by taking more extreme examples: what is the appropriate attitude toward the tubercle bacillus (a natural being) and toward radioactive waste (something relatively unnatural, insofar as it has been produced only under very recent and peculiar historical conditions and requires enormously complex technology)? I choose a naturally occurring being for the first example since it points up a problem with the appropriation of Taoist and Buddhist ideas by recent deep ecology, with its "ultimate norm" of "biocentric equality."[20] This seems a rather infelicitous name for an ultimate norm—surely "biotic equality" would be more appropriate—but it does point up the narrower focus of deep ecology as compared with Taoism or Zen, where the inorganic realm of mountains and streams is as important as the vegetal and animal realms.

The principle, or "intuition," of biocentric equality, as defined by Devall and Sessions, is that "all things in the biosphere have an equal right to live and blossom and to reach their own individual forms of unfolding and self-realization" (67), and deep ecology is also said to advocate "biospecies equality" as the idea that "all nature has intrinsic worth" (69). While the sentiment behind this ideal is commendable, the formulation is flawed: to adopt this idea as an ultimate norm would mean abandoning the work of human culture—and perhaps the human race—altogether. Imagine if, on discovering the tubercle bacillus, we had upheld its "equal right to live and blossom and to reach its own individual form of unfolding and self-realization": tuberculosis would have decimated our best poets, painters, and composers long ago. Nor would it take much effort to ensure the flourishing of the Ebola virus and thus bring the human race to a gruesome finish. The deep ecologists would do well to take a few other leaves out of the Taoist/Zen book—those emphasizing the importance of context and perspective and the problems that arise when one tries to universalize.

Kūkai and Dōgen Defended

Let us begin with Kūkai. Just because the tubercle bacillus is part of the reality embodiment of the cosmic Sun Buddha does not mean that Kūkai would have us worship it and celebrate its equal right to unimpeded flourishing. The image of embodiment is important here. Things can go wrong in a human body which can be put right by getting rid of the noxious element and taking steps to see that it doesn't recur (as in excising a cancerous tumor, for example).[21] Insofar as the blossoming of the tubercle bacillus would jeopardize the flourishing of good Buddhist practice (among other things), Kūkai would surely see it as a baneful element within the body of Dainichi and approve appropriate surgery to get rid of it. The important thing is to consider the body and to appraise its health, holistically. He would similarly regard the tubercle bacillus as a part of Dainichi's exposition of the *dharma* for his own enjoyment. But Buddhist deities generally have their wrathful as well as their compassionate aspects, and there is no guarantee that their teachings will always be pleasing to the human ear.

The fact that radioactive waste is produced by humans would probably not be a factor in Kūkai's readiness to recommend surgery to remove it from the *dharmakāya*. But in view of the centrality of *impermanence* in Buddhist teachings, and since the half-life of something like plutonium is measurable in *kalpa*s, one can imagine that the relative *non*-impermanence of radioactive waste would be a reason for Kūkai's wanting to get rid of it. And if radioactive waste is expounding the *dharma* in any way, it is probably by showing us that the farther things get from being impermanent, the more lethal they become.

What would Dōgen say about these causes of fatal disease and lethal pollution? Are deadly viruses and plutonium waste part of Buddha-nature? The former surely are, along with the tubercle bacillus, poisonous snakes, and other sentient beings that are deadly to humans. Dōgen naturally subscribes to the Buddhist view of the sacredness of life and the precept of not killing, but he (and a follower of his philosophy) would observe these precepts in the context of other features of his worldview, such as the "Buddha-nature of non-being" (*mu busshō*), the interfusion of life and death (*shōji*), and the functional interdependence (*engi*) of all things more generally.[22] And given the difference in the "*dharma* positions" (*hōi*) occupied by humans and bacilli, Dōgen would surely *not* condemn, in most circumstances, attempts to eradicate the tubercle bacillus as evil or as pernicious anthropocentrism. The "in most circumstances" is meant to suggest the importance, for Zen, of broadening one's perspective in order to see the total context.

These considerations demand a slight modification of my earlier formulation: a view of the world as the body of Dainichi or as Buddha-nature would naturally lead to reverence for and respectful treatment of the totality—but would not rule out destroying certain parts of it under certain circumstances.

The status of radioactive waste with respect to Buddha-nature would, I suspect, be somewhat problematic for Dōgen. There is no denying that his philosophy is distinguished by a radical expansion of the traditional concept of Buddha-nature:

> Since ancient times, foolish people have believed man's divine consciousness to be Buddha-nature—how ridiculous, how laughable! Do not try to define Buddha-nature, this just confuses. Rather, think of it as a wall, a tile, or a stone, or, better still, if you can, just accept that Buddha-nature is inconceivable to the rational mind.[23]

Here is another instance of Dōgen's superseding the distinction between sentient and nonsentient beings: he conversely claims in another passage that "walls and tiles, mountains, rivers, and the great earth" are all "mind-only."[24] He is also apparently contradicting a statement in the *sūtras* to the effect that "fences, walls, tiles, stones, and other nonsentient beings" do *not* have Buddha-nature.

Now, to ascribe Buddha-nature to stones is one thing, but to include walls and tiles is another, far more provocative thing. One reason for this is that the *shō* of *busshō* has important connotations of "birth," "life," and "growth"—such that it would be counterintuitive to apply the term to something constructed or fabricated by human beings.[25] It is doubtful whether the technology used in Dōgen's day to produce fences, walls, and roof tiles was environmentally destructive, but one might reasonably wonder whether Dōgen would be comfortable saying that even fences or roof tiles made of nonbiodegradable plastic are Buddha-nature. But again, as in the case of Kūkai's talk of the body of Dainichi, the important feature of Buddha-nature for Dōgen, exemplified in his identification of it as "total-being" (*shitsu-u*), is that it constitutes an organized totality. He would thus *not* be committed to celebrating the chemicals polluting a river (which render the resident fish more impermanent than they would otherwise be) or the radioactive waste stored all over the planet (which is capable of radicalizing the impermanence of all life to the point of extinction) as venerable manifestations of Buddha-nature.

Dōgen was influenced, as was Kūkai, by classical Taoist thinkers (Lao-tzu and Chuang-tzu), as evidenced by his frequent talk of the "Buddha Way" (*butsudō*, or Buddha *tao*)—not to mention his name (which means "source of the Way"). Throughout his writings Dōgen advocates paying close attention to the natural world, just as the Taoists recommend following *t'ien tao* (the Way of Heaven). And, just as the Taoist sage practices an enlightened "sorting" (*lun*) of things on the basis of the broadest possible perspective on their various *te* (powers, potencies), so Dōgen exhorts his readers to "total exertion" (*gūjin*) in attending to the different ways things "express the Way" (*dōtoku*) and occupy their special "*dharma* positions" (*hōi*) in the vast context of the cosmos.[26] By contrast with the radical-egalitarian deep-ecological picture of Taoism and Zen, whereby all living beings are to be encouraged to blossom and flourish, both

Chuang-tzu and Dōgen would want to take into account the effects of propagating tubercle bacilli or radioactive waste on the flourishing of human (and other) beings before deciding to let them bloom.

Practical Postscript

The crucial question concerning these Japanese Buddhist ideas about nature is to what extent they can contribute to the solution of our current ecological problems. It would clearly be difficult to convince most citizens in Western countries, or their political representatives, that the solution lies in the ideas of a ninth-century thaumaturge from Japan. But it is demonstrable that this Japanese Buddhist understanding of the relations between human beings and the natural world has close parallels in several (admittedly non-mainstream) currents of Western thinking. (In the United States, the relevant figures would range from the Native Americans to more intellectually "respectable" characters, such as Emerson, Thoreau, Aldo Leopold, and John Muir; in Germany, there would be Böhme, Goethe, Schelling, and Nietzsche; in France, Rousseau; and so on.) If one were to show the underlying harmony among these disparate worldviews, and how these ideas conduce to a fulfilling way of living that lets the natural environment flourish as well, there might be a chance of some progress.

The problem is how to bring about an experiential realization of the validity of such ideas on the part of the large numbers of inhabitants of postindustrial societies whose lives are fairly well insulated from nature. A few days away from watching television in a more or less hermetically sealed space, and spent in an unspoiled natural environment, would help immeasurably; but, since some kind of guidance is desirable, this is a labor-intensive project (already being undertaken at certain Zen centers, colleges, and universities) that can reach only small numbers of people at a time.

There is justified doubt as to whether the task could be well accomplished by publishing a book, since the people whose perspectives need to be changed (the politicians and general populace) do not read much anymore. But they do watch television—and so an optimal medium for the dissemination of these ideas would be film, which can show as forcefully as it can tell, and offers the alternatives of documentary (which can vividly present the dire situation we are in) and drama (which can make the problems and their potential solutions *personal*). A pioneer in this field, in the area of the art film, is John Daido Loori, whose Zen videography beautifully and forcefully conveys Dōgen's understanding of the natural world as a source of wisdom.[27]

With respect to film drama, it is by no means inconceivable, in view of the number of Hollywood stars and rock musicians who visibly promote environmental causes, that the right dramatic script(s) could attract the talents of some world-famous actors and actresses, with some well-known popular musicians for the soundtrack, and eventuate in a feature film with a salutary ecological message. We might then look forward to seeing, in worldwide distribution, the cosmic Buddha expounding the true teachings not only through mountains, trees, and rivers but also by way of celluloid and fiber-optic cable.

This little flourish of fantasy points up one of the more encouraging implications of the Japanese Buddhist outlook for our contemporary situation—insofar as that kind of philosophy resolves the tension between nature and culture. As the example of Dōgen (and of other figures in the Zen tradition) shows, there is no necessary contradiction between a simple life lived lightly on the earth and a life rich in refined culture. If Thoreau took his Homer to Walden, we can probably in good ecological conscience have our *sūtra*s on CD-ROM to complement the scriptures in mountains, rivers, and trees.

Notes

* The writing of this paper was supported by a research grant from the Japan Studies Endowment of the University of Hawaii, funded by a grant from the Japanese government.

1. Harold Bloom has remarked on the pronounced Gnostic strain in contemporary American religion, thanks to which believers understand themselves as being in essence separate from nature; see his *The American Religion: The Emergence of the Post-Christian Nation* (New York: Simon and Schuster, 1992), chaps. 1 and 2.

2. In the course of an attack on the "new fundamentalism" of deep ecology, the French philosopher Luc Ferry refers to its non-anthropocentric worldview as an "as yet unprecedented vision of the world" (*The New Ecological Order* [Chicago: University of Chicago Press, 1995], 60–61). Ferry is apparently unaware that similarly non-anthropocentric perspectives have informed sophisticated Taoist and Buddhist philosophies for centuries.

3. For an illuminating account of this debate, see William R. LaFleur, "Saigyō and the Buddhist Value of Nature," in *Nature in Asian Traditions of Thought: Essays in Environmental Philosophy*, ed. J. Baird Callicott and Roger T. Ames (Albany: State University of New York Press, 1989), 183–209. The author goes on to show how these ideas were subsequently elaborated by several prominent figures in the Japanese Tendai school, notably Ryōgen in the tenth century and Chūjin in the twelfth. In the same volume, see also David Edward Shaner, "The Japanese Experience of Nature," 163–82.

4. *Kūkai: Major Works*, translated with an account of his life and a study of his thought, by Yoshito S. Hakeda (New York: Columbia University Press, 1972) (hereafter cited as Hakeda, *Kūkai*), 254–55.

5. Kūkai, *Hizō ki* (Record of the secret treasury), in *Kōbō daishi zenshū* (*KDZ*), ed. Yoshitake Inage, 3rd ed. rev. (Tokyo: Mikkyō Bunka Kenkyūsha, 1965), 2:37; cited in LaFleur, "Saigyō and the Buddhist Value of Nature," 186.

6. Kūkai, *Sokushin jōbutsu gi* (Attaining enlightenment in this very body), in Hakeda, *Kūkai*, 229.

7. Kūkai, *KDZ*, 1:516; cited in Hakeda, *Kūkai*, 93.

8. For a fine explication of this idea, see Thomas P. Kasulis, "Reality as Embodiment: An Analysis of Kūkai's *Sokushinjōbutsu* and *Hosshin Seppō*," in *Religious Reflections on the Human Body*, ed. Jane Marie Law (Bloomington: Indiana University Press, 1995), 166–85. See also, by the same author, "Kūkai (774–835): Philosophizing in the Archaic," in *Myth and Philosophy*, ed. Frank E. Reynolds and David Tracy (Albany: State University of New York, 1990), 131–50.

9. Kūkai, *KDZ*, 3:402; cited in Hakeda, *Kūkai*, 91.

10. "We show you the revelation of the Godhead through nature. . . . how the Unground or Godhead reveals itself with this eternal generation, for God is

spirit. . .and nature is his corporeal being, as eternal nature. . . . For God did not give birth to creation in order thereby to become more perfect, but rather for his own self-revelation and so for the greatest joy and magnificence" (Böhme, *De Signatura Rerum*, 3.1, 3.7, 16.2).

11. Dōgen, *Shōbōgenzō*, "Busshō" (Buddha-nature). Subsequent references to Dōgen will be made simply by the title of the relevant chapter/fascicle of his major work, *Shōbōgenzō* (in vol. 1 of *Dōgen zenji zenshū*, ed. Ōkubo Dōshū [Tokyo, 1969–70]).

12. Dōgen, "Zenki" (Total working); cited in Hee-Jin Kim, *Dōgen Kigen— Mystical Realist* (Tucson: University of Arizona Press for the Association for Asian Studies, 1975), 164.

13. Dōgen, "Keiseisanshoku" (Sounds of the valley, color of the mountains), in *Shōbōgenzō*, trans. Kōsen Nishiyama and John Stevens, 4 vols. (Sendai: Daihokkaikaku, 1975–83), 1:92.

14. Dōgen, "Mujō-seppō" (Nonsentient beings expound the *dharma*); cited in Kim, *Dōgen Kigen*, 253–54.

15. Dōgen, "Keiseisanshoku"; cited in Kim, *Dōgen Kigen*, 256.

16. Hee-Jin Kim's formulation (*Dōgen Kigen*, 256). See his insightful account of Dōgen's understanding of nature and the force of the nature imagery in his texts, in the section entitled "Nature: The Mountains and Waters" (253–62).

17. Dōgen, "Jishō zammai" (The *samādhi* of self-enlightenment); cited in Kim, *Dōgen Kigen*, 97.

18. Yuriko Saito, "The Japanese Love of Nature: A Paradox," *Landscape* 31, no. 2 (1992):1–8.

19. Kūkai, *KDZ* 1:95; cited in Hakeda, *Kūkai*, 80.

20. Bill Devall and George Sessions, *Deep Ecology: Living As If Nature Mattered* (Salt Lake City: G. M. Smith, 1985), 66, where the norm is said to have been developed by Arne Naess. While there is no mention of Kūkai in this book, there are several references to Taoist ideas (which influenced Kūkai as well as Zen), as well as references to or quotations from Dōgen on 11 (where he is invoked as a representative of Taoism), 100–101, 112–13, and 232–34.

21. The analogy between the *dharmakāya* and a physical body or organism breaks down with the consideration that there can be nothing *outside* the *dharmakāya*, though this does not reduce the efficacy of the analogy in other respects.

22. Dōgen's idea of Buddha-nature—including "total-being Buddha-nature" (*shitsu-u busshō*), "non-being Buddha-nature" (*mu busshō*), and "emptiness Buddha-nature" (*kū busshō*)—is incredibly complex. See Kim's chapter "The Buddha-nature" (136–227) in his *Dōgen Kigen*, and Masao Abe, "Dōgen on Buddha-nature," in *A Study of Dōgen: His Philosophy and Religion* (Albany: State University of New York Press, 1992), 35–76.

23. Dōgen, "Busshō," in Nishiyama and Stevens, *Shōbōgenzō*, 4:140. It is significant that the term *garyaku* in *shōheki garyaku* ("fences, walls, tiles, stones") also has the connotation of useless, insignificant things.

24. Dōgen, "Sangai yuishin" (The three worlds are mind-only); cited in Kim, *Dōgen Kigen*, 157.

25. Similarly, *hsing*, the Chinese equivalent of *shō*, is derived from *sheng*, meaning "birth, life, growth." At the same time, interestingly, the radical in the graph *shō/hsing* is the *risshinben*—denoting "mind."

26. Hee-Jin Kim lays appropriate emphasis on the antiquietistic aspect of Dōgen's philosophy: "In his view things, events, relations were not the given (entities) but were possibilities, projects, and tasks that can be acted out, expressed, and understood as self-expressions and self-activities of the Buddha-nature. This did not imply a complacent acceptance of the given situation but required man's strenuous efforts to transform and transfigure it" (*Dōgen Kigen*, 183).

27. See—or, rather, view—the VHS tapes *Mountains and Rivers: An Audio-visual Experience of Zen's Mystical Realism* (Mt. Tremper, N.Y.: Dharma Communications, 1994) and *Sacred Wildness: Zen Teachings of Rock and Water* (Mt. Tremper, N.Y.: Dharma Communications, 1996).

Buddhism and Animals:
India and Japan

Animals and Environment
in the Buddhist Birth Stories

Christopher Key Chapple

Prologue: Animal Spirit

Thomas Berry, in his *Dream of the Earth*, lauds the importance of species protection. He notes that the disappearance of each endangered animal and plant from the planet results in the diminishment of human consciousness. He states:

> If we have powers of imagination these are activated by the magic display of color and sound, of form and movement, such as we observe in the clouds of the sky, the trees and bushes and flowers, the waters and the wind, the singing birds, and the movement of the great blue whale through the sea. If we have words with which to speak and think and commune, words for the inner experience of the divine, words for the intimacies of life, if we have words for telling stories to our children, words with which we can sing, it is again because of the impressions we have received from the variety of beings about us.[1]

Animals as described by Berry hold the potential for enriching human consciousness; by observing animals and the natural order we learn not only about their behavior but also acquire insights and metaphors that deepen our own experience as human beings.

Walt Whitman, a great observer of the natural and human world, once wrote in regard to animals that:

> I think I could turn and live with animals,
> They are so placid and self-contained

I stand and look at them long and long.
They do not sweat and whine about their condition,
They do not lie awake in the dark and weep for their sins,
They do not make me sick discussing their duty to God,
Not one is dissatisfied,
Not one is demented with the mania of owning things.[2]

For Whitman, the very being of animals in their seeming simplicity provided a moral example for humans to emulate.

Animals throughout the world's folklore have been used as metaphors and as inspiration, as prophetic and imaginative tools. From the Anansi spider tales of the Yoruba to the coyote stories of North America, from the Brer Rabbit tales of the American South to Aesop's Fables from ancient Greece, animal stories have provided amusement, delight, and wisdom for millennia. From the dawn of human history in the caves of Lascaux to the therioanthropic images of Pharaonic Egypt and Shang dynasty China, as well as India's Indus Valley civilization depictions of rhinoceri and various forms of cattle and cats, animals have been central to human self-orientation and definition, with humans seeking in various ways to capture the power of animals, to be safe from harm from animals, to feed upon animals, and, through ritual, to revere animals.

Animals demonstrate a wide range of behaviors. Although Aristotle and Descartes did not attribute cognition to animals, the versatility and profundity of animal consciousness has received positive attention from recent scientists, who now acknowledge an awareness in animals that includes intentionality, emotion, and, to a degree, logic. In addition to the pioneering scientific work of Donald R. Griffin, Carolyn A. Ristau, Frans de Waal, Dorothy Cheney,[3] Irene Pepperberg,[4] Donald Kroodsma,[5] and others, research in animal cognition has been popularized in such books as *When Elephants Weep: The Emotional Lives of Animals*.[6] A high level of intelligence is now recognized in chimpanzees, dolphins, and many other animals, and, in some ways, this paradigm shift in science makes the Buddhist attitude toward animals as exhibited in the Jātaka stories more interesting, if not more credible.

Animal Awareness in Buddhism

In the cosmology of the early renouncer traditions of Buddhism and Jainism, animals play a vital role. Not only do animals occupy their own important niche in the categorization of realms that also house humans, gods, hell beings, and ghosts, each animal can serve as host to life-forms involved in an ever-changing game of cosmic musical chairs. An animal in one birth might take the form of the same or a different animal in the next lifetime, might advance to human or godly status, or might descend to the hellish or ghostly realms. Unlike the *Ṛg Veda*, which regards animals as tools for human sustenance or sacrifice, the early literature of the Buddhist and Jaina Śramaṇical treatment of animals accords to them an important place in the hierarchy of life.

The status of animals in the early Buddhist tradition has been the topic of three recent studies: a chapter entitled "Nonviolence, Buddhism, and Animal Protection" in my *Nonviolence to Animals, Earth, and Self in Asian Traditions*,[7] which discusses the Buddhist precept against taking life and surveys the Aśokan materials; James P. McDermott's article "Animals and Humans in Early Buddhism," which examines materials in the Sutta and Vinaya texts regarding the treatment of animals;[8] and Padmanabh S. Jaini's "Indian Perspectives on the Spirituality of Animals," which cites a wide range of animal stories from the Hindu, Buddhist, and Jaina traditions that indicate belief in innate spiritual and ethical capacities within animals.[9]

In Buddhist countries, a genre of text has arisen known as the Jātaka or birth stories. Each of these stories tells the past lives of the Buddha and includes a moral lesson. This genre includes stories embedded in canonical texts, collections that isolate and embellish the birth tales, and regional stories in local languages.[10] The most comprehensive translation of Jātaka tales is included in a six-volume work titled *The Jātaka or Stories of the Buddha's Former Births*, published in 1895. This work includes 550 stories which were translated from Pali into English and spans over two thousand pages. According to E. B. Cowell, the editor of this massive project, the collection arose from Singhalese stories that were developed from much shorter Pali verses. Buddhaghosa, the great fourth-century Sri

Lankan Theravāda redactor, translated the stories into Pali.[11] In addition to the original Pali verse (*gāthā*), these tales include an introductory context story, a longer version of the verse story, and a brief mention of the identities of the animals and persons in the tale, including who in the tale was later to be reborn as the Buddha. Excerpts, anthologies, and children's books based on this massive work have appeared in English throughout the past century.[12]

There are additional "apocryphal" Jātaka tales,[13] as well as Arya Sūra's Sanskrit retelling of 34 Jātaka stories in the *Jātakamālā* (ca. 400 C.E.), which was recently newly translated by Peter Khoroche.[14] Other Jātaka stories occur in later Mahāyāna texts, particularly the *Mahāratnakūṭa Sūtra*.[15]

For the purposes of this study, I will focus on the 550 stories traditionally accepted within the Theravāda tradition. Of the 550 tales, a full half of them (225) mention animals, usually as the central characters. Seventy different types of animals are mentioned and 319 animals or groups of animals appear in the 225 stories. Monkeys lead the pack, being represented in 27 different tales, followed with elephants (24), jackals (20), lions (19), crows (17), deer (15), birds (15), fish (12), and parrots (11). Of special interest are 10 stories in which the Buddha and other beings take the form of tree spirits (see the table found at the end of this chapter).

In most instances, animals represent prior life-forms of persons living at the time of the Buddha. The actions of these animals in the past help explain present-day human behavior. In some cases, this animal behavior is auspicious and has laid the foundation for later auspicious human action; in other cases, the behavior is objectionable and helps account for heinous human behavior committed by the Buddha's contemporaries. Many stories of the latter type relate to Devadatta, the cousin of the Buddha who plotted his downfall and actually attempted to kill the Buddha by hurling rocks at him and sending a raging elephant in his path.

This study will focus specifically on the portrayal of animals (and plants) in select stories from the Pali collection of Jātaka stories. I have grouped these stories into examples that illustrate the wisdom and/or compassion exhibited by animals; karmic moral fables wherein animals are punished for their folly or cruelty; stories pertaining to vegetarianism and meat-eating; stories designed to discourage animal sacrifice; and tales that contain what seems to be an inherently ecological message.

Compassionate and Wise Animals

The first representative sampling I have chosen under this category tells the story of a time when the Buddha lived in a prior birth as a woodpecker (the *Javasakuṇa Jātaka*).[16] One day he noticed that a lion was in great discomfort, due to a bone being lodged in his throat. After propping the lion's jaws open with a stick, the woodpecker enters the mouth of the lion and dislodges the stick, enabling the lion once again to breathe and eat easily. At a later time, the bird comes near the lion while the lion is devouring a wild buffalo. To test the lion, the woodpecker flies near to him and asks of him a favor. The lion haughtily replies that he had spared the woodpecker's life once, and that was enough of a favor. The bird chastises the ungrateful lion and hastens on his way. After telling this tale, the Buddha states that the rude lion was Devadatta in a prior life and that he, the Buddha, was the helpful woodpecker.

In the *Suvaṇṇamiga Jātaka*, a golden stag became trapped in a snare.[17] Despite his strong efforts and the encouragement of his wife, he could not free himself. His devoted wife then confronted the hunter who had come to collect his catch. She offered her own life in place of her husband's life. Stunned, the hunter freed both of them. In thankfulness for the hunter's change of heart, the stag later presented the hunter with a "jewel he had found in their feeding ground" and implored the hunter to abstain from all killing, to establish a household, and to become involved with good works. Following the story, Buddha notes that he himself was the royal stag.

In both of these stories, animals exhibit meritorious behavior and set an example for the humans listening to each tale. In the first instance, the Buddha teaches the importance of generosity and gratitude. In the second, he teaches the power of self-sacrifice and devotion. This latter example also includes an animal advocating for abstention from hunting, a reflection of the Buddhist precept of non-injury to life.

Foolish Animals

The *Kokālika Jātaka* tells that many years ago in Banaras, the king
had a bad habit of talking too much. A wise and valued minister
decided to teach the king a lesson. A cuckoo (like the North
American cowbird), rather than rearing her own young, had laid an
egg in a crow's nest. The mother crow, thinking the egg to be one
of her own, watched over the egg until it hatched and then fed the
young infant bird. Unfortunately, one day, while not yet grown, the
small intruder uttered the distinct call of the cuckoo. The mother
crow grew alarmed, pecked the young cuckoo with her beak, and
tossed it from her nest. It landed at the feet of the king, who turned
to his minister. "What is the meaning of this?" he asked. The wise
minister (the future Buddha) replied that

> They that with speech inopportune offend
> Like the young cuckoo meet untimely end.
> No deadly poison, nor sharp-whetted sword
> Is half so fatal as ill-spoken word.

The king, having learned his lesson, tempered his speech, and
avoided a possible overthrow of his rule. In his commentary, the
Buddha notes that he was the wise minister and the talkative king
one of his garrulous monks, Kokālika.[18]

In the *Laṭukika Jātaka*, the Buddha tells of two elephants, one
the regal leader of the herd, the other a rogue marauder.[19] The head
elephant one day comes upon a mother quail whose youngsters had
just hatched. The quail implores the head elephant to protect her
children, and he arranges for all eighty thousand of his followers
to step carefully around the birds. He warns the mother quail of a
rogue elephant that might come by and advises her likewise to
implore him to spare her children. Despite her entreaties, the rogue
elephant nastily ignores the mother quail and crushes the young
quail with his left foot. The mother quail, angered by the cruel
murder of her brood, sets out in search of revenge. She meets with
a crow, a fly, and a frog, who agree to help her. The crow pecks
out the eyes of the elephant. The fly lays its eggs in the empty
sockets. After the fly eggs turn into maggots and cause a frenzy of
itchiness in the elephant's head, he blindly seeks out water to give
him some relief. Under the guidance of the quail, the frog croaks

first at the top of the mountain, leading the elephant to a precipice. He then jumps to the bottom of the cliff and croaks again. The elephant, following the sound of the frog in search of water, plunges into the chasm, and rolls to his death at the foot of the mountain. After telling this story, the Buddha states, "Brethren, one ought not to incur the hostility of anyone." He then notes that he was the friendly head elephant and that Devadatta was the rogue elephant.

The first story warns against taking on negative habits associated with particular animals, in this case, excessive loquaciousness. The second story, albeit its gruesome nature, warns that one must not commit random acts of destruction.

The Question of Vegetarianism in the Jātaka Tales

One issue that arises in the discussion of Theravāda Buddhism is the question of whether to eat flesh foods. As D. Seyfort Ruegg has pointed out, this policy varies from country to country, according to the customs of the host country.[20] Although it has been somewhat disputed whether the Buddha himself ate meat, it clearly has been acceptable for monks in Southeast Asia to receive meat as part of their alms, as long as meat dishes have not been prepared especially for them. At variance with the dietary laws of Śramaṇic groups in India, particularly the Jainas, we find the Buddha proclaiming that it is acceptable to eat meat as long as one did not directly kill the animal. This is also at variance with Vyāsa's proclamation in his commentary on the *Yoga Sūtras* that not only is direct violence to be avoided, but the practitioner of *ahiṃsā* must abstain from assent to violence.[21] The Buddha suggests that as long as one does not become entangled in violence, it is acceptable to allow oneself to be used for another's evil purposes to avoid harm.

In the *Telovāda Jātaka*, the Buddha directly criticizes a Jaina ascetic by the name of Nāthaputta, who ridicules the Buddha for accepting food with meat.[22] The Buddha tells the story of a time when he was a Brahmin living in Banaras. In his old age he had retired to the Himalayas to pursue a religious life. During a periodic visit to the city to get salt and seasoning, a wealthy man intentionally prepared for him a meal with fish and then ridiculed the holy man for eating it. In reply, the Brahmin retorted, "If the holy eat, no sin is done," affirming that aspect of the monastic code that states "my priests have permission to eat whatever food is customary to eat in any place or country, so that it be done without the indulgence of the appetite, or evil desire."[23] In closing, the Buddha remarks that he had been the Brahmin, and that the Jaina monk Nāthaputta had been the wealthy man.

In yet another story based in Banaras, the *Tittira Jātaka* (no. 319), the Buddha tells of a time when he lived as a Brahmin ascetic of great spiritual accomplishment. During this time, a fowler had trained a partridge to serve as a decoy, attracting other partridges into the fowler's snare. At first the decoy partridge resisted his task, but the fowler beat him on the head with bamboo until the partridge learned to be submissive. In his conscience, the partridge suffered greatly, wondering if he accrued great sin through his complicity. One day, the fowler brought his partridge down to the river, near the hut of the accomplished ascetic. While the fowler slept, the partridge asked the ascetic if in fact his life as a decoy was in error. The Brahmin replied:

> If no evil in thy heart
> Prompts to deed of villainy,
> Shouldst thou play a passive part,
> Guilt attaches not to thee.
> If not sin lurks in the heart,
> Innocent the deed will be.
> He who plays a passive part
> From all guilt is counted free.

Freed from remorse, the partridge is carried off again by the fowler. After telling this tale, the Buddha announces that he was the ascetic and his son, Rahūla, the partridge.[24] Rather than using this tale as an opportunity to denounce all forms of hunting, the Buddha acknowledges that circumstance sometimes forces compromise.

Animal Sacrifice

In one birth long ago, as told in the *Dummedha Jātaka*, the Buddha, a prince of Banaras, was appalled by the sacrificial massacre of sheep, goats, poultry, pigs, and other animals, in accordance with Vedic ritual. Each year, until the death of his father, he performed his own rituals—without killing animals—to the spirit of a special banyan tree. After the death of his father, he ascended the throne and revealed to his subjects the nature of his worship at the tree, announcing that he had promised to offer to the tree the lives of one thousand humans who violate the precept of nonviolence. Once this proclamation had been made, all the townfolk forever renounced the practice of animal sacrifice. Thus, "without harming a single one of his subjects, the *bodhisattva* made them observe the precepts."[25] This underscores the Buddhist commitment to giving up the *dummedha*, or evil sacrifice, in order to spare the lives of innocent animal victims.

The theme of campaigning against the bloody Brahmanical sacrifice of animals to placate the gods continues in the *Lohakumbi Jātaka*. While the Buddha was dwelling at Jetavana, he told a story about the king of Kosala. One evening the king heard four terrible wails. He consulted with a group of Brahmins, who advised him that the calls in the darkness indicated imminent destruction and that to propitiate the gods the king must offer a fourfold sacrifice and kill men, bulls, horses, elephants, quails, and other birds in sets of four. The Brahmins happily set about building a fire pit and collecting their sacrificial victims and became "highly excited at the thought of the dainties they were to eat and the wealth they would gain."[26] Queen Mallika, skeptical of the goings-on, urged the king to consult the Buddha. The king traveled to Jetavana and told the Buddha of his anxiety regarding the four screams in the night. The Buddha assured him that this had happened long ago. In the tale that follows, the Buddha repeats the story, but includes among the Brahmins one priest who questions the need to kill so many beings. The priest encounters an ascetic in the garden, who explains that the king is mistaken about the true cause of the noises in the night. The young priest escorts the ascetic to the king, where the ascetic explains that the four cries were uttered by four men who long ago committed adultery and, as a result of their sin, were condemned to be reborn in the Four Iron Caldrons, where they dwelt for thirty thousand years, periodically boiling to the top and uttering their sickening moans. It was these moans that the king of long ago heard, and these same moans that the king of Kosala heard as well. The Buddha, both in his life long ago as an ascetic, and in Jetavana, assured the respective kings that no harm would befall them due to the four cries. Consequently, both kings canceled the sacrifice and released all the numerous victims. This story ridicules the Brahmanical sacrificial process, carrying the message that misguided notions and greed lie at the heart of such behavior. This story also emphasizes the Buddhist teachings on the inevitability of karmic punishment for wrongdoings and hence undermines the notion that Brahmanical sacrifice can be expiatory. Both stories invoke the Buddhist precept that the lives of animals must be protected.

Jātaka Tales and Ecology

In the *Rukkhadhamma Jātaka*, a quarrel had arisen regarding water rights. In response, the Buddha told a tale of his past life as the spirit of a sal tree in the Himalayas. During the reign of King Vessava, the trees, shrubs, bushes, and plants were all invited to choose a new abode. The future Buddha-tree advised all his kinsfolk to "shun trees that stood alone in the open and to take up their abodes" in the forest. The wise vegetative spirits followed his advice, but the proud and foolish ones instead chose to dwell outside the villages and towns, to reap the benefits offered by townspeople who worship such trees. They left the forest and came to inhabit "giant trees which grow in an open space." One day a mighty storm swept over the countryside. The solitary trees, despite their years of growth deep into the rich farmland, suffered greatly: their branches snapped, their trunks collapsed, and they were uprooted, "flung to the earth by the tempest." But when the storm hit the sal forest of interlacing trees where the future Buddha dwelt, "its fury was in vain. . .not a tree could it overthrow." In a touching conclusion, the Buddha narrates that the "forlorn fairies whose dwellings were destroyed took their children in their arms and journeyed to the Himalayas." The future Buddha responded with the verse:

> United, forest-like, should kinsfolk stand;
> The storm o'erthrows the solitary tree.[27]

This was later repeated by the Buddha when he addressed the villagers during a dispute over water, reminding them to work in unity toward a common goal. This story could be interpreted as a call to heed the lessons of the forest, to acknowledge the strength of the interconnectedness of life.

In the *Kusanjāli Jātaka*,[28] the future Buddha dwelt as a clump of kusa grass near a beautiful wishing tree (*Mukkhaka*) with a strong trunk and spreading branches. The spirit of this tree had once been a mighty queen. The grass was an intimate friend of this noble tree. Nearby, the palace of King Brahmadatta in Banaras had only one main pillar, which had become shaky. The king sent his carpenters to find wood with which to replace the pillar, and they came upon the wishing tree. They resisted cutting it down, and yet could find no other suitable candidate for the job. When they told the king of their troubles, he told them to cut the wishing tree to make his roof secure. The carpenters went and made a sacrifice to the tree, asking for its forgiveness and announcing that they would return the next day to execute their deadly deed. The tree burst into tears, and the various spirits of the forest came to console her, yet none could think of a way to thwart the carpenters. Finally, the kusa grass Buddha called up to her and assured her that he had a plan.

The next day, the kusa grass took on the personality of a chameleon and worked his way up from the roots of the tree through the branches, making the tree appear as if it were full of holes. When the carpenters came, the leader exclaimed that the tree was rotten and that they had not properly inspected it the day before. Consequently, the tree was saved. The noble tree rejoiced and lauded the lowly clump of grass for saving her life. She assembled the forest spirits and announced that they "must make friends of the wise whatever their station in life."[29] After telling this story, the Buddha explains that Ānanda, his loyal follower, was the tree sprite, and that he, the Buddha, was the kusa sprite.

Although this story was used as a moral tale to encourage people to support one another and accept one another regardless of social rank, this also can be seen as an environmental fable wherein the salvation of the tree stands for the preservation of both remarkable trees and the larger ecosystem in which they thrive.

The last story I have chosen with an underlying ecological theme is the *Vyaddha Jātaka*, a tale reminiscent of Aldo Leopold's concept of "thinking like a mountain."[30] In this story, the Buddha dwelt in a forest as a tree spirit. In this particular forest also lived a lion and a tiger, who used to "kill and eat all manner of creatures," leaving behind their offal to fester and decay. Because of the ferociousness of these predators, no humans dared to enter the forest, let alone cut down even a single tree. However, one of the tree spirits could not stand the stench generated by the lion's and tiger's rotting victims. One day, against the advice of the Buddha-tree, the spirit assumed an awful shape and scared off the killers. The people of a nearby village noticed that they no longer saw the tracks of either the lion or the tiger and began to chop down part of the forest. Despite the entreaties of the foolish tree spirit, the animals would not return, and after a few days the men "cut down all the wood, made fields, and brought them under cultivation,"[31] thus driving out the spirits of the forest.

The moral given by the Buddha was that one should recognize that one's peace sometimes depends upon being able to stave off the incursion of others, and that one should not disturb such a state of affairs. From an environmental perspective, the presence of predators maintained an acceptable balance within the ecosystem, a balance that could not be restored after the predators were driven off, opening the land for clear-cutting and agricultural use.

Each of these three stories exhibits a continuity of life-forms illustrative of the integrated nature of Buddhist cosmology. Human consciousness has been shaped and informed by the observation of animals and trees. According to the Buddha, we can learn from animals and trees because we were once animals and trees ourselves. In the time of the Buddha, in a time when agriculture and the building of cities and towns threatened nature, it was recognized that trees were not readily able to advocate for themselves. By telling the tale, in this third instance, of the foolish destruction of a forest, the Buddha has provided a lasting fable that can likewise help contemporary persons acknowledge the shortsightedness of such actions and thus, hopefully, avoid future destruction of life systems.

Conclusion

The animals stories of the Jātaka tales include simple moral tales advising Buddha's followers to avoid hurting people through physical violence or slander, using examples of rogue or rascal animals and their exploits as a didactic tool. In other fables, the Buddha tells of meritorious actions performed by animals, including remarkable acts of charity and compassion. He uses examples of the suffering of animals to sermonize against the ritual use of animals in sacrifice, and he speaks with high praise of animals who have sacrificed their own lives to save others.

In the Jātaka tales, animals can be seen to represent human qualities. However, to the extent that the animals are personified, it can also be argued that humans themselves more often than not exhibit qualities easily recognizable in animals, whether manifested as moral exemplars or as fools. The animal tales of Buddhism illustrate and underscore the position that life from one form to the next is continuous. The Buddhist doctrine of reincarnation supports this theory in two ways. First, according to reincarnation theory, present life will continue in some future form. Second, because lives have endured so many incarnations, a familial link may be assumed. The *Laṅkāvatāra Sūtra*, a Mahāyāna Buddhist text, states:

> In the long course of *saṃsāra* [reincarnation], there is not one among living beings with form who has not been mother, father, brother, sister, son, or daughter, or some other relative. Being connected with the process of taking birth, one is kin to all wild and domestic animals, birds, and beings born from the womb [32]

Repeated birth generates an interconnected web of life which, according to the Buddhist precept of harmlessness, must be respected.

Animals in Buddhism, however, are not universally lauded, romanticized, or idealized. The foibles of animals are often presented. Animals are depicted as being cruel to other animals. Furthermore, human treatment of animals is not always kind. The Buddha-to-be kills a tortoise,[33] and a recurring theme involves human destruction of animal habitats.

It may be said that animals in the Jātaka tales are seen not so much as animals but as potential humans or as animals that can teach humans a lesson. However, it should also be noted that the Buddha was very familiar with animals. He lived during a time and in a place where the boundaries between humans and animals were far more fluid than in contemporary industrialized societies. In fact, his descriptions of monkeys, elephants, quail, cuckoos, and the rest are presented with remarkable detail and accuracy. His insights into both animal and human behavior combine to make the Jātaka stories very effective didactic tools.

Reflecting back upon the opening statements quoted from Thomas Berry, the varied beings of the natural world shaped the consciousness and imagination of the Buddha and early Buddhists as they repeated and shared hundreds of animal fables. Through direct observation of the natural world, a wisdom arose, communicated through a medium accessible to adults and children alike.

Likewise, when Walt Whitman proclaims that he looks at animals "long and long," he also takes from them an understanding of their simplicity, their innate sense of purpose, and their self-dignity. Without complaining, without speculating, without committing the sin of possessiveness, animals move through their existence with seeming grace and ease. Both Berry and Whitman suggest that animals can provide a moral example for humans and also deepen the threads of human experience. The Jātaka tales affirm this interpretation of animal worth.

Human consciousness can be shaped by its experiences of animals. The depth and profundity of human experience can be enriched and improved through contact with and observation of the ways of animals. Contrary to Aristotelean and Cartesian attitudes toward animals, animals possess cognition, will, emotion, and reason. As noted by the Buddha, animals, like ourselves, make choices that govern both this immediate life and future experiences.

NUMBER OF ANIMALS APPEARING IN THE JĀTAKA TALES
(THERAVĀDA TRADITION)

IN ALPHABETICAL ORDER		IN NUMERICAL ORDER	
antelope	2	monkey	27
bear	1	elephant	24
beetle	1	jackal	20
bird	15	lion	19
boar	5	crow	17
buffalo	1	bird	15
bull	4	deer, stag, doe	15
cat	1	fish	12
chameleon	2	parrot	11
chicken	2	snake	10
cow	3	tree spirit	10
crab	3	horse	8
crane	3	goose	7
crocodile	4	tiger	7
crow	17	tortoise	7
cuckoo	3	boar	5
deer, stag, doe	15	goat	5
dog	4	quail	5
donkey	2	bull	4
duck	1	crocodile	4
eagle	1	dog	4
elephant	24	ox	4
elk	1	partridge	4
falcon	2	peacock	4
fish	12	rat	4
fly	1	vulture	4
fox	1	cow	3
frog	1	crab	3
goat	5	crane	3
goose	7	cuckoo	3
grass spirit	1	lizard	3
hawk	1	pig	3
horse	8	pigeon	3
hound	1	serpent	3
iguana	1	woodpecker	3
jackal	20	antelope	2
jay	1	chameleon	2

NUMBER OF ANIMALS APPEARING IN THE JĀTAKA TALES
(THERAVĀDA TRADITION), CONTINUED

IN ALPHABETICAL ORDER		IN NUMERICAL ORDER	
lion	19	chicken	2
lizard	3	donkey	2
mongoose	1	falcon	2
monkey	27	osprey	2
mosquito	1	owl	2
mouse	1	rabbit	2
osprey	2	rooster	2
otter	1	viper	2
owl	2	water spirit	2
ox	4	bear	1
panther	1	beetle	1
parrot	11	buffalo	1
partridge	4	cat	1
peacock	4	duck	1
pig	3	eagle	1
pigeon	3	elk	1
quail	5	fly	1
rabbit	2	fox	1
rat	4	frog	1
rooster	2	grass spirit	1
rhinoceros	1	hawk	1
serpent	3	hound	1
shrew	1	iguana	1
snake	10	jay	1
tiger	7	mongoose	1
tortoise	7	mosquito	1
tree spirit	10	mouse	1
viper	2	otter	1
vulture	4	panter	1
water spirit	2	rhinoceros	1
wolf	1	shrew	1
woodpecker	3	wolf	1

Notes

1. Thomas Berry, *The Dream of the Earth* (San Francisco: Sierra Club Books, 1988), 11.

2. As quoted in John Lockwood Kipling, *Beast and Man in India: A Popular Sketch of Indian Animals in Their Relations with the People* (London: MacMillan, 1892), i.

3. Donald R. Griffin, *Animal Thinking* (Cambridge, Mass.: Harvard University Press, 1984), and *Animal Minds* (Chicago: University of Chicago Press, 1992); Carolyn R. Ristau, ed., *Cognitive Ethology: The Minds of Other Animals: Essays in Honor of Donald R. Griffen* (Hillsdale, N.J.: L. Erlbaum Associates, 1991); Frans de Waal, *Good Natured: The Origins of Right and Wrong in Humans and Other Animals* (Cambridge, Mass.: Harvard University Press, 1996); Dorothy Cheney and Robert M. Seyfarth, *How Monkeys See the World: Inside the Mind of Another Species* (Chicago: University of Chicago Press, 1990).

4. See Michael Haederle, "Talking and Reasoning? It's for the Birds," *Los Angeles Times*, 14 April 1996, p. E1.

5. Donald E. Kroodsma, *Acoustic Communication in Birds* (New York: Academic Press, 1982).

6. J. Moussaieff Masson and Susan McCarthy, *When Elephants Weep: The Emotional Lives of Animals* (New York: Delacorte Press, 1995).

7. Christopher Key Chapple, *Nonviolence to Animals, Earth, and Self in Asian Traditions* (Albany: State University of New York Press, 1993), 21–48.

8. James P. McDermott, "Animals and Humans in Early Buddhism," *Indo-Iranian Journal* 32, no. 2 (1989).

9. Padmanabh S. Jaini, "Indian Perspectives on the Spirituality of Animals," in *Buddhist Philosophy and Culture: Essays in Honour of N. A. Jayawickrema*, ed. David J. Kalupahana and W. G. Weeraratne (Columbo, Sri Lanka: N. A. Jayawickrema Felicitation Volume Committee, 1987), 169–78.

10. Donald Swearer is currently working on the Jātaka tales of Thailand.

11. E. B. Cowell, ed., *The Jātaka or Stories of the Buddha's Former Births*, 6 vols. (London: Pali Text Society, 1895–1907), x.

12. These include Ellen C. Babbit, *Jataka Tales* (New York: Century, 1912); Noor Inayat Khan, *Twenty Jātaka Tales* (The Hague: East West Publications, 1939); and many retellings by American Buddhist storyteller Rafe Martin and numerous picture books published by Shambhala Press.

13. I. B. Horner and Padmanabh S. Jaini, *Aprocryphal Birth-Stories* (London: Pali Text Society, 1985).

14. Peter Khoroche, *Once the Buddha Was a Monkey: Arya Sura's Jatakamala* (Chicago: University of Chicago Press, 1989). This was earlier translated by J. S. Speyer, *The Jatakamala, or Garland of Birth Stories by Arya Sura* (Oxford University Press, 1895).

15. Garma C. C. Chang, ed., *A Treasury of Mahāyāna Sūtras: Selections from the Mahāratnakūṭa Sūtra* (University Park: Pennsylvania State University Press, 1983).

16. Cowell, *The Jātaka*, story 308, 4:17–18.

17. Cowell, *The Jātaka*, story 359, 5:120–23.

18. Cowell, *The Jātaka*, story 331, 4:68–69.

19. Cowell, *The Jātaka*, 5:115–17.

20. D. Seyfort Ruegg, "Ahimsa and Vegetarianism in the History of Buddhism," In *Buddhist Studies in Honour of Walpola Rahula*, ed. Somaratna Balasooriya et al. (London: Gordon Fraser, 1980), 234–241.

21. Christopher Chapple and Yogi Anand Viraj (Eugene P. Kelly, Jr.), *The Yoga Sūtras of Patañjali* (Delhi: Sri Satguru Publications, 1990), 70–71.

22. Cowell, *The Jātaka*, story 246, 2:182–83.

23. Hardy, *Manual*, 327, as cited in Cowell, *The Jātaka*, story 246, 2:182.

24. Cowell, *The Jātaka*, story 319, 4:43–44.

25. Cowell, *The Jātaka*, story 50, 1:126–28,

26. Cowell, *The Jātaka*, story 314, 4:29–32.

27. Cowell, *The Jātaka*, story 74, 1:181–82.

28. Cowell, *The Jātaka*, story 121, 1:267–69.

29. Ibid., 269.

30. See Susan L. Flader, *Thinking Like a Mountain: Aldo Leopold and the Evolution of an Ecological Attitude toward Deer, Wolves, and Forests* (Columbia: University of Missouri Press, 1974).

31. Cowell, *The Jātaka*, story 272, 3:244–46.

32. D. T. Suzuki, trans., *The Laṅkāvatāra Sūtra* (London: Routledge and Kegan Paul, 1932).

33. Cowell, *The Jātaka*, story 178, 2:55–56.

Animal Liberation, Death, and the State: Rites to Release Animals in Medieval Japan

Duncan Ryūken Williams

Introduction

> The prohibition on the taking of life must be observed in the period just before the Iwashimizu *hōjō-e*.
> —Shogunal order, 1280

This order was sent from both the shogunal and imperial governments to various provinces in Japan in the year 1280 C.E. (Kōan 2).[1] The provinces were to observe this rule on not taking life (that is, not killing animals) during the two-week period which preceded the *hōjō-e* ceremony held annually on the fifteenth of the Eighth Month at the Iwashimizu Hachiman Shrine in the city of Yawata in present-day Kyoto Prefecture.

The *hōjō-e*,[2] a ceremony of releasing living beings (most usually birds, fish, or other animals) into the wild, is a Buddhist ritual which can be seen across a number of Buddhist countries, particularly in East Asia. This study outlines how this ritual, based on the principle of compassionate action toward animals and merit-making therefrom, developed in Japan. There were two peculiarly Japanese ways in which this ceremony was transformed. First, the direct involvement of the medieval Japanese state in promoting and supporting this Buddhist ritual and the concurrent enforcement of a ban on the taking of life (*sessho kindan*) made this ritual into a state rite as opposed to simply a Buddhist ritual. Second, the most well known medieval site for the *hōjō-e*, the Iwashimizu Hachiman Shrine, which will be taken up as a case study, was not a Buddhist temple per se, but a "Shintō" shrine with a Buddhist component.

Thus, rather than a purely Buddhist ritual, the *hōjō-e* in Japan can be identified as a "Shintō-Buddhist" combinative ritual.

In addition to documenting these two developments of the *hōjō-e* in Japan, the rite is used as a case study to reflect on critiques of Buddhism as not necessarily environmentalist.[3] Although as an ideal, the *hōjō-e* seems to represent a Buddhist view of animals that is sympathetic, there are a number of ways in which the *hōjō-e* as a ritual practice in medieval Japan is problematic if seen as having only a positive assessment of animals.

The Textual Basis of the *Hōjō-e* in Japan

The Japanese *hōjō-e* can be traced to two Buddhist canonical sources: the *Bommyōkyō* (Sanskrit, *Brahmajāla Sūtra*) and the *Kongōmyōkyō* (Sanskrit, *Suvarṇaprabhāsa Sūtra*). The *Bommyōkyō* states:

As a [child] of the Buddha, one must with a compassionate heart practice the work of liberating living beings. All men are our fathers. All women are our mothers. All our existences have taken birth from them. Therefore, all the living beings of the rokudō [six realms] are our parents, and if we kill them, we kill our parents and also our former bodies; for all earth and water are our former bodies, and all fire and wind are our original substance. Therefore, you must always practice liberation of living beings (since to produce and receive life is the eternal law), and cause others to do so; and if one sees a worldly person kill animals, one must by proper means save and protect them from misery and danger.[4]

This portion of the *sūtra* has been interpreted by such Buddhist monks as Keishu, in his commentary the *Hōjō jissai katsuma giki*, as meaning that sentient beings in the six realms were once one's parents and to kill sentient beings is tantamount to killing one's parents, which provided the rationale for releasing living beings from suffering.

The other canonical source, the *Kongōmyōkyō*, includes a section (the "Rūsui chōja shijin") which relates more directly to the practice of releasing fish and other animals. The basic story involves the Buddha in a previous life (as a rich man named Rūsui) coming across ten thousand fish that were about to die because the pond in which they were dwelling was about to dry up. Developing the mind of compassion, he had elephants help carry enough water to the pond for the fish to survive. Thereafter a banquet was hosted for the fish, at which the future Buddha preached the *Dharma* (particularly the doctrine of dependent origination). Unfortunately, soon thereafter an earthquake hit the region and all the fish died. They were reborn in Tōriten (Thāyashimsat heaven), and out of gratitude to the man who saved them, the fish offered him precious jewels and other treasures.[5] It is this *sūtra* which is most often cited in ritual documents—for example, in Kōfukuji Temple's *Hōjō-e hōsoku*—as the source for the *hōjō-e* ritual.

Yamamoto Haruki has argued that the *sūtra*s are interpreted in two different ways in relation to the Japanese performance of the *hōjō-e*. On the one hand, the *Bommyōkyō*'s emphasis on other sentient beings as one's parents becomes related to the development of ancestral worship (*sosen kaikō*) as part of the rite in Japan. On the other hand, the *Kongōmyōkyō*'s emphasis on the merit derived from helping animals ("treasures bestowed" on helper) becomes related to the notion of performing the *hōjō-e* for this-worldly benefits (*riyaku shinkō*).[6]

In terms of their respective views on animals, it is possible to see, on the one hand, the *Bommyōkyō* holding a position that the boundary between the human and animal worlds is like a semipermeable membrane, as either oneself or one's parent can be an animal in a past or future life. This view might be understood as parallel to the deep ecological worldview in which the natural world or the animal world is seen as part of a "deeper ecological self."[7] On the other hand, while this view is not absent from the *Kongōmyōkyō*, the emphasis there is rather on the altruistic act itself that comes simply from seeing animals, as animals, suffering. This view might be said to be more akin to animal rights perspectives regarding the sentience of animals and their standing, or rights, independent from human beings.[8]

The Nature of the Ritual

The ritual itself has both a broad and a narrow meaning. From the shrine/temple's point of view, the entire day of ritual observance was termed the *hōjō-e*. As such, the *hōjō-e* was a festival day, dancing, music, horse riding, processing the shrine *kami* (Shintō deity) out on a portable shrine (*mikoshi*), and wrestling performances, among other activities.[9] But from a narrower point of view, the *hōjō-e* can be considered a ritual activity centered on the release of birds, fish, or other animals.

Because the rite was classified as an observance of the "non-killing precept," both the Buddhist monks who oversaw the ritual[10] and the Shintō priests were to observe abstinence from meat and fish during the festival period.[11] On the actual day of the rite, at the Iwashimizu Hachiman Shrine, fish and clams were released in the *hōjō* river (*hōjōgawa*), which was on the south side of the shrine compound. While the fish were being released, the priest (*dōshi*) chanted Buddhist scriptures, particularly the *Kongōmyōkyō*. He also announced the performances of all the other *hōjō-e* ceremonies that were held across Japan in the past year.[12] At Kōfukuji, the headquarters of the Hossō (Yogācāra) school, *hōjō-e* was held on the seventeenth of the Fourth Month at the temple's Tōtōkōzen-in Ichigonkannondō. After a *dharma* meeting at that building, the lay members (*sankeisha*) would take carp to be released in a pond (*hōjō-ike/enketsuchi*) specifically designated for the protection of animals released for this rite. When the carp were released, the lay members would also release pieces of paper upon which they had written wishes, in the belief that the wishes would be granted because of the merit accrued from releasing animals from captivity.[13]

The Transformation of the *Hōjō-e* at Hachiman Shrines in Japan

> Every fifteenth of the Eighth Month, around the country is the Hachiman *hōjō-e*.
>
> —*Genpei Seisuiki*[14]

One of the key characteristics of the Japanese tranformation of the *hōjō-e* was its performance at so-called Shintō shrines which had the deity Hachiman as their cultic center.[15] The first account of the *hōjō-e* ritual in Japan is recorded in the *Jūji enjishō* as being held in 710 (Yōrō 4) at the Usa Hachiman Shrine.[16] Archaeological digs at Iba Iseki and Chigasakishi Motomura have given us evidence of early performances of the *hōjō-e* at Buddhist temples.[17] Yet, what is most peculiar about the development of this ritual in Japan is the concentration of the performance of this ceremony at Hachiman shrines,[18] which are generally categorized as "Shintō" shrines, not Buddhist temples. Because *hōjō-e* were primarily held at Hachiman shrines, the rite, while officially classified as a Buddhist ritual, was often understood as a "Shintō-Buddhist" or a "shrine-temple" ritual.[19] Indeed, Nakano Hatayoshi has argued that the Usa *hōjō-e* is a mixture of an older "Shintō" ritual of installing a bronze mirror at the shrine and the 710 Buddhist ceremony.[20] This Usa Shrine ritual was then transmitted to other Hachiman shrines and Buddhist temples during the Heian period (for example, in 859 by the Buddhist monk Gyōkyō from the temple Daianji to the Iwashimizu Hachiman Shrine). By the Kamakura and Muromachi periods, the Iwashimizu Hachiman Shrine's *hōjō-e*[21] on the fifteenth of the Eighth Month of each year became the best-known and most elaborate example of this ritual.[22]

The *Hōjō-e* and the State: Especially the Case of Iwashimizu Hachiman

By the late medieval period, the ritual life of the state consisted of three important state rituals (*sanchokusai*): the Kamo Festival, the Kasuga Festival, and the Iwashimizu *hōjō-e*.[23] In the case of the Iwashimizu *hōjō-e*, both the imperial court and the Kamakura and Muromachi shogunates observed this state ritual by sending envoys and monetary offerings[24] to Iwashimizu on the appointed day. Court and shogunal representatives (*jōkei/chokushi*) observed a strict abstention from any fish or meat (*shōjin kessai*) during the period prior to their visit to the shrine.[25] By the Muromachi period, the importance of state attendance at this rite was so great that all four Ashikaga shoguns (Yoshimitsu, Yoshimochi, Yoshinori, and Yoshimasa) went to Iwashimizu in person as representatives (*jōkei*) of the state. Once at the shrine, the envoys made offerings and attended the various stages of the ceremony, including the release of fish and clams into the river.[26]

While state support of a rite to release animals may seem at first glance to be a positive development in terms of ecological activity, there were in fact a number of nonaltruistic factors in the state's interest in this rite. Particularly in the case of the Iwashimizu Hachiman Shrine, there are three major reasons for the state's involvement with this ritual: 1) Hachiman was an *ujigami* (clan deity) of the Minamoto clan and their descendants. To support the most important rite (the *hōjō-e*) at the most significant Hachiman shrine was, then, considered a familial obligation by the members of the Kamakura and Ashikaga governments, many of whom were connected to this clan.[27] 2) Iwashimizu had a huge military force that rivaled the government's forces,[28] and thus the state needed to appease the shrine and its quasi-military (*shinjin*) by providing the funds for the *hōjō-e*. 3) Iwashimizu occupied a strategic geo-military position, and any medieval political power had to negotiate and curry favor with the shrine by supporting its rites.

Furthermore, as quoted at the beginning of this essay, both shogunal and court governments sent out orders to the provinces to observe the rule on not taking life (not killing animals) during the two-week period between the first and fifteenth of the Eighth Month. This has been termed by certain Japanese medieval historians as the "ideology of nonkilling" (*sessho kindan* ideology).[29] What ties the *hōjō-e* and "ideology of nonkilling" is, of course, the first of the Buddhist precepts: "do not kill." This "ideology" was a part of the system of the twenty-two shrine-temple complexes in which the Japanese state appropriated Shintō shrine and Buddhist temple rites and doctrines during the medieval period.[30] Iwashimizu Hachiman, at the height of its political power, stood at the head of this system of shrine-temple complexes. This kind of tie between the state and the *hōjō-e* can be highlighted as a Japanese innovation to the character of this ritual.

Problematic Issues of the *Hōjō-e* in Japan and Challenges to the Image of Buddhism as Environmentalist

The two characteristic features of the Japanese transformation of the *hōjō-e* outlined above—namely, the "Shintō-Buddhist" nature of the majority of the *hōjō-e* and the identification of the *hōjō-e* as a *state rite* in the medieval period—allow us the opportunity to examine several problematics associated with the ritual in terms of the perspective on animals that the rite may reflect.

While it is tempting to suggest, as a number of scholars and Buddhist practitioners have done recently,[31] that the *hōjō-e* demonstrates a Buddhist view of animals which is sympathetic and positive, this view needs to be qualified in a number of ways. First, as we have shown in the case of "Shintō-Buddhist" complexes such as the one at Iwashimizu Hachiman, the rite was a very elaborate affair because of the importance attached to it by the state. Taira Masayuki's research has shown that in the medieval period, the shrine was extremely concerned about having enough fish and clams to release (usually in the range of one to three thousand). Thus, more than triple the number were captured several weeks ahead of time to ensure that enough animals were available by the time the state envoy arrived. In other words, if three thousand fish were to be released at the *hōjō-e*, a total of nine thousand would need to be captured and purchased by the shrine with the understanding that two-thirds of them might die before they could be released.[32] The display of power was more important than the lives of the animals themselves. The release of animals, then, that occurred at these major medieval shrines and temples was more often a matter of displaying political power or appeasing various deities.[33]

Another problem concerning the *hōjō-e*'s intimate connection with the state was the use of the so-called ideology of not killing promoted by the state to gain control of land. Itō Seirō has argued persuasively that local lords of estates (*shōen*) built smaller Hachiman shrines on their lands. In doing so they made use of the fact that sites where the *hōjō-e* rituals are performed were to be deemed sacred, and thereby they controlled hunting, fishing, and agricultural activities on the feudal estates. In other words, the ritual and ideological basis of *hōjō-e* was also sometimes used to control new lands won through war in medieval Japan.[34] On the one hand, the ceremony of the release of animals was seen as a way to atone for the blood spilled during warfare, but, on the other hand, the rite was used to justify warfare and the continued control over the lands won in war.

There is thus a paradox built into the medieval Japanese *hōjō-e*. The importance placed on this ritual and the notion of "nonkilling" was precisely what caused shrines to go to great lengths (even "sacrificing" two-thirds of the fish) to perform this rite as a grand state ritual involving the release of thousands of fish. The idea that sites where *hōjō-e* were held could be designated as places where people could be prohibited from hunting and fishing and from engaging in other agricultural activities was what made the ritual so attractive to provincial lords who had, ironically, just taken the land through force and bloodshed. The paradox is also inherent in the broader practice of memorializing animals that one has killed. For example, traditionally in fishing villages, memorial rites for fish just captured were performed.[35] Likewise in more contemporary Japan, some Buddhist priests have joined conservative politicians for "whale banquets" (eating illegally caught whales), explaining their activity as being one of "memorializing" whales.

Conclusion

In 1017, Fujiwara no Sanesuke, one of the leading courtiers of the day, sent the governmental envoy off to the Iwashimizu *hōjō-e* at the Kamo River. As he was bidding the messenger farewell, he saw two men fishing on the banks of the river. As it was the day of the *hōjō-e*, he bought the fish the fishermen had caught and released them.[36] I end with this story as a counterpoint to the ways in which I have shown the *hōjō-e* to be problematic, to be, indeed, other than an environmentally friendly act. Just as the *hōjō-e* functioned in a number of environmentally unfriendly ways, the story of Sanesuke reveals a significant example of the way in which the *hōjō-e* and the notion of nonkilling entered the world of medieval Japanese society. There are most probably many more unrecorded private acts of relieving the suffering of animals which were generated through contact with the rite of *hōjō-e* or the idea of nonkilling.

At the same time, the need for careful reflection of idealized notions of Buddhism as environmentalist is clear. When one reviews the history of the interface of Buddhism and environmentalism,[37] the overwhelming tendency has been to define the Buddhist contribution to environmentalism in terms of the most idealized notions of what Buddhism is. Though my tendency is to emphasize the more practical dimensions of Buddhist contributions to ecology, the principle of taking the best ideals of a tradition for constructive theology or philosophy is, in itself, not a problem. What is troubling, however, is the tendency to define Buddhist ecological worldviews in contradistinction to other religious traditions, such that the worst actual practices of Christianity and other traditions are contrasted with the best, most ideal components of Buddhism.[38] My hope is that this paper has provided a useful survey of the Japanese development of the *hōjō-e* and a balanced, critical reflection on the ways in which this Buddhist rite might be considered environmentally friendly.

Notes

1. A shogunal order from the *kansenji* official was sent on the fifteenth of the Twelfth Month, 1280 (Kōan 2), to the five provinces closest to the capital (*gokinai shokoku*). The same order went out as an imperial edict from Emperor Go-Uda three days later to all provinces (*kyōi shokoku*). This order is quoted in full and discussed in detail in Itō Seirō, "Iwashimizu hōjō-e no kokkateki ichi ni tsuite no ikkōsatsu" (A consideration of the state-like aspect of the Iwashimizu Shrine's *hōjō-e*), *Nihonshi Kenkyu* 188 (April 1978):36–37.

2. The term literally means release, living (beings), meeting/ceremony.

3. For a very interesting critique of "ecoBuddhism," see Ian Harris, "How Environmentalist Is Buddhism?" *Religion* 21 (April 1991):101–14, and "Buddhist Environmental Ethics and Detraditionalization: The Case of EcoBuddhism," *Religion* 25, no. 3 (July 1995):199–211.

4. This passage was originally translated by M. W. de Visser and later revised by Jane Marie Law in "Violence, Ritual Reenactment, and Ideology: The *Hōjō-e* (Rite for Release of Sentient Beings) of the Usa Hachiman Shrine in Japan," *History of Religions* 33, no. 4 (May 1994):325–26. The *Bommyōkyō* section above can be found in the Taishō edition of the Buddhist canon (T. 1484, 24:997A–1003A).

5. A more elaborated account of the story can be found in Law, "Violence, Ritual Reenactment, and Ideology," 326, in English; or in Haruki Yamamoto, "Hōjō-e ni tsuite" (On the *hōjō-e*), *Shukyō Kenkyū* 56, no. 4 (1983):294, in Japanese.

6. Yamamoto, "Hōjō-e ni tsuite," 294–95.

7. For more on deep ecological views regarding the notion of an "ecological self" which includes animals, see Bill Devall and George Sessions, *Deep Ecology: Living As If Nature Mattered* (Salt Lake City: Peregrine Smith Books, 1985); Bill Devall, "Ecocentric Sangha," in *Dharma Gaia: A Harvest of Essays in Buddhism and Ecology*, ed. Allan Hunt Badiner (Berkeley: Parallax Press, 1990), 155–64; and Joanna Macy's works: "Our Life as Gaia," in *Thinking Like a Mountain: Towards a Council of All Beings*, ed. John Seed et al. (Philadelphia: New Society Publishers, 1988), 59–65; "The Ecological Self: Postmodern Ground for Right Action," in *Sacred Interconnectedness: Postmodern Spirituality, Political Economy, and Art*, ed. David Ray Griffin (Albany: State University of New York Press, 1990), 35–48; and "The Greening of the Self," in *Dharma Gaia: A Harvest of Essays in Buddhism and Ecology*, ed. Allan Hunt Badiner (Berkeley: Parallax Press, 1990), 53–63.

8. Perhaps the deep ecology versus animal rights positions of the present day have some precedent in medieval Japanese Buddhism.

9. For a detailed description of the shrine and temple activities, such as the movement of the *mikoshi*, the procession of shrine-temple officials, horse racing,

and wrestling, see documents such as the *Kashiwagashū* or the *Jūji enjishō busshinji shidai,* which can be found in Itō, "Iwashimizu hōjō-e no kokkateki ichi ni tsuite no ikkōsatsu," 33.

10. At Iwashimizu, although the institution was generally considered to be a "Shintō" shrine, Buddhist monks at the nearby temple, Zenpōji, controlled much of the administration of the shrine. For example, the monk Kenshū was appointed by Takauji to fill the *bakufu* position of *bugyō* for Iwashimizu's administration, just as similiar posts were created by the shogunate to administer Ise, Mt. Hiei, Tōdaiji, and Kōfukuji. See Ken'ichi Futaki, "Iwashimizu hōjō-e to Muromachi bakufu: shōgun jōkei sankō o megut'te" (The Iwashimizu *hōjō-e* and the Muromachi *bakufu*: The visit by the shōgun as the jōkei [envoy]), *Kokugakuin Nihonbunka Kenkyūjo Kiyō* 30 (September 1972):101.

11. Ibid., 102.

12. Itō, "Iwashimizu hōjō-e no kokkateki ichi ni tsuite no ikkōsatsu," 33.

13. Yamamoto, "Hōjō-e ni tsuite," 294.

14. This can be found in the *Yamaki yauchi goto* section of the *Genpei Seisuiki,* which is quoted in Itō, "Iwashimizu hōjō-e no kokkateki ichi ni tsuite no ikkōsatsu," 39.

15. One of the difficult aspects of studying the *hōjō-e,* because of its "combinative" Shintō-Buddhist character, is the relative lack of documents on the ritual at Hachiman shrines due to the destruction of these texts during the Meiji period (1868–1912). That the Meiji government's policy of *haibutsu kishaku* (abolition of Buddhism and destruction of Śākyamuni) and *shimbutsu bunri* (separation of Shintō and Buddhism) helped to destroy documents and practices of Shintō-Buddhist combinative character has been well documented in English by Martin Collcutt, "Buddhism: The Threat of Eradication," in *Japan in Transition: From Tokugawa to Meiji,* ed. Marius B. Jansen and Gilbert Rozman (Princeton: Princeton University Press, 1986), 143–67; Allan G. Grapard, "Japan's Ignored Cultural Revolution: The Separation of Shinto and Buddhist Divinities in Meiji *(shimbutsu bunri)* and a Case Study: Tōnomine," *History of Religions* 23, no. 3 (February 1984):240–65; and James Ketelaar, *Of Heretics and Martyrs in Meiji Japan: Buddhism and Its Persecution* (Princeton: Princeton University Press, 1990). The destruction of *hōjō-e* documents in particular is taken up in Japanese in Futaki, "Iwashimizu hōjō-e to Muromachi bakufu," 99.

At the Iwashimizu Hachiman, the *hōjō-e* was canceled at the beginning of the Meiji period but later restored under the name Iwashimizusai, which is now performed on 15 September. The rite one would see today must be considered to be quite different from the medieval *hōjō-e,* as all "Buddhist" elements were purged in the Meiji period to make it conform to the state directive.

16. This date is cited in Itō, "Iwashimizu hōjō-e no kokkateki ichi ni tsuite no ikkōsatsu," 32. Jane Marie Law, however, in the best article in English on the *hōjō-e* ("Violence, Ritual Reenactment, and Ideology," 326), gives 745 as the date

of the first occurrence of the rite. The Yōrō 4 (710) is a date that most scholars of the *hōjō-e* cite (see Nakano Hatayoshi, *Hachiman Shinkōshi no Kenkyū* [Studies on the cult of Hachiman] [Tokyo: Yūzankaku, 1976], upon whom Law relies).

17. A minority view, held by Okada, indicates that the first *hōjō-e* were held in 676 (Temmu 5), as recorded in the Nihonshoki and for which archaeological digs provide evidence: see Sōji Okada, "Iwashimizu hōjō-e no kōsaika" (The development of the Iwashimizu *hōjō-e* as a state ritual), *Kokugakuin Daigaku Daigakuin Kiyō* 24 (1992):3. For a study on the *hōjō-e* at the three Buddhist institutions of Kōfukuji, Konbu-in, and Yoshidadera, see Yamamoto, "Hōjō-e ni tsuite."

18. There are three major Hachiman shrines (Usa Hachiman, in Kyūshū; Iwashimizu Hachiman, in Kyoto; and Tsurugaoka Hachiman, in Kamakura), which have numerous sub-shrines (*massha*). In addition, there are numerous Buddhist temples which have Hachiman as part of the temple's cultic life (for example, Tōdaiji, Tōji, and Yakushiji). See Christine Guth, *Shinzō: Hachiman Imagery and Its Development* (Cambridge, Mass.: Council on East Asian Studies, Harvard University, 1985), for more general information in English.

19. Okada Sōji goes so far as to argue that, especially in the case of the Iwashimizu *hōjō-e*, the *hōjō-e* rite itself is fundamentally a "Shintō" ritual with Buddhist overlays ("Iwashimizu hōjō-e no kōsaika," 16).

20. There has been, in addition, a very popular theory that the Usa *hōjō-e* began as a "ceremony to appease spirits" (*chinkonsai*), "to appease the malevolent spirits of the Hayato tribe defeated and slaughtered by forces of the centralized government in a bloody battle in 720" (Law, "Violence, Ritual Reenactment, and Ideology," 327). This is clearly a possibility, though Nakano had seen this explanation of the origins of the rite as a later addition (see in Itō, "Iwashimizu hōjō-e no kokkateki ichi ni tsuite no ikkōsatsu," 32).

21. Though Gyōkyō brought the "deity-body" of the Hachiman from Usa to Iwashimizu in 859, the most commonly held view is that the *hōjō-e* rite itself was not held until 864 (Jōkan 5); the minority views include the dates 861 (Jōkan 3) or 877 (Jōkan 18) (see Okada, "Iwashimizu hōjō-e no kōsaika," 3).

22. By the Kamakura period, the *hōjō-e* ceremony was held on the fifteenth of the Eighth Month at the following Hachiman shrines: Tsurugaoka, Usa, Usa gosho bekkyū (Hakozaki, Senguri, Fujisaki, Nittagū, kuma kakugū), and Sakuharagū. Only Nittagū's *hōjō-e* had a different date (the fifteenth of the Ninth Month). The Kamakura shogunate, however, naturally centered most of its attention on the geographically closer Tsurugaoka Hachiman Shrine, which became classified as a *nenju gyōji* (an official annual observance of the state).

23. Futaki, "Iwashimizu hōjō-e to Muromachi bakufu," 99; and Itō, "Iwashimizu hōjō-e no kokkateki ichi ni tsuite no ikkōsatsu," 33.

24. These monetary offerings were often in the form of large quantities of salt or rice. The rice came from the Inayama estate in Yamashiro province, and the

salt came from the Bizen, Iyo, and Yamashiro estates. The sending of envoys was also a major expenditure. Itō Seirō has analyzed governmental budgets during the Kamakura period in reference to the costs of state support of the Kasuga Festival, the Kamo Festival, and the Iwashimizu *hōjō-e* and the orders sent from the court or the shogunate to lands they controlled to provide for these supplies (Itō, "Iwashimizu hōjō-e no kokkateki ichi ni tsuite no ikkōsatsu," 35–36).

25. We should note here, though, that during the Kamakura period, it was the imperial court which had the stronger connections to Iwashimizu, while the Kamakura *bakufu* had close ties to the Tsurugaoka Hachiman Shrine's *hōjō-e* ceremony. By the Muromachi period, the Ashikaga shogunate and the imperial court turned their attention solely to the Iwashimizu *hōjō-e*, as the first Ashikaga shogun, Yoshimitsu, shifted the political center back to Kyoto and had a personal interest in the fusion of warrior (*bushi*) culture and aristocratic (*kuge*) culture (Futaki, "Iwashimizu hōjō-e to Muromachi bakufu," 103–12).

26. The exception to this was Yoshimitsu, who left before the release of the fish and was seen to be somewhat irreligious for doing so. His successor, Yoshimochi, on the other hand, stayed for the release of the animals and personally observed extra days of abstention from fish or meat (*shōjin kessai*) before the day of the ritual at the nearby lodgings (*shukubō*) belonging to the temple Zenpōji.

We should also note here how the changing state structures coincided with changes in the performance of the *hōjō-e*. Changes in the order of one's position during the ceremony or in the parade, the wearing of swords by government officials, and the various procedures in the appointment of Buddhist priests to preside over the ceremony all reflected changes in the political structure from the Heian to the Kamakura/Muromachi periods.

27. Futaki, "Iwashimizu hōjō-e to Muromachi bakufu," 100.

28. During the medieval period, in addition to the government military forces, there were three main nongovernmental military forces: 1) the private armies of the *shōen* (estates) of local lords; 2) the armed monks (*sōhei*) associated with Buddhist temples, such as Kōfukuji, Mt. Hiei, Mt. Kōya, Negoroji, Daigoji, Tōdaiji, and Enryakuji; and 3) the armed shrine affiliates (*shinjin*) of the Iwashimizu Hachiman Shrine and several other shrines.

29. Scholars such as Koyama Yasunori and Itō Seirō have argued that the "nonkilling ideology" was used by the medieval state to control the peasant class. Particularly, so-called peasant activities such as hunting, fishing, forestry, irrigation, and slash-and-burn agriculture were periodically prohibited by local provincial governments, using the "nonkilling" order from the central government as a pretext (for more on this particular argument, see Itō, "Iwashimizu hōjō-e no kokkateki ichi ni tsuite no ikkōsatsu," 39–40). However, the evidence for this is somewhat unconvincing given Itō's usually meticulous documentation, although I will return to the problem of the state appropriation of the *hōjō-e* and the ideology of nonkilling below.

30. For more on the medieval state and the system of twenty-two shrine-temple complexes, see Allan G. Grapard, "Institution, Ritual, and Ideology: The Twenty-Two Shrine-Temple Multiplexes of Heian Japan," *History of Religions* 27, no. 3 (February 1988):246–69.

31. This has been particularly often cited in the United States by Philip Kapleau (*To Cherish All Life: A Buddhist Case for Becoming Vegetarian* [San Francisco: Harper and Row, 1982], among others) and the journal *Buddhists Concerned for Animals*.

32. Masayuki Taira, "Debating the Buddhist Prohibitions against the Taking of Life," talk given in Japanese at "New Directions in the Study of Social History, Status, Discrimination, Popular Culture in Premodern Japan" workshop, Princeton University, 26–28 October 1995.

33. The custom of *kessai* (abstention) from meat, sexual activity, and such, was a common short-term practice followed by political elites for accumulating the favors of a particular deity or, in the case of emperors performing *kessai*, for the protection of the nation.

34. This partly explains why Hachiman was one of the most popular deities among the emerging warrior class and the provincial warlords of medieval Japan.

35. Eiki Hoshino, with Dōshō Takeda, "*Mizuko Kuyō* and Abortion in Contemporary Japan," in *Religion and Society in Modern Japan*, ed. Mark Mullins et al. (Berkeley: Asian Humanities Press, 1993), 171–90.

36. From the 1017 (Kan'in 1) entry in Fujiwara no Sanesuke's diary, *Shōyūki*, as found in Okada, "Iwashimizu hōjō-e no kōsaika," 6.

37. See Duncan Williams, "The Interface of Buddhism and Environmentalism in North America" (B.A. thesis, Reed College, 1991).

38. This, of course, refers not only to Buddhism but also to the rhetoric of Western exploitation of nature and Eastern harmony and oneness with nature. For early versions of this view, see Masao Abe, "Man and Nature in Christianity and Buddhism," *Japanese Religions* 7, no. 1 (July 1971):1–10; Hwa Yol Jung, "Ecology, Zen, and Western Religious Thought," *Christian Century*, 15 November 1972, 1153–56; and Lynn White, Jr., "The Historical Roots of Our Ecologic Crisis," *Science* 155 (March 1967). For more recent manifestations of this idea, see Stephen R. Kellert, "Concepts of Nature East and West," in *Reinventing Nature? Responses to Postmodern Deconstruction*, ed. Michael E. Soulé and Gary Lease (Washington, D.C.: Island Press, 1995), 103–21; and Yuriko Saito, "The Aesthetic Appreciation of Nature: Western and Japanese Perspectives and Their Ethical Implications" (Ph.D. diss., University of Wisconsin, 1983).

Zen Buddhism:
Problems and Prospects

Mountains and Rivers and the Great Earth: Zen and Ecology

Ruben L. F. Habito

The question I address in this essay is this: does Zen practice and teaching support and foster an active engagement toward the earth's well-being and an ecologically viable way of life and vision? Rather than writing of Zen in a generic and idealized way, here I refer mainly, though not exclusively, to the Zen practice and teaching offered in the Sanbō Kyōdan community, a direct continuation of what is known as the Harada-Yasutani lineage, which has had considerable influence in North America and Europe in the last two or three decades.[1]

The first section will note attitudes that appear to serve as obstacles to a commitment to our ecological well-being on the part of those who practice Zen. The second section will describe three fruits that manifest themselves in the life of the Zen practitioner, which may enable one to overcome those attitudes discussed in the first. The third section will then look at possible Zen contributions to our ecological well-being, considering the connection between Zen practice and ecologically oriented life and action.

Some Pitfalls in Zen Practice

To our central question of whether Zen practice and teaching support and foster active engagement in the ecological well-being of the earth community and an ecologically viable way of life and vision, a first-impression answer would be, "it appears not" (*videtur quod non*), on at least two counts. First, many Zen practitioners are on a journey of self-discovery, having taken an "inward turn" that de-emphasizes their engagement with events in the "outside world." Second, the Zen dictum of "living in the present moment" can foster an attitude of indifference toward the future—not only the individual practitioner's own future, but also the communal future of living beings on earth.

On the first count, it is a fact that many individuals begin their Zen practice as their entry into a journey within. This tradition, which focuses on meditative practice, itself encourages the inward turn that enables the individual to disengage him- or herself from distracting and secondary "worldly" preoccupations and to focus on "the one thing necessary"—the awakening to one's true self, understood to be the basis of true inner peace and fulfillment.

There are, of course, those who begin Zen practice out of mere curiosity or out of a desire to partake in the benefits it offers to one's physical well-being, such as improved posture, the cure of certain ailments, and so on. There are also those who are already engaged in social and ecological issues when they begin their Zen practice and turn to it precisely in order to derive nourishment and energy for their tasks in that arena. It must be noted however that a good number of those who turn to Zen do so spurred on by a felt inner need to set their lives in order, to find their "ground" or "center" amid the vicissitudes of life, to solve some fundamental questions on the meaning of one's presence on this earth, or just to find inner peace and serenity, in a practice centered on the awareness of one's breath and seated meditation. The Zen journey undoubtedly is an interior-oriented one that involves rigorous and continual practice. It is a journey that takes up one's full attention and energy over a long period of time, perhaps one's whole lifetime.

This inward turn of the Zen practitioner can militate against a commitment toward an ecologically viable way of life in this way: the emphasis on "listening within" may lead to a dichotomous view of the "within" and the "without," to the extent that the practitioner disengages from the concerns of the rest of society, diminishing the individual's interest in and engagement with events in the world "outside."

Thus, the toxic wastes that are wreaking havoc on our natural habitat are not considered as great a threat to one's being as are the three poisons of greed, anger, and selfish ignorance, which the serious practitioner feels one must first battle with and attempt to uproot from within, before being able to address the issue of the toxic wastes "outside." The "mountains and rivers" that appear in Zen discourse are often merely idealized images in the practitioner's mind, with no connection at all to the actual mountains in many parts of the world that are being denuded because of indiscriminate logging practices or to the rivers reeking with chemical pollutants.

On the second count, the emphasis in Zen writings and teachings on "living in the present moment" may give practitioners the misguided impression that Zen practice discourages thinking about or has nothing to do with one's individual or the earth's communal future. It may even lead to an irresponsible attitude that constantly seeks to "seize the day" (*carpe diem*) and forgets or ignores the consequences of one's actions, passions, or omissions for one's own or others' future. This attitude admittedly is an erroneous one based on a misunderstanding of the Zen dictum, but it is one that must be dealt with nevertheless. This type of one-sided emphasis on the present moment thus would tend to diminish the concern that many species on earth are becoming extinct and that, because of this, the whole earth community is heading toward a bleak future.

In sum, these two points—the preoccupation with the "within" that stands in opposition to or excludes the "without"; and the preoccupation with the present that excludes the past and the future—would incline us to give a negative response to the initial question of whether Zen practice and teaching supports an ecologically viable way of life and vision.

However, an examination of the actual fruits of Zen practice in the lives of practitioners may offer a perspective that can overcome the aspects that militate against or diminish practitioners' engagement with the ecological well-being of the whole earth community.

Fruits of Zen Practice

The three fruits that are made manifest in the life of the Zen practitioner as she or he deepens in *zazen,* or seated meditation, and the cultivation of awareness in one's daily life are as follows: 1) the deepening of one's mindfulness (*jōriki* in Japanese; literally "the power of *samādhi*"); 2) the experience of awakening to one's true self (*kenshō-godō,* or "the way of enlightenment through seeing one's true nature"); and 3) the realization and personalization of this true self in one's ordinary life (*mujōdō no taigen*, literally "the bodily manifestation of the peerless way").[2]

First, with the deepening of one's mindfulness, the Zen practitioner is able to gather together the disparate elements of one's life and achieve ever greater integration. The practitioner comes to be *fully there* at every moment, alive in the here and now. The practice of *just sitting* (Japanese, *shikan-taza*), with one's legs folded and one's back straight, with one's whole being fully at attention in the *here and now,* relishing the freshness of each breath as it comes and goes, has this natural effect of bringing about a greater sense of wholeness and at-homeness in ordinary life.

Just sitting in this way invites one to live at the core of one's being, to *do* nothing and to *have* nothing, but simply to *be*. Focused on *be*-ing, rather than on doing or having, one is able to celebrate and relish all things in the universe, just as they are.

This first fruit opens the way for the second, namely, the experience of awakening to one's true self. This experience involves a revolution in one's way of seeing and relating to everything in the universe. One way to describe the experience is the arrival at what can be called a *zero-point,* wherein opposing concepts of being and nonbeing, doing and nondoing, having and nonhaving, plus and minus, and so on, converge and cancel each other out. At this zero-point, the separation between subject and object, between the "I" and the world, is overcome, and the practitioner is opened to an entirely new way of seeing and way of being.[3]

The thirteenth-century Japanese Zen master Dōgen gives expression to this experience of the disintegration of the boundary between subject and object:

> I came to realize clearly, that mind is no other than mountains and rivers and the great earth.[4]

This second fruit of Zen practice, the experience of awakening to one's true nature, is seen as a pivotal point in an individual person's Zen journey, but it is still regarded only as the practitioner's entry-point into the Zen way of life. A person who has been confirmed in this initial awakening experience is led deeper into the life of Zen with the continued practice of selected *kōan*s, numbering around five to six hundred in the Sanbō Kyōdan lineage, under the guidance of an authorized teacher.[5]

The experience of Zen awakening enables a practitioner to overcome the dichotomy in one's consciousness between subject and object and to bridge the gap between the "I" and the whole universe. An initial experience of this sort, incidentally, is usually accompanied by a deep joy that may be manifested in bursts of laughter and also in tears and convulsions. Arriving at a standpoint totally different from ordinary consciousness (characterized by the subject-object polarity), the practitioner experiences profound emotions of exhilaration, inner peace, and gratitude.

The emotional impact can be like a "pink cloud" that lasts for days, or even longer. But the emotions eventually subside, and the practitioner comes back to the "ground" of ordinary life with its ups and downs and with its concomitant tasks. The integration of the vision of nonseparateness, glimpsed in the initial awakening experience, with the rest of one's life is the third, and most significant, fruit of Zen. This is the fruit described as the "embodiment of enlightenment in one's daily life" and is a process which takes a whole lifetime.

As one continues practice in this direction, one is enabled to live in ever deeper awareness of the mystery of each present moment as one goes about daily activities, from washing one's face in the morning to preparing for bed at night.

Kōan practice becomes a powerful way of embodying the enlightenment experience in one's daily life. Each *kōan* is a renewed invitation to return to the primordial experience of awakening, with a new and fresh angle offered by the particular *kōan* in question.

An example of such a *kōan* given to a practitioner in this context is the following, from the collection entitled *The Book of Serenity*:

> Officer Lu Geng said to Nanquan,"Teaching Master Zhao was quite extraordinary: he was able to say, 'Heaven and earth have the same root, myriad things are one body.' " Nanquan pointed to a peony in the garden and said, "People today see this flower as in a dream."[6]

In this *kōan*, the Zen practitioner is invited by Teaching Master Zhao to experience this zero-point as the dynamic ground of all that exists: "Heaven and earth have the same root, myriad things are one body." In other words, this experience of zero-point is presupposed in this expression, and the practitioner is enjoined to demonstrate her understanding of it as coming from that experience, in the one-to-one encounter with the Zen teacher.

The last line, then, is taken up to call our attention to how our ordinary perceptions, which presuppose a subject-object duality, are based on an illusion: "People today see this flower as in a dream." That is, they are not able to "see" the real flower as they remain trapped in the ordinary consciousness that separates the "object" (flower) from the "subject" ("I" as seer).

The three fruits thus can be summed up as an ever-deepening process of integration of one's whole life, involving a constant return to that primordial experience of awakening to one's true self in the ordinary events of life, such as looking at a flower or chopping wood or carrying water.

The question to be addressed, then, is this: how does the realization of these three fruits of Zen practice enable one to overcome the aspects that militate against active engagement in issues involving our ecological well-being, as noted in the first section?

First, as one continues practice, enabling these three fruits to mature in one's daily life, one overcomes the dichotomy of the "inward" versus the "outward." In rediscovering that one's true self is not separate from "the mountains and rivers and the great earth" and all sentient beings, there is no longer anything in the universe that is outside of one's concerns.

Such a perspective transforms one's fundamental attitude toward the natural world and all sentient beings. Mountains, rivers, and the great earth are experienced as manifestations of one's own true self; they are no longer seen as "out there," entities separate from oneself. One is enabled to feel and see things *from the perspective* of the mountains, the rivers, the great wide earth, of everything that lives and breathes—pelicans and dolphins, dragonflies and ladybugs, and, of course, other human persons.

Another passage from Dōgen comes to mind here:

Delusion is seeing all things from the perspective of the self. Enlightenment is seeing the self from the perspective of the myriad things of the universe.[7]

To see everything "from the perspective of the myriad things of the universe" is also to experience that each element in this universe is interconnected with everything else. This vision of interconnectedness is described with rich imagery in the *Flower Garland Sūtra*.[8] One key image in this *sūtra* is the jeweled net of Indra. This is a wondrous net which stretches out infinitely in all directions, and a single bright jewel is in each eye of the net. Each jewel, in its marvelous transparency and uniqueness, reflects all the other jewels in this infinite net. And conversely, each unique jewel is likewise reflected on every other in this wondrous net.[9]

The experiential appropriation of this image grounds one in a transformative process that encompasses one's whole life. One comes to deepen one's awareness in daily life, enlightened by the wisdom that sees all things "as they are," that is, as not separate from one's true self. This wisdom flows out into a life of compassion, wherein one literally "feels in with" other beings, suffering with them in their suffering, being joyful with them in their joy.

Thus, with the maturation of the three fruits of practice, the Zen dictum of "living in the present moment" is understood no longer as an exclusion of the past and the future, but precisely as a recognition of one's past and one's future as contained in the fullness of this present moment. In other words, the present moment understood in the context of Zen practice is not a point in linear time but a dimension of fullness that enables one to embrace one's past and all its consequences and to take responsibility for one's future as the natural unfolding of this present moment. With such an understanding of living the present moment, one lives life and makes decisions in the present in a way that is open to the future and is thereby responsible for it.

In sum, the three fruits of Zen practice thus enable one to overcome the dichotomies of the "inward" as against the "outward," the "present moment" as against past and future. In particular, the maturation of the third fruit, which comes with the continuation of *zazen* and *kōan* practice, deepens one's awareness of interconnectedness with all beings, and this awareness comes to ground one's every thought, word, and action in one's daily life.

Zen Practice and Ecological Action: Prospects

The person wherein the three fruits of Zen practice are in the process of maturation sees oneself as not separate from mountains, rivers, and the great wide earth. To see one's true self *as* the mountains, rivers, and forests, and as the birds, dolphins, and all the inhabitants of the great wide earth, constitutes a solid basis for living an ecologically sound way of life. This way of seeing everything as one's true self leads to actions that would not destroy but would protect, revere, and celebrate the mountains and rivers and the great wide earth as one's own body. It is this living sense of oneness with the mountains, rivers, the great wide earth lived and felt as one's own body which can provide us humans with a key to the way out of our critical ecological situation.

From this vision, nonseparation, opened to the practitioner in the initial awakening experience and cultivated in continued *zazen* and *kōan* practice, enables one to feel, as one's very own, the pangs of hunger of those who are deprived of the basic necessities of life, the pain of the victims of violence and discrimination and injustice, in their different forms.

Further, one is enabled to feel as one's very own the pain of the whole earth being destroyed by human selfishness and greed and shortsightedness: the mountains being denuded, the rivers being polluted, the species of life-forms being decimated. In all this, one feels one's own body racked in pain.

Such a sensitivity to the pain of the earth may thus become the source of the energy that can lead to the transformation of the way we live and relate to one another and to the earth.

The task, then, is one of translating this experiential realization of oneness with mountains, rivers, and the great earth into a mode of life and mode of action that addresses the concrete issues we face in our contemporary world. This task invites one to a deeper experiential appropriation of the wisdom of nondiscrimination, that is, the vision of reality that has overcome the dualistic walls separating subject and object, oneself and the natural world. But further, it calls for the activation of skillful means (*upāya*) that will enable one to respond, grounded in compassion, to different situations, based on the needs of sentient beings. It is in this activation of the various "skillful means" necessary to address our contemporary ecological crisis that Zen practitioners may be able to contribute to the common task of healing the earth's wounds.

There are now many groups and communities bonded together in the practice of Zen, spread out in different parts of the world. These communities have the potential of becoming centers of ecological awareness. In addition to promoting various ways of living a more simple, sound ecological life on the individual and family level—the natural outflow of their communal Zen practice as described above—these communities could also be matrices of support for socioecologically oriented action programs undertaken in solidarity with other groups already engaged in various ecological issues.

Concretely, participation by Zen practitioners in events or in group action organized to call attention to specific local issues of ecological import (such as sit-ins to protest logging practices in certain areas, information campaigns against a certain company's waste disposal habits, or the denouncement of development projects that would threaten the ecological well-being of a certain local area) could be seen as the "activation of skillful means" called for in particular situations. Such participation in concrete modes of action would be seen as the natural outflow of the vision and the experience that is nurtured and deepened by Zen practice.

Over and above the particular forms of action Zen practitioners can engage in—needless to say, in solidarity with many groups already engaged in various types of ecological concerns—the most significant contribution Zen can make toward supporting and fostering the earth's well-being and promoting an ecologically viable way of life is in offering a fundamental vision of reality that invites human beings to an experiential oneness with mountains and rivers and the great earth.

Notes

1. For a current lineage chart of the Harada-Yasutani lineage of Zen, which includes the Sanbō Kyōdan line transmitted to Yamada Kōun (1907–1989), the White Plum line of the late Maezumi Taizan (1931–1995), another *dharma* heir of Yasutani Hakuun (1885–1973), as well as the Diamond Saṅgha established by Robert Aitken (b. 1917), a *dharma* heir of Yamada Kōun, see the World Wide Web page prepared by Matthew Ciolek of the Australian National University (http://coombs.anu.edu.au/WWWVLPages/-BuddhPages/HaradaYasutani.html).

2. Kapleau, Philip, ed., *The Three Pillars of Zen: Teaching, Practice, and Enlightenment* (Boston: Beacon Press, 1967), 76–49; Ruben Habito, *Healing Breath: Zen Spirituality for a Wounded Earth* (Maryknoll, N.Y.: Orbis Books, 1993), 32–37. Kapleau describes the particular elements involved in the practice and teaching of the Sanbō Kyōdan Zen tradition, based in Kamakura, Japan. On this, see also Robert Aitken, *Taking the Path of Zen* (San Francisco: North Point Press, 1982) and Habito, *Healing Breath*.

3. Ruben Habito, *Total Liberation: Zen Spirituality and the Social Dimension* (Maryknoll, N.Y.: Orbis Books, 1989), 29–32.

4. Dōgen Zenji, *Sokushin Ze-Butsu*, in *Shōbōgenzō*, 3 vols. (Tokyo: Iwanami Bunko, 1939), 98. Incidentally, this noted passage has served as the trigger for an enlightenment experience of a twentieth-century Japanese Zen master whose authorized disciples now continue the lineage in Europe and North America: Kapleau *Three Pillars of Zen*, 202–4. See also "*In Memoriam*: A Tribute to Yamada Kōun," *Buddhist Christian Studies* 10 (1990):231–37, for an account of Yamada's life and Zen teaching and for a partial list of his *dharma* heirs now continuing his teaching in different countries throughout the world. This account of Yamada's Zen enlightenment experience is also quoted in Ruben Habito, "The Ecological Implications of Zen Buddhism," in James Veitch, ed., *Can Humanity Survive? The World's Religions and the Environment* (Auckland, New Zealand: Awareness Book Company, 1996), 137–38, a related article also addressing the ecological implications of Zen Buddhism.

5. After a practitioner is confirmed in the initial experience, she or he is guided through a set of twenty-two miscellaneous *kōan*s, after which follow the *kōan* collections entitled *Wu-Men Kuan* (*Mumon Kan* in Japanese pronunciation), or "Gateless Gate," the *Pi-yen Lu* (*Hekigan Roku*), or "Blue Cliff Records," the *Tsung-jung Lu* (*Shōyō Roku*), or "Book of Serenity," and the *Denkō-roku* or "Transmission of Light." In addition to these collections, the practitioner goes through *The Five Ranks, The Three Treasures, The Threefold Pure Precepts*, and *The Ten Grave Prohibitions*.

6. Thomas Cleary, trans., *The Book of Serenity* (New York: Lindisfarne Press, 1990), 390.

7. Dōgen, *Genjō kōan*, in *Shōbōgenzō*, 77.

8. Francis Cook, *Hua-Yen Buddhism: The Jewel Net of Indra* (University Park and London: Pennsylvania State University Press, 1981).

9. Ibid., 1–19, 56–74.

The Precepts and the Environment

John Daido Loori

Imagine, if you will, a universe in which all things have a mutual identity. They all have a codependent origination: when one thing arises, all things arise simultaneously. And everything has a mutual causality: what happens to one thing happens to the entire universe. Imagine a universe that is a self-creating, self-maintaining, and self-defining organism—a universe in which all the parts and the totality are a single entity, all of the pieces and the whole thing at once are one thing.

This description of reality is not a holistic hypothesis or an all-encompassing idealistic dream. It is your life and my life. The life of the mountain and the life of the river. The life of a blade of grass, a spiderweb, the Brooklyn Bridge. These things are not related to each other. They are not part of the same thing. They are not similar. Rather, they are *identical* to each other in every respect.

But the way we live our lives is as if this were not so. We live our lives in a way that separates the pieces, alienates, and hurts. The Buddhist Precepts are a teaching on how to live our lives in harmony with the facts described above. When we look at the Precepts, we normally think of them in terms of people. Indeed, most of the moral and ethical teachings of the great religions address relationships among people. But these Precepts do not exclusively pertain to the human realm. They are talking about the whole universe, and we need to see them from that perspective if we are to benefit from what they have to offer, and if we are to begin healing the rift between ourselves and the universe.

First among the sixteen Precepts are the Three Treasures. We take refuge in the Three Treasures—the Buddha, the *Dharma*, and the

Saṅgha. Understood from three different perspectives, the Three Treasures present different virtues. The first perspective is called the One-Bodied Three Treasures; the second is called the Realized Three Treasures; and the third is called the Maintained Three Treasures.

From the perspective of the One-Bodied Three Treasures, *anuttara-samyaksambodhi*—supreme enlightenment—is the Buddha Treasure. Master Dōgen taught, "Being pure, genuine, apart from the dust is the *Dharma* Treasure." The reason it is apart from the dust is that it *is* the dust. That is what the virtue of purity is about. There is nothing outside of it. The merits of harmony are the *Saṅgha* Treasure. Together, these are the One-Bodied Three Treasures.

To realize and actualize *bodhi*, or enlightenment, is the Buddha Treasure of the Realized Three Treasures. The realization of Buddha is the *Dharma* Treasure, and to penetrate into the Buddhadharma is the *Saṅgha* Treasure. These are the Realized Three Treasures.

Among the Maintained Three Treasures, their manifestation in the world, "guiding the heavens and guiding people, sometimes appearing in vast emptiness, sometimes appearing in dust, is the Buddha Treasure. Sometimes revolving *sūtra*s and sometimes revolving the oceanic storehouse, guiding inanimate things and guiding animate things, is the *Dharma* Treasure. And freed from all suffering and liberated from the house of the Three Worlds is the *Saṅgha* Treasure." This is what we take refuge in. These Three Treasures are the universe itself. They are the totality of the environment and oneself.

Next are the Three Pure Precepts. The first of the Three Pure Precepts is *not creating evil*. This is based on the assumption that there is an inherent purity and goodness in the universe. Actually, there is neither goodness nor badness, neither good nor evil. These polarities do not exist until we create them. This precept is saying that *not creating evil* is the abiding place of all Buddhas, the source of all Buddhas.

The second of the Three Pure Precepts is *practicing good*. Not to create evil means not to become involved in any activity that will give rise to evil. Although from the absolute perspective, there is neither good nor evil, every activity is going to create some consequence in the world of phenomena. The minute there is action, either good or evil appears. So, do not let evil appear but, rather, practice good. This is the *dharma* of *samyaksambodhi*, the way of all beings.

The third of the Three Pure Precepts is *actualizing good for others*. This is to transcend the profane and go beyond the holy, to liberate oneself and others.

The Three Pure Precepts are a definition of harmony in an inherently perfect universe, a universe that is totally interpenetrated, codependent, and mutually arising. But the question is, how do we accomplish that perfection? The Ten Grave Precepts point that out. Looking at the Ten Grave Precepts in terms of how we relate to our environment is a step in the direction of appreciating the continuous, subtle, and vital role we play in the well-being of this planet. It is the beginning of taking responsibility for the whole catastrophe.

The First Grave Precept is *affirm life—do not kill*. What does it mean to kill the environment? It is the worst kind of killing. We are decimating many species. There is no way that these life-forms can ever return to the earth. The vacuum their absence creates cannot be filled in any other way, and such a vacuum affects everything else in the ecosystem, no matter how infinitesimally small it is. We are losing species by the thousands every year—the last of their kind on the face of this great earth. And because someone in South America is doing it, that does not mean we are not responsible. We are as responsible as if we are the ones clubbing an infant seal or burning a hectare of tropical forest. It is as if we were squeezing the life out of ourselves: killing the lakes with acid rain; dumping chemicals into the rivers so that they cannot support any life; polluting our skies so our children choke on the air they breath. Life is nonkilling. The seed of the Buddha grows continuously. Maintain the wisdom life of Buddha and do not kill life.

The Second Grave Precept is *be giving—do not steal*. Do not steal means not to rape the earth. To take away from the insentient is stealing. The mountain suffers when you clear cut it. Clear cutting is stealing the habitat of the animals that live on the mountain. When we overcut, streams become congested with the sediments that wash off the mountain slopes. This is stealing the life of the fish that live in the river, of the birds that come to feed on the fish, of the mammals that come to feed on the birds. Be giving, do not steal. The mind and externals are just thus, the gate of liberation is open.

The Third Grave Precept is *honor the body—do not misuse sexuality*. Honor the body of nature. When we begin to interfere with the natural order of things, when we begin to engineer the genetics of viruses and bacteria, plants and animals, we throw off the whole ecological balance. Our technological meddling affects the totality of the universe and there are karmic consequences to that. The three wheels—body, mind, and mouth, or, greed, anger, and ignorance—are pure and clean. Nothing is desired. Go the same way as the Buddha, do not misuse sexuality.

The Fourth Grave Precept is *manifest truth—do not lie*. One of the very common kinds of lying that is currently popular is called greenwashing. Greenwashing is like whitewashing: it pretends to be ecologically sound and politically correct. Monsanto Chemical Company tells us how wonderful they are and how sensitive they are to the environment. Exxon tells us the same thing. The plastic manufacturers tell us the same thing. Part of what they say is true: without plastic there could be no special pump for failing hearts; without plastic there could be no oxygen tent. But plastic cups and plates that are not biodegradable and are filling up the dumps continue to be made. Another kind of lying is the lying that we do to ourselves about our own actions. We go into the woods and, rather than take the pains to haul out the nonbiodegradable stuff that we haul in, we hide it. We sink the beer cans, bury the cellophane wrappings under a root. We know we have done it, but we act as though we have behaved differently. Gain the essence and realize the truth. Manifest the truth and do not lie.

The Fifth Grave Precept is *proceed clearly—do not cloud the mind*. Do not cloud the mind with greed; do not cloud the mind with denial. It is greed that is one of the major underlying causes of pollution. We can solve all the problems; we have all the resources to do so. We can deal with our garbage, we can deal with world hunger, we can deal with the pollution that comes out of the smokestacks. We have the technology, but the solutions will cost a lot of money, which means that there will be less profit. If there is less profit, people will have to make do with a little bit less. Our greed prevents us from accepting this. Proceed clearly, do not cloud the mind with greed.

The Sixth Grave Precept is *see the perfection—do not speak of others' errors and faults*. For years we have manicured nature because in our opinion nature does not know how to do things. That manicuring may continue, for example, in the way we view the shifting shores of a river. We conclude that the river is wrong. It erodes the banks and floods the lowlands. It needs to be controlled. So, we take all the curves out of it, line the banks with stone, and turn it into a pipeline. This effectively removes all the protective space that the waterbirds use for nesting and the places where the fish go to find shelter when the water rises. Then, the first time there is a spring storm, the ducks' eggs and the fish wash downstream and the river is left barren. Or, we think there are too many deer, so we perform controlled genocide. The wolves kill all the livestock, so we kill the wolves. Each time we get rid of one species, we create an incomprehensible impact and traumatize the whole environment. The scenario changes and we come up with another solution. We call this process wildlife management. What is this notion of wildlife management? See the perfection, do not speak of nature's errors and faults.

The Seventh Grave Precept is *realize self and other as one—do not elevate the self and put down others.* Do not elevate the self and put down nature. We hold a human-centered notion of the nature of the universe and the nature of the environment. We believe God put us in charge, and we live out that belief. The Bible confirms this for us. We live as though the universe were spinning around us, with humans at the center of the whole picture. We are convinced that the multitude of things are there to serve us, and so we take without any sense of giving. This is elevating the self and putting down nature. In this universe, where everything is interpenetrated, codependent, and mutually arising, nothing stands out above anything else. We are inextricably linked and nobody is in charge. The universe is self-maintaining. Buddhas and ancestors realize the absolute emptiness and realize the great earth. When the great body is manifested, there is neither inside nor outside. When the *Dharma* body is manifested there is not even a single square inch of earth on which to stand. It swallows it. Realize self and other as one. Do not elevate the self and put down nature.

The Eighth Grave Precept is *give generously—do not be withholding.* We should understand that giving and receiving are one. If we really need something from nature, we should vow to return something to nature. We are, without question, dependent on nature.

There is a vast difference between recognizing dependency, and entering it consciously and gratefully, and being greedy. Native Americans lived amidst the plenty of nature for thousands of years. They fed on the buffalo when they needed that type of sustenance. We nearly brought that species to extinction in two short decades. And it was not because we needed the food. Tens of thousands of carcasses rotted while we took the skins. It is the same with our relationship to elephants, seals, alligators, and countless others. Our killing has nothing to do with survival. It has nothing to do with need. It has to do with greed. Give generously, do not be withholding.

The Ninth Grave Precept is *actualize harmony—do not be angry.* Assertive, pointed action can be free of anger. We can fence the deer out of our garden and prevent them from eating our vegetables without hating the deer. Also, by simply being patient and observing the natural cycles we can avoid unnecessary headaches and emotional outbreaks. Usually we will discover that the things we believe to be in the way are really not. When the gypsy moths descended in swarms one year and ate all the leaves off the trees so that in the middle of June the mountain looked like it was late fall, the local community of Woodstock, New York, became hysterical. We made an all-out attack. Planes came daily and sprayed the slopes with chemicals. People put tar on the bases of trees to trap the caterpillars. The gypsy moths simply climbed up, got stuck in the tar and piled up so others could crawl across the backs of the dead ones and went up the trees to do what they needed to do. Amidst all of these "disasters," with the leaves gone and the shrubbery out of the shade, the mountain laurel bloomed like it had never bloomed before. I had no idea we had so much mountain laurel on this mountain. It is true the gypsy moths damaged the trees. The weak trees died. But by July, there were new leaves on the trees and the mountain was green again. Yet, the anger and the hate we felt during those spring months was debilitating and amazing. The air was filled with it.

In another incident, the fellow who owned the house that is now the monastery abbacy had beavers on his property. They were eating up his trees so he decided to exterminate them. A neighbor told him beavers were protected, so he called the Department of Environmental Conservation. The rangers trapped and removed the animals. When we moved into the house, however, a pair of beavers showed up and immediately started taking down the trees again. In fact, they toppled a beautiful weeping willow that my students had presented to me as a gift. I was supposed to sit under it in my old age, but now it was stuck in a beaver dam, blocking the stream. With the stream dammed, the water rose, the pond grew and filled with fish. With the abundance of fish, ducks arrived, followed by the fox and the osprey. Suddenly the whole environment came alive because of those two beavers. Of course, they didn't stay too long because we didn't have that much wood, and after two seasons they moved on. The dam disintegrated, the water leaked out, and the pond shrank. It will remain that way until the trees grow back and the next pair of beavers arrives. If we can just keep our fingers out of it and let things unfold, nature knows how to maintain itself. It creates itself and defines itself, as does the universe. And, by the way, the weeping willow came back, sprouted again right from the stump. It leans over the pond watching me go through my own cycles.

The Tenth Grave Precept is *experience the intimacy of things—do not defile the Three Treasures.* To defile is to separate. The Three Treasures are this body and the body of the universe, and when we separate ourselves from ourselves, and from the universe, we defile the Three Treasures.

To practice the Precepts is to be in harmony with your life and the universe. To practice the Precepts means to be conscious of what they are about—not just on the surface but on many levels, plumbing the depths of the Precepts. It means being deeply honest with yourself. When you become aware that you have drifted away from the Precepts, simply acknowledge that fact. Acknowledgment means to take responsibility for your life; taking responsibility plays a key role in our practice. If you do not practice taking responsibility, you are not practicing. It is as simple as that. There is nobody checking when you are doing *zazen* whether you are letting go of your thoughts or sticking with them. It has to do with your own honesty and integrity. Only you know what you are doing with your mind. It is the same with the Precepts. Only you know when you have actually violated a precept. And only you can be at one with that violation, can atone. To be at one with it means to take responsibility. To take responsibility means to acknowledge yourself as the master of your life.

To take responsibility empowers you to do something about whatever is hindering you. As long as we blame, as long as we avoid or deny, we are removed from the realm of possibility and power to do something about our lives. We become totally dependent upon the ups and downs that we create around us. There is no reason that we should be subjected to anything when we have the power to see that we create and we destroy all things. To acknowledge that simple fact is to take possession of the Precepts. It is to make the Precepts your own. It is to give life to the Buddha, to the environment and all beings, and to the universe itself.

American Buddhism:
Creating Ecological Communities

Great Earth *Sangha*: Gary Snyder's View of Nature as Community

David Landis Barnhill

In the poem "O Waters" (*TI* 73),[1] Gary Snyder presents the following image:

> great
> earth
> sangha.

Sangha, of course, is the Buddhist term for religious community, one of the "three jewels" along with *dharma* (truth or teachings) and Buddha. Traditionally, *sangha* refers to the community of monks, people who have devoted their lives to spiritual practice separated from normal society. Snyder has clearly departed from that notion here: the *sangha* is the ecosphere of the planet. In this one image is suggested two fundamental characteristics of his thought: a creative extension of both Buddhism and ecology by seeing each in terms of the other, and an overriding concern with community.

The notion of community is one of the central ideas in both ecological science and environmental philosophy, and its general significance is worth reviewing. Seeing nature as community is a "radical" perspective: it changes at the root level our view of nature. We can see some implications of this perspective by considering how it opposes the traditional view of nature as "Other." The concept of Other is complex, but for our purposes here we can focus on three aspects: our relation to the Other, its value, and our obligations to it. When we think of something as Other, we hold that there is a profound *split* between "us" and "them." Certainly, that is how Western culture at least has tended to see our relationship to nature.

But if nature is our community, then it is not separate from us but rather is the fundamental existential context of our lives. Similarly, when we think of something as Other, then we *devalue* it: any value it may have is instrumental. But if nature is considered a community we are part of, then its value is intrinsic: both the individual beings and the system as a whole have their own integrity. And when we treat something as Other, there is little if any sense of *obligation* to it. But if nature is our community, then our obligation is to preserve it. In Aldo Leopold's famous words, "A thing is right when it tends to preserve the integrity, stability, and beauty of the biotic community. It is wrong when it tends otherwise."[2] But even more: if nature is truly a community we belong to, then there is a responsibility to *participate in it* as community.[3]

But while the idea of nature as community has these basic implications, it can be developed in a number of different ways. We need to pay close attention to its distinctive uses by each thinker, refining our sense of the various meanings and functions it has in a person's ecological thought. Surely one of the most complex and significant presentations of nature as community is by Gary Snyder.

The Ecological Community of Indra's Net

One principal aspect of Snyder's view of community involves the basic cycles of nature, in particular the food web and the cycle of production by plants, consumption by animals, and decomposition by fungi and other organisms. Early in his writings Snyder asks the question, "Just where am I in the food chain?" (*EHH* 32). For Snyder that is a religious question, and the answer points to our essential place—our niche—in the community of life.

For Snyder, the food web does not suggest that nature is "red in tooth and claw" but is instead a community that consists of "a gift-exchange, a potluck banquet, and there is no death that is not somebody's food, no life that is not somebody's death. . . . The shimmering food chain, food-web, is the scary, beautiful condition of the biosphere" (*G* 1). The intimacy of this gift exchange leads Snyder to speak in anthropomorphic terms. "Looking closer at this world of oneness, we see all these beings as our flesh, as our children, our lovers. We see ourselves too as an offering to the continuation of life" (*G* 1). This community of mutual gift exchange leads him to exclaim "What a big potlatch we are all members of!" (*PofW* 18–19).

This food-web community is sacramental: "To acknowledge that each of us at the table will eventually be part of the meal is not just being 'realistic.' It is allowing the sacred to enter and accepting the sacramental aspect of our shaky temporary personal being" (*PofW* 18–19). Involved, then, is a particular kind of community consciousness, "the sacramental food-chain mutual-sharing consciousness. . ." (*PIS* 95–96). Such a consciousness enables us to see that the sacramental community is fundamentally one of love. Turning the conventional attitude of survival of the fittest on its head, Snyder can ask rhetorically: "if we eat each other, is it not a giant act of love we live within?" (*G* 1). This love is clearly an extended form: "What are we going to do with this planet? It's a problem of love; not the human love of the West—but a love that extends to animals, rocks, dirt, all of it" (*TRW* 4). This love creates communion, found in the "sacramental energy-exchange, evolutionary mutual sharing aspect of life. . . . And that's what communion is" (*TRW* 89).

In articulating this sacred food-web community, Snyder refers to the traditional Buddhist idea of interpenetration and specifically refers to the image of Indra's net found in the Avataṃsaka school of Buddhism (Hua-yen in Chinese and Kegon in Japanese). In this image,

> the universe is considered to be a vast web of many-sided and highly polished jewels, each one acting as a multiple mirror. In one sense each jewel is a single entity. But when we look at a jewel, we see nothing but the reflections of other jewels, and so on in an endless system of mirroring. Thus in each jewel is the image of the entire net.[4]

For Snyder that mirroring is found in the interdependencies of nature's web. He has taken a Buddhist idea and applied it to ecology—or we could say that he has applied ecology to Hua-yen Buddhism. He has, in effect, "ecologized" the Buddhist notion of interpenetration and the image of Indra's net and "Buddhacized" the notion of ecosystem. "The web of relationships in an ecosystem

makes one think of the Hua-yen Buddhist image of Indra's net. . ."
(*PIS* 67). Snyder cites a Buddhist text to suggest the ecology of
Buddhist metaphysics: "If you can understand this blade of rice you
can understand the laws of interdependence and origination. . .[and]
you know the Buddha" (*TRW* 35).

One of the principal activities of any ecosystem, of course, is
eating. The implication is that the ecological net of Indra is made
not of jewels but of flesh: that of plants, animals, our own. This
seems at first to be at odds with the "ecstatic" quality that is
characteristic of traditional discussions of Indra's net. But as we
have seen, the enfleshed Indra's net of the "gift exchange" is
something viewed as positive, as well as something we must actively
participate in. "Everything that breathes is hungry. But not to flee
such a world! Join in Indra's net!" (*PIS* 70).[5]

Snyder thus sees our relationship to nature as being a part of a
communion of beings which constitutes Indra's net of the food web.
This view has important implications for the notion of the self as
well as the issue of the one and the many, the whole and parts.
Snyder's view is not a monism in which differentiation is lost or
individuality is denied or devalued. As Snyder says, "all is one and
at the same time all is many" (*OW* 9). Speaking of both art and life,
Snyder has said "A poem, like a life, is. . .a uniqueness in the
oneness" (*PIS* 115). To emphasize his point, he has cited a Chinese
Buddhist saying: "Easy to reach nirvana, / Hard to enter difference"
(*PIS* 212). The Buddhist notion of interpenetration and the image
of Indra's net makes clear that the ecological self is not indistin-
guishable from the whole. The self is *both* the individual *and* the
whole. Snyder presents the following image to suggest this point:
"We are many selves looking at each other, through the same eye"
(*OW* 62). He specifically cautions against a simplistic notion of
oneness that would deny individuality. "The work of art has always
been to demonstrate and celebrate the interconnectedness: not to
make everything 'one' but to make the 'many' authentic, to help
illuminate it all" (*PIS* 90). This retention of a nonmonist but
nondualistic sense of difference allows for a vital sense of a
community of beings.

It is important to realize that Snyder's view of a sacred community of love is both descriptive and normative. It is not simply
that we are physically interrelated by the food web and we ought

to embody it with a feeling of love. Snyder claims that, even though we don't realize it, we *do* exist in a web of love. To see the food web as simply a necessity of survival is to fail to see that it is also an interaction of love. Thus his statement cited above: "if we eat each other, is it not a giant act of love we live within?" (*G* 1). A Buddhist parallel here can be found in the notion of our original Buddha-nature, especially as interpreted by Dōgen. We already *are* Buddhas, we simply don't realize it. Similarly, nature's ecological relations, including the food web, *are* the functioning of love. This is not to assert that pain and loss are unreal or that "everything turns out for the best." It is instead to extend the notion of love and to make our vision more subtle. But, as with Dōgen, it is not enough to be descriptive. We need to realize, to make real, this already existing condition: we need to recognize its true character and live in a way that authentically embodies it. Thus, the normative is implied in the descriptive. The "practice of the wild" is to realize in practice the essential condition of the community of life. For Snyder, as for Dōgen, practice is itself the goal.

The Bioregional Community

As we have seen, Snyder "ecologizes" the Buddhist notion of Indra's net and "Buddhacizes" science's view of the ecology of the food web. What implications does such a view have on a broader understanding of the interrelationship between human culture and the rest of nature? For Snyder, bioregionalism has been the principal framework for articulating a philosophy of culture and nature, and it is central to his view of a new, extended form of Buddhist community. The goal of bioregionalism can be put in simple terms: the creation of a society in which "A people and a place become one" (*PIS* 95). The focus here is not some abstract or generalized oneness but a concrete unity with a particular place. It is not realized in some aloof mystical state but in the very physical practice of "reinhabitation," dwelling fully at home and in place. Reinhabitation involves substantial bioregional education: where the water comes from and where the waste goes, what species of birds and bugs are part of our local community, what kind and quantity of food and housing the bioregion naturally supports, the myths and practices of the native peoples (they too are part of the community, even if they are no longer here). And reinhabitation calls for long-term commitment to live and work in the place, "to become people who are learning to live and think 'as if' they were totally engaged with their place for the long future" (*PIS* 247). To live this way develops community. "To restore the land one must live and work in a place. To work in a place is to work with others. People who work together in a place become a community, and a community, in time, grows a culture" (*PIS* 250).

For Snyder community is a spiritual path which centers on having a deep sense of place.

> Because by being in place, we get the largest sense of community. We learn that community is of spiritual benefit and health for everyone, that ongoing working relationships and shared concerns, music, poetry, and stories all evolve into the shared practice of a set of values, visions, and quests. That's what the spiritual path really is (*TRW* 141).

The bioregional community is "the largest sense of community" in part because it includes all species.

> Human beings who are planning on living together in the same place will wish to include the non-human in their sense of community. This also is new, to say our community does not end at the human boundaries; we are in a community with certain trees, plants, birds, animals. The conversation is with the whole thing. That's community political life (*TT* 18).

In a Snyderesque statement of deep ecology's[6] principle of ecocentric egalitarianism, he says we must "take ourselves as no more and no less than another being in the Big Watershed. We can accept each other all as barefoot equals sleeping on the same ground" (*PofW* 24).

Bioregionalism has been accused of leading toward a provincialism that ignores planetary issues such as global warming as well as concern for peoples and bioregions remote from one's local place. Snyder's particular development of bioregionalism answers this criticism in a way that reflects his Hua-yen vision of the nonduality of holism and individualism: the local bioregion interpenetrates with the planetary. This is seen in the title of his recent autobiographical reflections on his life at Kitkitdizze, his name for his home in the Sierra Nevada foothills. The title of these reflections, "Kitkitdizze: A Node in the Net" (the concluding essay of *PIS*), suggests that his local bioregion is one distinct part of the vast, single, whole of Indra's net. In promoting a balanced view that integrates the local and the global (in addition to the individual and the whole), Snyder is concerned with the predominance of an excessively global perspective.

> Continuing a dialogue between cosmopolitanism and the matter of being deeply local is crucial. To be merely cosmopolitan, merely international is not interesting. . . . So the check that is imposed upon the tendency toward centralization is the actual diversity of the world (*TT* 14).

Note, however, that the local does not exclude the global. We should recognize that ultimately we live on one planet, while acknowledging that such holism consists of diversity. "We should be dubious of fantasies that would lead toward centralizing world political power, but we do need to nourish interactive playful diversity on this one-planet watershed" (*PIS* 212). As such, the whole can be known through the parts.

> I'm not saying that the continent as a whole, or even the planet as a whole, cannot be, in some sense, grasped and understood, and indeed it should be, but for the time, especially in North America, we are extremely deficient in regional knowledge—what's going on within a given region at any given time of year. Rather than being limiting, that gives you a lot of insight into understanding the whole thing, the larger system (*TRW* 27).

Bioregionalism, then, "implies an engagement with community and a search for the sustainable sophisticated mix of economic practices that would enable people to live regionally and yet learn from and contribute to a planetary society" (*PIS* 247). In fact, such an Indra's net version of bioregionalism suggests a new perspective on the common phrase "think globally, act locally." The split seems unnecessary. Ultimately thinking globally and thinking locally go hand in hand; to act locally is to act globally.[7]

Indra's net is not the only Buddhist image that applies to Snyder's view of the unity-in-diversity of bioregionalism and the intrinsic value of every member of the community.

> One of the models I use now is how an ecosystem resembles a mandala. A big Tibetan mandala has many small figures as well as central figures, and each of them has a key role in the picture: they're all essential. . . . Every creature, even the little worms and insects, has value. Everything is valuable—that's the measure of the system (*WM* 23).

Snyder relates this *mandala* vision of nature with the view of the Ainu of northern Japan. "Each type of ecological system is a different mandala, a different imagination. Again the Ainu term *iworu*, field-of-beings, comes to mind" (*PofW* 107). In discussing the "field of beings," Snyder seems to suggest another combination of the descriptive and the normative: ". . . how totally and uniquely *at home* each life-form must be in its own unique 'buddha-field'" (*PofW* 108). Perhaps we too are essentially at home, even though we do not realize it and act contrary to it. If so, a deeper sense of how all things are at home in this *mandala* of life will help us see how we are as well.

The Mythological, Shamanistic Community

Snyder's scientifically based but Buddhistically developed notion of the ecological/bioregional community is complemented by a different sense of community, arising principally from his study of Native American cultures. We can call this the "mythological, shamanistic community," in which plants, animals, and humans are seen as part of an interactive social community. As Bert Almon has said: "Many of Snyder's 'people' are birds. . .even plants. . . . His problem as a poet of the whole range of living beings is to create poems in which animals and plants appear as autonomous presences. . . . The aim is. . .to see all beings as co-citizens in a community of life."[8] The result is a "True Communionism" that differs from the ideals of both capitalism and communism. As Hwa Yol Jung and Petee Jung have written: "Communionism is first and foremost the way of seeking a deep sense of communion with myriads of natural things on earth, who are also called 'peoples' without any facile dualism and unnecessary hierarchism of any kind."[9] This ideal, as true of his essays and interviews as his poems, is founded on the view of primal cultures that animals are people who coexist with us as part of nature. "People of primitive cultures appreciate animals as other people off on various trips" (*EHH* 121).

Snyder's perception that animals are our fellow creatures is extended in one of his "Little Songs for Gaia" in *Axe Handles* (50). Once Snyder was sitting with fellow poet Lew Welch in the mountains. Welch asked him whether he thought the rocks were paying attention to the trees. Snyder said he didn't know and wondered what Welch was driving at. Welch replied: "The trees are just passing through."[10] That idea inspired the following poem.

> As the crickets' soft autumn hum
> is to us,
> so are we to the trees
>
> as are they
>
> to the rocks and the hills.

Our fellow "creatures" include plants and even rocks. Note how this poem brings time into his presentation of community. The community is not just now but is part of the entire geological process. One might wonder if this sequential "equation" can be extended. After all, the mountains too are just passing through, so perhaps it is appropriate to add: "as the rocks and hills are to the ocean and air." But then they too are just passing through.

For Snyder, plants and animals are not just our fellows, they are our elders. Describing a time he was in an old growth forest, Snyder has said, "For hours we were in the company of elders" (*PofW* 135). As elders they bear nature's information: "The old stands of hoary trees. . .are the grandparents and information-holders of their communities" (*PofW* 139). And these elders are our teachers: "I suspect that I was to some extent instructed by the ghosts of those ancient trees as they hovered near their stumps" (*PofW* 118).

Community involves some sense of communication or communion, and part of Snyder's view of the interactive character of nature's community concerns interspecies communication. Animals can speak to us. One morning as he awoke in his sleeping bag on a long trip by car through the West, a magpie came close to him and gave him the following song (*TI* 69).[11]

MAGPIE'S SONG

Six A.M.,
Sat down on excavation gravel
by juniper and desert S.P. tracks
interstate 80 not far off
   between trucks
Coyotes—maybe three
      howling and yapping from a rise.

Magpie on a bough
Tipped his head and said,

   "Here in the mind, brother
   Turquoise blue.
   I wouldn't fool you.
   Smell the breeze
   It came through all the trees
   No need to fear
   What's ahead
   Snow up on the hills west
   Will be there every year
   be at rest.
   A feather on the ground—
   The wind sound—

Here in the Mind, Brother,
Turquoise Blue"

Such interspecies communication also allows another kind of link between humans and the rest of nature. We can speak *for* animals and plants in the sense of political representation. Snyder sees this as one of his roles, "an occasional voice for the nonhuman rising within the human realm. . ." (*TRW* 159). Such a possibility, in fact, becomes a moral imperative. Plants and animals are a part of our *political* community and their voice needs to be heard in the chambers of government as well as in the books of poetry (*TI* 106).

This mythic, anthropomorphic perspective is also part of Snyder's presentation of the food web as community. Snyder notes that in traditional Native American belief, the animal offers itself to the worthy hunter, expecting gratitude and conscientiousness in return.[12] "The world is not only watching, it is listening too. . . . Other beings (the instructors from the old ways tell us) do not mind being killed and eaten as food, but they expect us to say please, and thank you, and they hate to see themselves wasted" (*PofW* 20–21). Animals, then, are our helpers, giving us sustenance.

Commenting on the poem "Soy Sauce" (*AH* 30–31), Woody Rehanek has stated that Snyder presents himself as "identifying with, representing, and finally *becoming* a totem animal. This experience transcends intellectual rapport and becomes a total affinity with the nonhuman. . . . A vital aspect of shamanism is this ability to become one with the animal."[13] But the term "oneness" can refer to several kinds of states and relationships. Snyder is concerned with the oneness of a community, not some monistic unity achieved by a solitary mystic. There is, for instance, the oneness of interspecies transformation which allows transhuman community to occur.

> We are all capable of extraordinary transformations. In myth and story these changes are animal-to-human, human-to-animal, animal-to-animal, or even farther leaps. The essential nature remains clear and steady through these changes. So the animal icons of the Inupiaq people ("Eskimos") of the Bering Sea (here's the reverse!) have a tiny human face sewn into the fur, or under the feathers, or carved on the back or breast or even inside the eye, peeping out (*PofW* 20).

Such a view may at first seem very different from Buddhism, but only if we rather artificially separate an "elite" Buddhism from the broader context of popular religion. Such a separation has not been characteristic of East Asian Buddhism, and there are some interesting parallels between Snyder's mythic/shamanistic views and Japanese poets. As Snyder sees ancient trees as elders, the Japanese Buddhist poet Saigyō (1118–1190) saw cherry trees and the moon as companions and models in his spiritual journeys.[14] The Buddhist-influenced poet Bashō (1644–1694) suggested that he heard the voice of his parents in the cry of a bird, and he spoke of the true poet as someone who is able to enter into a bamboo and speak its subtle feelings.[15] In general, transformations between the human and animal worlds is a common theme in Japanese Shintō and shamanistic East Asian folk Buddhism. Snyder, in fact, gives Native American stories of transformations a Buddhist interpretation, as he continues the passage cited above.

> This is the *inua*, which is often called "spirit" but could just as well be termed the "essential nature" of that creature. It remains the same face regardless of the playful temporary changes. . . . This is not the same as an anthropocentrism or human arrogance. It is a way of saying that each creature is a spirit with an intelligence as brilliant as our own. The Buddhist iconographers hide a little animal face in the hair of the human to remind us that we see with archetypal wilderness eyes as well (*PofW* 20).

But what does all this mythic discourse amount to? Does Snyder actually *believe* in interspecies transformation and the rest? I think such a positivist question is the wrong one to ask. The fundamental function of myth is not to state what is "objectively real," which opens the door to arguments about what is "really true." An animal rights advocate, for instance, once complained to me after I delivered a paper on Snyder's view of hunting that "animals don't really give up themselves to the hunter, that's just a rationalization." Snyder's presentation of hunting as gift and communion certainly could be used as a rationalization for needless killing, but we should avoid rejecting out of hand the traditional views of Native American hunters. I would prefer to begin with the hypothesis that there is some important wisdom involved in such mythic thinking which cannot be captured by our modern notions of objective reality. Myth, after all, articulates what is *psychologically* and *spiritually* real, what is essential in our relationships with the world. Snyder's mythological community suggests the multidimensional intimacy of our connection to and communion with the rest of nature, our fundamental similarity to all other beings, and our co-participation in the community of nature. And it does so in a way that can promote a fuller realization of our deep interrelationship with all of life. As Murphy has noted concerning Snyder's retelling of the Native American story of "The Woman Who Married a Bear" (in *PofW* 155–74), "What is revealed here. . .is the power that myth can carry in the present day and the ways by which it can help bridge the gap between animal and human. . . ."[16]

Snyder is well aware of—and critical of—the tendency in Western Buddhism to reject the "popular" elements of Buddhist belief and practice.

> There's a big tendency right now in Western Buddhism to psychologize it—to try and take the superstition, the magic, the irrationality out of it and make it into a kind of therapy. You see that a lot. Let me say that I'm grateful for the fact that I lived in Asia for so long and hung out with Asian Buddhists. I appreciate that Buddhism is a whole practice and isn't just limited to the lecture side of it; that it has stories and superstition and ritual and goofiness like that. I love that aspect of it more and more (*WM* 25–26).

Part of Snyder's Buddhistic totalism is to embrace the long-standing tradition of merging sophisticated philosophy and advanced mystical disciplines with "popular" beliefs and rituals. His admixture of Buddhism and Native American culture is in line with blendings of Buddhism and popular/shamanistic religions in China and Japan.[17] The obvious difference is that Snyder has not turned to the popular religion of the majority of Americans (our "masses") but to the minority tradition of Native Americans. He has done so in part because the majority tradition of Protestantism has rejected "magic" (including the Catholic Eucharist) and emphasized the "fallenness" of the natural world and our separation from it. The shamanistic aspects of popular religions in East Asia are largely absent from our "popular" religion, while they are strong in the native but minority traditions of America.

Shaman as Ecologist

Snyder has associated shamanism and ecology since his earliest writings. Recently David Abram has discussed that association in a way that clarifies Snyder's view.

> The traditional shaman, as I came to discern in the course of my twelve months in Asia, is in many ways the "ecologist" of a tribal society. He or she acts as intermediary between the human community and the larger ecological field. . . . By his or her constant rituals, trances, ecstasies, and "journeys," the shaman ensures that

the relation between human society and the larger society of beings
is balanced and reciprocal.[18]

We have failed to recognize this ecological role of the shaman,
Abram says, because of our assumptions about nature and the
supernatural.

> Countless anthropologists have managed to overlook the ecological
> dimension of the shaman's craft, while writing at length of the
> shaman's rapport with "supernatural" entities. We must attribute
> much of this oversight to the modern assumption that nonhuman
> nature is largely determinate and mechanical, and that that which
> is regarded as mysterious, powerful, and beyond human ken must
> therefore be of some other, nonphysical realm outside of nature—
> "supernatural."[19]

Abram discovered that for the shaman/ecologist, the natural *is* the
supernatural, the supernatural is nature.

This general point, of course, is true of Mahāyāna Buddhism as
well: "form is emptiness, emptiness is form." But philosophical/
mystical Mahāyāna tends to see the ultimate reality in terms of an
impersonal, single Mind, rather than a community of beings.
Shamanism, on the other hand, sees nature as characterized not by
Mind but by minds. As a result, Abram notes, human intelligence
is considered "simply one form of awareness among many others,"[20]
a theme found throughout Snyder's writings.

> Magic, then, in its perhaps more primordial sense, is the experience
> of living in a world made up of multiple intelligences, the intuition
> that every natural form one perceives—from the swallow swooping
> overhead to the fly on a blade of grass and indeed the blade of grass
> itself—is an experiencing form, an entity with its own predilections
> and sensations, albeit sensations that are very different from our
> own.[21]

Here the Buddhist parallel is with the karmic cosmology of the six
realms, which includes animals and four other kinds of supernatural
(or supranatural) beings: hell-dwellers, hungry ghosts, titans, and
heavenly beings. Both Abram and Snyder differ from this model by
locating all transhuman intelligence in the palpable, sensuous world

of nature and also by including plants and (at least in Snyder's thought) ecosystems from watersheds to Gaia. Nature, and all of nature, is the compass of community.

For both Abram and Snyder, the shaman is one who has developed special expertise in establishing a communion with transhuman intelligences. "The shaman's magic is precisely this heightened receptivity to the meaningful solicitations—songs, cries, gestures—of the larger, more-than-human field."[22] In such a view, nature is a "sentient landscape,"[23] and spiritual communion requires a sensuous acuity to nature's voices. As nature and the "supernatural" are inextricably linked, so too our physical senses are intimately tied to our awareness of the spiritual dimensions of life. As a result, the cultivation of a sensuous communion with nature is critical for the health of individuals and for a culture as a whole.

> To shut ourselves off from these other voices, to continue by our lifestyles to condemn these other sensibilities to the oblivion of extinction, is to rob our own senses of their integrity, and to rob our minds of their coherence. We are human only in contact and conviviality with what is not human. Only in reciprocity with what is Other will we begin to heal ourselves.[24]

This reciprocity leads Abram into a new kind of dialogue.

> I found myself caught in a nonverbal conversation with this Other, a gestural duet with which my reflective awareness had very little to do. It was as if my body were suddenly being motivated by a wisdom older than my thinking mind, as though it were held and moved by a logos—deeper than words—spoken by the Other's body, the trees, the air, and the stony ground on which we stood.[25]

There is a problem with Abram's use of the terminology of the Other. As we noted in the introduction, the notion of Other usually implies separation, devaluing, and a lack of obligation, aspects of imperialist and patriarchal views toward other beings. Abram clearly is pointing to a different relation: interdependence and mutual implication, a community. One of the characteristics of a true community (as opposed to a social collectivity) is that other members are part of the definition of who and what we are. In order to clarify Abram's comments and Snyder's view of community, we need to be able to articulate the kind of relation with others that affirms difference but avoids separation and alienation. Mikhail Bakhtin's thought is helpful in this task.

Anotherness and Dialogics: Bakhtin and Interspecies Community

Mikhail Bakhtin, the Russian literary critic, has argued for an alternative to an alienational and oppressive sense of Otherness.

> Russian distinguishes between *drugoi* (another, other person) and *chuzhoi* (alien; strange; also, the other). The English pair 'I/other,' with its intonations of alienation and opposition, has specifically been avoided here. The *another* Bakhtin has in mind is not hostile to the *I* but a necessary component of it, a friendly other, a living factor in the attempts of the *I* toward self-definition.[26]

Patrick D. Murphy has made an important contribution to the study of community and ecology by extending Bakhtin's thought to include nature as well as humans. Although the Western tradition has traditionally conceived of both nature and foreign cultures as Other, Murphy asks: "What if instead of alienation we posited *relation* as the primary mode of human-human and human-nature interaction without conflating difference, particularity and other specificities? What if we worked from a concept of relational difference and *anotherness* rather than Otherness?"[27] Clearly Abram conceives of the natural world as "another" rather than an Other. Murphy correctly sees the same vision of relation in Snyder, who (at one time a Ph.D. student of anthropology) "argues for each of us to turn from being 'ethnologist' to being 'informant,' to move from objectifying detachment from the other to subjectivity-sharing engagement with the other as another."[28]

Such a view of anotherness rejects a strict dichotomy of self and other, as does Buddhism and ecological thought. As Murphy notes, Bakhtin's thought implies "recognizing the concepts of both self and other as interdependent, mutually determinable, constructs. . . ."[29] It is appropriate here to recall Bert Almon's account of Snyder's poetry "in which animals and plants appear as autonomous presences."[30] By autonomous Almon is arguing that animals and plants are not Others and not mere background or symbol. Yet in a Bakhtinian view—and in Snyder's ecological-shamanist-Buddhist view—nothing is autonomous. It would be better to argue that for Snyder animals and plants have their own *integrity*—not in being autonomous but by being *integrated* in the interdependent web as "anothers."[31]

It is important to recognize that this view of self and Other as interpenetrating functions both normatively and descriptively.[32] It articulates an ideal relationship to nature and argues for its possibility: we are not necessarily relegated to an alienated relationship with nature as Other. But it also implies a descriptive affirmation of our essential condition: whether we realize it or not, we exist in a web of interrelationships. Such a view entails a Hua-yen-like perception of relation, "Conceptualizing self/other as interpenetrating part/part and part/whole relationships rather than dichotomy. . . ."[33] Murphy comments on the works of Ursula Le Guin in a way that applies to Snyder's Hua-yen vision of community: "self and other, and individual and community, are complementarities that when unified produce a sense of wholeness, although not necessarily completedness."[34]

Community thus conceived is not simply interexistent but also interactive, and Bakhtin's theory of dialogics is useful in understanding the interactive quality of Snyder's—and Abram's—view of community. Complex and ambiguous, Bakhtin's notion of dialogue has been interpreted in various ways, with conflicting schools of thought claiming him as their own.[35] But the central point relevant to our discussion is that anotherness enables dialogical interchange. To be another instead of an other is to be a speaking subject rather than an analyzed or utilized object. Although Bakhtin was anthropocentric in his dialogical theory, Abram, Snyder, and Murphy all emphasize that transhuman nature also can have such a voice. "Such a perception of interconnectedness not only enables one to move from the self/other as dichotomy to viewing both terms as mutually constitutive forms of being another, but also enables one to listen to others, whether human or not, as speaking subjects, sentient and creative. . . .[36] Since transhuman nature does not speak in a human voice, there is a need for someone to render that voice: a shaman or a poet. In a statement that clearly applies to Snyder's view of the shaman poet, Murphy has said that "The implications of this other as speaking subject need to be conceptualized as including more than humans, and as potentially being constituted by a speaker/ author who is not the speaking subject but a renderer of the other as speaking subject. . . ."[37]

A central issue raised by the notion of Other is the possibility of true communication. Tim Dean, approaching the notion of Other from a Lacanian viewpoint, states that "The paradox of the Other is that it both enables relation and disables relation, rendering communication always imperfect and effectively disharmonising connection."[38] While it is true that communication is imperfect and relation is never totally harmonious and transparent, to claim that Otherness always and inevitably implies a disharmonizing connection is to reject the possibility of anotherness. Such a view may fit Lacan, but Snyder and shamanistic traditions (and Bakhtin) clearly disagree: relationship with "another" may be imperfect but it can become a true *inter*relationship in which the integrity of each is maintained, true learning occurs, and communication can be effectively (though not perfectly) harmonizing.

But, for the traditional shaman the ultimate goal is not one's personal communion with the transhuman. It is healing. In traditional shamanistic societies, personal illness is seen not simply as a function of a body but in the person's relationship with the larger society of beings. "Disease, in most such cultures, is conceptualized as a disequilibrium within the sick person, or as the intrusion of a demonic or malevolent presence into his or her body. . . . Yet such influences are commonly traceable to an imbalance between the human community and the larger field of forces in which it is embedded."[39] Abram's view of "physical" illness is echoed in Patricia Clark Smith's view of social conflict among Native American women. There is, she says, "a tendency to see conflict between women as not totally a personal matter but, rather, as part of a larger whole, as a sign that one of the pair has lost touch not with just a single individual but with a complex web of relationship and reciprocities."[40] The etymological root of the term "to heal," Snyder often notes, is "to make whole." This is the shaman's task— to bring wholeness to society by bringing it in to wholeness with nature.

A Community of Practice

One of the reasons Snyder has been drawn to the notion of the shaman is because a shaman is a religious *practitioner*. For all his reputation as a "nature poet," Snyder does not fit the conventional mode of the contemplative. True to his Zen roots, Snyder emphasizes a path of practice. And, his focus is not on the shaman or monk as individual practitioner but on the community as the context for interdependent religious practice. Since the beginning of his poetic career, participation in a community of religious practice has been a central goal for Snyder. He was drawn early in life to Native American spirituality but turned instead to Buddhism because he found it a more accessible community. Snyder recalls that he "saw that American Indian spiritual practice is very remote and extremely difficult to enter, even though in one sense right next door, because it is a practice one has to be born into. Its intent is not cosmopolitan. Its content, perhaps, is universal, but you must be a Hopi to follow the Hopi way" (*TRW* 94). He found in Japanese Zen a community of practice that he could participate in, and he was attracted to its discipline. "Its community life and discipline is rather like an apprenticeship program in a traditional craft. The arts and crafts have long admired Zen training as a model of hard, clean, worthy schooling" (*PofW* 148).

By the time of his return from Japan in the 1960s, however, he began to articulate his view of the limitations of the traditional Buddhist *sangha*. In 1969 he stated that his ideal was an expanded community of spiritual practice, one which would retain the universality and intellectual sophistication of Buddhism but be a broader, nonmonastic community like those found in tribal societies.

> The Buddhist and Hindu traditions. . .lost something which the primitives did have, and that was a total integrated life style. . . . Certain primitive cultures that are functioning on a high level actually amount to what would be considered a spiritual training path in which everyone in the culture is involved and there are no separations between the priest and layman or between the men who become enlightened and those who can't. What we need to do now is to take the great intellectual achievement of the Mahayana Buddhists and bring it back to a community style of life. . . (*TRW* 15–16).

More recently (1990) he has repeated this ideal, emphasizing the necessity of staying involved in the sometimes unpleasant aspects of domestic and ecological relationships.

> There are additional insights that come only from the nonmonastic experience of work, family, loss, love, failure. And there are all the ecological-economical connections of humans with other living beings, which cannot be ignored for long, pushing us toward a profound consideration of planting and harvesting, breeding and slaughtering. All of us are apprenticed to the same teacher that the religious institutions originally worked with: reality (*PofW* 152).

As a result, Snyder has emphasized the importance of family as a context for spiritual practice. "To me, the natural unit of practice is the family. The natural unit of the play of practice is the community. A *saṅgha* should mean the community, just as the real Mahayana includes all living beings" (*TRW* 136).

Such a view of the *saṅgha* as a family-based community departs from the traditional monastic notions. In fact, it recalls instead basic Confucian ideals. But Snyder's impetus is clearly rooted in Mahāyāna Buddhism. Mahāyāna arose in part as a more inclusive branch of Buddhism, loosening the strict separation of monk and laity and opening spiritual aspiration and practice to those outside the monastery walls. Mahāyāna philosophy, especially Hua-yen, has consistently critiqued dualities and taken a totalistic view of a comprehensive interrelated reality. For Snyder, the spiritual community and its practice must reflect such a totalistic view.

Snyder's Buddhist notion of a totalistic community leads to a view of social revolution that departs from traditional Buddhism. If the *saṅgha* is all beings, and morality is the central aspect of the path, then social morality is necessary and leads to political radicalism.

> The mercy of the West has been social revolution; the mercy of the East has been individual insight into the basic self/void. We need both. They are both contained in the traditional three aspects of the Dharma path: wisdom (prajñā), meditation (dhyāna), and morality (śīla). . . . Morality is bringing it back out in the way you live, through personal example and responsible action, ultimately toward

the true community (saṅgha) of "all beings." This last aspect means, for me, supporting any cultural and economic revolution that moves clearly toward a free, international, classless world. . . . Working on one's own responsibility, but willing to work with a group. "Forming the new society within the shell of the old"—the I.W.W. slogan of fifty years ago (*EHH* 92).

There is an ongoing—and creative—tension in Snyder's expanded view of the *saṅgha*. Social morality combined with a totalistic metaphysic calls for an international and planetary political radicalism. But his bioregional and tribal leanings point toward a local focus; the practice of a bioregional community is the practice of place.

> Ultimately we can all lay claim to the term *native* and the songs and dances, the beads and feathers, and the profound responsibilities that go with it. . . . Part of that responsibility is to choose a place. To restore the land one must live and work in a place. To work in a place is to work with others. People who work together in a place become a community, and a community, in time, grows a culture (*PIS* 250).

Such practice may be local political work, "the tiresome but tangible work of school boards, county supervisors, local foresters, local politics. . ." (*WM* 23–24). It also includes the work of being a family. "There's a fatherly responsibility there, and a warm, cooperative sense of interaction, of family as extended family, one that moves imperceptibly toward community and a community-values sense" (*WM* 24). Family leads into community (again, a very Confucian idea), and neighborhood community ties to ecological community. "Neighborhood values are ecosystem values, because they include all the beings" (*WM* 24). While all of nature is included in the *saṅgha*, it is the ecological neighborhood of the watershed that is the place of practice.

> The watershed is our only local Buddha mandala, one that gives us all, human and non-human, a territory to interact in. That is the beginning of dharma citizenship: not membership in a social or a national sphere, but in a larger community citizenship. In other

words, a *sangha*; a local dharma community. All of that is in there, like Dōgen when he says, "When you find your place, practice begins" (*WM* 24).

In place, and in community, one can begin the real work of Buddhist-ecological practice. This is what true revolution involves.

Conclusion: Building a Community

It is appropriate to conclude by examining a recent poem that touches on a number of aspects of Snyder's view of community, some of which we have discussed and others that we can only note briefly here. The poem, "Building" (*NN* 366–67), includes a narrative account of the construction of buildings in his Sierra Nevada community. It is worth quoting in full here.

We started our house midway through the Cultural Revolution,
The Vietnam war, Cambodia, in our ears,
    tear gas in Berkeley,
Boys in overalls with frightened eyes, long matted hair, ran
    from the police.
We peeled trees, drilled boulders, dug sumps, took sweat baths
    together.
That house finished we went on
Built a schoolhouse, with a hundred wheelbarrows,
    held seminars on California paleo-indians during lunch.
We brazed the Chou dynasty form of the character "Mu"
    on the blacksmithed brackets of the ceiling of the lodge,
Buried a five-prong vajra between the schoolbuildings
    while praying and offering tobacco.
Those buildings were destroyed by a fire, a pale copy rebuilt
    by insurance.

Ten years later we gathered at the edge of a meadow.
The cultural revolution is over, hair is short,
    the industry calls the shots in the Peoples Forests,
Single mothers go back to college to become lawyers.

Blowing the conch, shaking the staff-rings
    we opened work on a Hall.
Forty people, women carpenters, child labor, pounding nails,
Screw down the corten roofing and shape the beams
    with a planer,

The building is done in three weeks.
We fill it with flowers and friends and open it up.

Now in the year of the Persian Gulf,
Of Lies and Crimes in the Government held up as Virtues,
    this dance with Matter
Goes on: our buildings are solid, to live, to teach, to sit,
To sit, to know for sure the sound of a bell—
This is history. This is outside of history.
Buildings are built in the moment,
    they are constantly wet from the pool
        that renews all things
    naked and gleaming.

The moon moves
Through her twenty-eight nights.
Wet years and dry years pass;
Sharp tools, good design.

The poem has strong allusions to the *Hōjōki* ("An Account of My Ten Foot Square Hut"), a famous prose piece by the Japanese Buddhist writer Kamo no Chōmei (1153–1216). Both works focus on buildings, place the reflections in a historical and political context of great disturbance and destruction, and are concerned about the continuity of culture in an impermanent world. In both works the failures of conventional society are assumed to be ongoing, though Snyder presents more of a political critique of the cause of the problem than Chōmei, who exemplifies the Buddhist belief that the world had entered into an irreversible historical era of the "decline of the Law" (*mappō*). Chōmei reflects on the impermanence of buildings and our relationship to them by stating near the end of his work that "Only in a hut built for the moment can one live without fears,"[41] a phrase Snyder echoes in the second to last stanza.

But the differences between the works are striking. Chōmei begins the *Hōjōki* with a poignant depiction of the essential insubstantiality of life: "The flow of the river is ceaseless and its water is never the same. The bubbles that float in the pools, now vanishing, now forming, are not of long duration: so in the world are man and his dwellings."[42] Snyder reveals instead a metaphysical optimism: near the end of the poem he alludes to the *Hōjōki* passage but shifts the emphasis from the passing away of things to their renewal: "constantly wet from the pool / that renews all things /

naked and gleaming." He also counters his political pessimism with a social optimism that is totally foreign to Chōmei. The Japanese writer left the capital city after its destruction and built a solitary hut in the foothills where he pursued the arts and Buddhist devotions. Snyder, too, leaves the city for the foothills, but he goes there to begin a new community. Unlike Chōmei's hut, Snyder's buildings are communal, built by and for the entire community: the schoolhouse was built "with a hundred wheelbarrows" while the Hall was constructed by "Forty people, women carpenters, child labor, pounding nails."

This community is presented in opposition not only to Chōmei's pessimistic reclusion but also to the violence and alienation of American society. His community is revolutionary not in having the intention of overthrowing the political establishment but in creating an alternative community that can be the basis for a post-civilization society, "forming the new society within the shell of the old." An essential characteristic of this vision of community is work.[43] It is physical and mundane rather than industrial or technologically sophisticated: "screw down the corten roofing and shape the beams with a planer." And work is a source of ritual that celebrates and reinforces social solidarity as it symbolically represents ecological interdependence with the use of natural materials: "Blowing the conch, shaking the staff-rings / we opened work on a Hall"; "The building is done in three weeks / We fill it with flowers and friends and open it up." The work of community is also religious, both in communal rituals and the practice of meditation. The repetition of "to sit" emphasizes the importance of this practice and suggests the necessity of ongoing practice: practice ultimately is not the means but the end. The result of such practice is the ability to participate fully in this dance with Matter, including the sensual acuity and decisive awareness to hear fully the sound of a bell.

Community is also the context and medium of cultural transmission. Snyder has always been concerned with the handing down of culture, though in earlier writings his view was more anthropological and historical, arguing for the continuity of shamanistic culture from the Paleolithic to the present. Especially with *Axe Handles* (1983), Snyder's writings discuss cultural transmission on a more personal level,[44] seen in the concluding lines of the title poem from that collection. He refers there to an ancient Chinese image of cultural continuity, an axe which functions both as a model for a new axe handle and as a tool in its construction.

> And I see: Pound was an axe,
> Chen was an axe, I am an axe
> And my son a handle, soon
> To be shaping again, model
> And tool, craft of culture,
> How we go on (*AH* 6).[45]

Similarly, "Buildings" concludes with the image of tools, objects of the human cycle of cultural transmission located in the context of nature's cycles. The concern with tools helps us connect with other cultures as well. In his early poem "Above Pate Valley" (*RR* 9), Snyder reflects on his work building trails in the Sierra Nevada. During a break from work he discovers thousands of arrowheads, which are both tools and products of tools. He concludes with a reference to his contemporary tools in the context of the sweep of time.

> . . Picked up the cold-drill,
> Pick, singlejack, and sack
> Of dynamite.
> Ten thousand years.

As Lionel Basney states, Snyder's tools "join with the obsidian flakes he has discovered earlier to form a bridge for sympathy, or at least contact, across the intervening ten thousand years. . . . The objects their work implies, both tools and implied products, such as arrowheads, identify meaning with an immediate practical use, and thus suggest a culture functioning in both pragmatic and spiritual terms."[46] Tools (cultural artifacts) and their design (cultural codes) are basic components of those fundamental cultural goals, "to live, to teach. . . . ," which is how, in fact, we go on. Snyder begins the poem "What Have I Learned" (*AH* 85) with the lines:

> What have I learned but
> the proper use for several tools?

Charles Molesworth has discussed the importance of cultural transmission in both *Turtle Island* and *Axe Handles*. Speaking of the latter, he states that "the central tension here is the same that animated *Turtle Island* (1974): how can we carry on the meaningful transmission of community and culture against the threatening background of ecological perversity and vast geological and cosmic processes."[47] Murphy claims, however, that "it does not seem entirely accurate to speak of 'tension' so much as of continuing concern."[48] But the transmission of true community is very much at tension with the ecological perversity of our times, not only because the destruction of nature is a symptom of a deep illness of society but because nature *is* our community. As the coyotes are exterminated from his California hills, Snyder says, "My sons will lose this / Music they have just started / To love" (*TI* 21). The coyotes are part of the community, and their music is part of the cultural transmission. On the other hand, the transmission of community and culture is not in tension with the geological and cosmic processes, or, in Snyder's terms, "the weathering land / The wheeling sky" (*RR* 8). They are the context of culture, and a healthy community is in harmony with those cycles. Thus the necessity to

> teach the children about the cycles.
> The life cycles. All the other cycles.
> That's what it's all about, and it's all forgot (*AH* 7).

As one would expect, the community of "Buildings" involves a connection to archaic shamanistic cultures and to Asian religions. During the building of the schoolhouse, they "Buried a five-prong vajra between the schoolbuildings / while praying and offering tobacco." Years later, "Blowing the conch, shaking the staff-rings / we opened work on a Hall." The community thus draws from both local/bioregional culture and global culture, the ancient California and early China.

Snyder's community exists within a more comprehensive and subtle context: the ongoing "dance with Matter." Ultimate reality is material; it is this very world of form. This world consists not of discrete and abiding substances but, rather, of things which, while solid and distinct, exist in an ongoing process of harmonious interaction such that all things are new each moment. This dance with matter is nondualistic: buildings are "solid" yet "wet from the pool / that renews all things. . . ." They, like all matter, are differentiated yet integrated. Community, then, is a dance within the dance of Matter. And, for Snyder, it is clearly a dance that matters ultimately.

This dance is both of history and beyond it. Snyder has spoken elsewhere of two modes of time by combining the indigenous Australian idea of "dreamtime" with Dōgen's notion of "being-time." One mode of time is "the eternal moment of creating, of being, as contrasted with the mode of cause and effect in time. . .where people mainly live. . ." (*PofW* 84–85). For Snyder, we need to see our buildings, our community, and culture itself as part of history, a response to a particular historical era, but we also need to recognize that buildings, and community, "are built in the moment," the timeless moment of renewal.

"Building" exemplifies Snyder's vision of an alternative community that is physically and metaphysically integrated into nature. Sherman Paul has noted the importance of community in Snyder's life by responding to Jack Kerouac's prophecy in *Dharma Bums* about Japhy Ryder (the novel's main character, based on Gary Snyder): "I think he'll end up like Han Shan living alone in the mountains and writing poems on the walls of cliffs." Paul corrects the prophecy by observing that Snyder "lives now in the mountains, but with his family, in community."[49] Paul is right to point to Snyder's combination of nature and community, but Snyder does not really live in the mountains in the way the semi-legendary recluse Han Shan did.[50] He dwells neither in the lowlands of American culture nor on the ascetic peaks of a cold mountain but in the foothills. At this intersection, he can pursue "solitude and community, *vajra* and *garbha*," thus embodying the "tension between the solitary eye and the nourishing kitchen [which] is at the root of the strength and magic of the Old Ways."[51] He is neither the Buddha achieving enlightenment on the mountain nor the Buddha descending the mountain to preach to the people. He is a re-inhabitant, dwelling in a bioregional community which combines Buddhism with the Old Ways. From that place, in place, he is able to cultivate his local community while staying interconnected with both mountain and city. Such an emplacement is both physically convenient and symbolically significant, for Snyder sees his community as limited to neither mountain nor city. Kitkitdizze is one node in the net of the great earth *saṅgha*.

Notes

1. Quotations from Gary Snyder's works are cited in the text with the abbreviations listed below:

AH   *Axe Handles: Poems*. San Francisco: North Point Press, 1983.

EHH   *Earth House Hold: Technical Notes and Queries to Fellow Dharma Revolutionaries*. New York: New Directions Press, 1969.

G   "Grace." *CoEvolution Quarterly* 43 (fall 1984):1.

NN   *No Nature: New and Selected Poems*. New York: Pantheon Books, 1992.

OW   *The Old Ways: Six Essays*. San Francisco: City Lights Books, 1977.

PIS   *A Place in Space: Ethics, Aesthetics, and Watersheds: New and Selected Prose*. Washington, D.C.: Counterpoint, 1995.

PofW   *The Practice of the Wild*. San Francisco: North Point Press, 1990.

RR   *Riprap and Cold Mountain Poems*. San Francisco: Four Seasons Foundation, 1965.

TRW   *The Real Work: Interviews and Talks 1964–1979*. New York: New Directions, 1980.

TI   *Turtle Island*. New York: New Directions, 1974.

"This Is Our Body." Audio tape from Watershed Tapes. 1989.

TT   *Turtle Talk: Voices for a Sustainable Future*. Ed. Christopher Plant and Judith Plant. *The New Catalyst* Bioregional Series. Philadelphia: New Society Publishers, 1990.

WM   "The Wild Mind of Gary Snyder." Includes quotations from an interview with Trevor Carolan. *Shambhala Sun* 4, no. 5 (May 1996):18–26.

2. Aldo Leopold, *A Sand County Almanac* (New York: Oxford University Press, 1949), 224–25.

3. For a further application of the idea of Other in the context of Mikhail Bakhtin's thought, see below, pp. 202–4.

4. David Landis Barnhill, "Indra's Net as Food Chain: Gary Snyder's Ecological Vision," *Ten Directions* 11, no. 1 (1990):20; quoted by Snyder in *PIS* 67. For other discussions of Snyder's Hua-yen vision with analyses of relevant poems, see Shu-chun Huang,"A Hua-yen Buddhist Perspective of Gary Snyder," *Tamkang Review* 20, no. 2 (winter 1989):195–216; and Christopher Parr, "Living Interdependence: Gary Snyder's Kegon and Zen Views of Work, Hunting and Place," paper delivered at the Midwest meeting of the American Academy of Religion, 23 March 1996.

5. For Snyder's views of eating, see Barnhill, "Indra's Net," and Snyder's "Nets of Beads, Webs of Cells" in *PIS,* 65–73.

6. Deep ecology is a contemporary movement in environmental philosophy first articulated by Arne Naess. Distinguishing itself from the "shallow ecology" of resource conservation and reformism, it emphasizes a deep questioning of the

philosophical and spiritual foundations of environmental issues. It draws on wisdom traditions of Asia and indigenous cultures in emphasizing the intrinsic value of the natural world and the absence of an ontological split between humans and nature.

7. In an interview with James Kraus, Snyder spoke of the need to extend knowledge of the local and its limits to the planetary scale. He also stated that his poem "Mountains and Rivers" "points in that direction, it tries to leap from the local to the planetary level and back again." James W. Kraus, "Gary Snyder's Biopoetics: A Study of the Poet as Ecologist" (Ph.D. diss., University of Hawaii, 1986), 182.

8. Bert Almon, "Buddhism and Energy in the Recent Poetry of Gary Snyder," *Mosaic* 11 (1977):121.

9. Hwa Yol Jung and Petee Jung, "Gary Snyder's Ecopiety," *Environmental History Review* 41, no. 3 (1990):84. Snyder speaks of True Communionism in "Revolution in the Revolution in the Revolution" (*RW* 49). Julia Martin discusses that idea, but focuses on Snyder's adaption of Christian terminology (e.g., communion) rather than his vision of community. See Julia Martin, "True Communionism: Gary Snyder's Transvaluation of Some Christian Terminology," *Journal for the Study of Religion* 1, no. 1 (1988):71–73.

10. Snyder gives this narrative account on the audio tape "This Is Our Body."

11. The narrative account of the origin of the poem is presented in "This Is Our Body." In performance, Snyder sings the section of the Magpie's song (in italics).

12. See Barnhill, "Indra's Net."

13. Woody Rehanek, "The Shaman Songs of Gary Snyder," *Okanogan Natural News* 19 (summer 1984):9; cited in Patrick D. Murphy, *Understanding Gary Snyder* (Columbia: University of South Carolina Press, 1992), 140.

14. William R. LaFleur, "Saigyō and the Buddhist Value of Nature," parts 1 and 2, *History of Religions* 13, no. 2 (November 1973):93–127; no. 3 (February 1974):227–47.

15. For a discussion of these aspects in Bashō's works, see David Landis Barnhill, "Folk Religion and Shintō in the Ecosystem of Bashō's Religious World," paper presented at the annual meeting of the American Academy of Religion, Chicago, Illinois, November 1994.

16. Patrick D. Murphy, *Understanding Gary Snyder* (Columbia: University of South Carolina Press, 1992), 164.

17. For an extensive discussion of the relationship between Ch'an (Zen) Buddhism and popular religions in China, see Bernard Faure, *The Rhetoric of Immediacy: A Cultural Critique of Chan/Zen Buddhism* (Princeton: Princeton University Press, 1991.

18. David Abram, "The Ecology of Magic," in *Finding Home: Writing on Nature and Culture from Orion Magazine*, ed. Peter Sauer (Boston: Beacon Press, 1992), 178–79.

19. Ibid., 180.

20. Ibid., 182.

21. Ibid., 183.

22. Ibid.

23. Ibid., 198.

24. Ibid., 201.

25. Ibid., 196–97.

26. Caryl Emerson, quoted in Patrick D. Murphy, *Literature, Nature, and Other: Ecofeminist Critiques* (Albany: State University of New York Press, 1995), 178, n. 4.

27. Murphy, *Literature, Nature, and Other*, 35.

28. Ibid., 115. For discussions of Snyder as ethnographic "informant," see David Robbins, "Gary Snyder: The Poet as Informant," *Dialectical Anthropology* 11, no. 2-4 (1986):203–10; and Nathaniel Tarn, "From Anthropologist to Informant: A Field Record of Gary Snyder," *Alcheringa: A Journal of Ethnopoetics* 4 (autumn 1972).

29. Murphy, *Literature, Nature, and Other*, 5.

30. Almon, "Buddhism and Energy in the Recent Poetry of Gary Snyder," 121.

31. See Roger T. Ames, "Putting the *Te* Back in Taoism," in *Nature in Asian Traditions of Thought: Essays in Environmental Philosophy*, ed. J. Baird Callicott and Roger T. Ames (Albany: State University of New York Press, 1989), 113–44, for an important discussion of integrity as integration in Taoist thought.

32. Ken Hirschkop has discussed the ambiguity between the descriptive and normative quality of Bakhtin's thought. See his "Introduction: Bakhtin and Cultural Theory," in *Bakhtin and Cultural Theory*, ed. Ken Hirschkop and David Shepherd (Manchester and New York: Manchester University Press, 1989), 1–38.

33. Murphy, *Literature, Nature, and Other*, 9.

34. Ibid., 118.

35. For a review of the diversity of interpretation of Bakhtin's thought, see Hirschkop, "Introduction: Bakhtin and Cultural Theory."

36. Murphy, *Literature, Nature, and Other*, 114.

37. Ibid., 9.

38. Tim Dean, *Gary Snyder and the American Unconscious: Inhabiting the Ground* (Houndmills, England: Macmillan, 1991), 151.

39. Abram, "The Ecology of Magic," 179.

40. Murphy, *Literature, Nature, and Other*, 178, n. 3.

41. Donald Keene, trans., "An Account of My Hut," in *Anthology of Japanese Literature from the Earliest Era to the Mid-Nineteenth Century*, ed. Donald Keene (New York: Grove Press, 1955), 209.

42. Ibid., 197.

43. On the importance of work in Snyder's writings, see Lionel Basney, "Having Your Meaning at Hand: Work in Snyder and Berry," in *World, Self, Poem:*

Essays on Contemporary Poetry from the "Jubilation of Poets," ed. Leonard M. Trawick (Kent, Ohio: The Kent State University Press, 1990); and Parr, "Living Interdependence."

44. This is not to say that the personal dimension was absent in earlier writings. *Earth House Hold* concludes with an account of the wedding of Snyder and Masa: "Standing on the edge of the crater, blowing the conch horn and chanting a mantra; offering shochu to the gods of the volcano, the ocean, and the sky; then Masa and I exchanged the traditional three sips. . ." (*EHH* 142). The image of the conch ties this piece to "Buildings," with the later poem suggesting a fuller realization of community as the context for the transmission of a culture that brings together primal/archaic and Buddhist elements.

45. Shih-hsiang Chen was Snyder's teacher of classical Chinese. He translated Lu Ji's *Wen fu* (Essay on literature), which refers to the axe image.

46. Basney, "Having Your Meaning at Hand," 135.

47. Charles Molesworth, "Getting a Handle on It," *American Book Review* 6, no. 5-6 (1984):15.

48. Murphy, *Understanding Gary Snyder*, 134.

49. Sherman Paul, *In Search of the Primitive: Rereading David Antin, Jerome Rothenberg, and Gary Snyder* (Baton Rouge: Louisiana State University Press, 1986), 188.

50. The legendary Han Shan, living alone in the high mountains scribbling poems on the wall of his cave, should not be confused with the historical poet whose identity and biography is little known but more complex than the legend. For the most comprehensive edition of his work, see Robert G. Henricks, *The Poetry of Han-shan: A Complete, Annotated Translation of* Cold Mountain (Albany: State University of New York Press, 1990).

51. From Snyder's response to Paul in Paul, *In Search of the Primitive*, 298, 299.

American Buddhist Response to the Land: Ecological Practice at Two West Coast Retreat Centers[1]

Stephanie Kaza

From a theoretical perspective, Buddhist philosophy appears to be highly congruent with an ecological worldview. Respected Buddhist teachers such as His Holiness the Dalai Lama and Vietnamese Zen master Thich Nhat Hanh frequently point to the interdependence of human life and the environment.[2] American Buddhist scholars, including many of those in this volume, show the bases in text and principle for a Buddhist environmental philosophy.[3] But how do these links translate into actual practice? Do American Buddhists "walk their talk"?

In this article I look at two American Buddhist centers to assess the extent of ecological practice at an institutional level. Retreat centers act as focal points for transmitting Buddhist values both to committed Buddhist practitioners and to the visiting public. To the extent that practice places reinforce ecological caretaking with spiritual principles, they provide a foundation for moral commitment to the environment. It is clear to many leading environmental thinkers that science, technology, and economics alone will not solve the environmental crisis.[4] Instead, they call for cultural transformation based on religious, moral, or spiritual values of deep care of and concern for the earth. How do American Buddhist centers contribute to this cultural shift? What in their efforts is distinctly Buddhist and what reflects the existing culture or reaction to it? Where are the points of tension around ecological practice in Buddhist centers? And on what institutional elements do these practices depend?

This article is a preliminary report of work in progress assessing environmental practices at diverse American Buddhist centers in the United States. The first two centers I have looked at are Green Gulch Zen Center, north of San Francisco, and Spirit Rock Meditation Center near San Rafael, in Marin County, California. Both are rural centers responsible for sizable portions of land. Though each has been established relatively recently, each has made some efforts toward appropriate land stewardship practices. I provide a brief land history of each center and a comparison of their similarities and differences. Information is drawn from center newsletters and journals, site visits, and interviews with staff members. I review the centers' current land practices in the context of Gary Snyder's core ethical guidelines for reinhabitation. I describe some points of tension and arenas for further ethical exploration. Much of what is reported here represents a dialogue unfolding. This paper itself may prompt further discussion and commitment toward turning the *Dharma* wheel another round.

Land Histories

Green Gulch Zen Center lies in a beautiful coastal valley in the narrow flood plain of Green Gulch Creek, just north of San Francisco. The land extends almost to the Pacific Ocean at Muir Beach and is surrounded by the public open space of Golden Gate National Recreation Area; nearby lands are protected by Mount Tamalpais State Park and Marin County Water District. The valley is flanked on the north and south by open, grass-covered ridges; remnants of redwood forest understory line the side canyons. In the next valley over lies Muir Woods National Monument, home to some of the tallest coast redwoods in the San Francisco Bay area.

Green Gulch Farm was purchased in 1972 from owner and rancher George Wheelwright ten years after San Francisco Zen Center was formally incorporated. Bay area Zen students had begun sitting with Shunryu Suzuki Rōshi in 1959 when he arrived at Sokoji Temple on Bush Street in Japantown. By 1966 Zen Center had become a stable practice community and Suzuki Rōshi was interested in finding rural land for a retreat center. With exuberant fundraising efforts (including generous rock and roll benefits), in

1967 Zen Center bought Tassajara Mountain Center, a former hot springs resort in the Big Sur area. Soon after, Zen Center moved from Sokoji to a new city facility on Page Street, which Suzuki named Hoshinji, Beginners' Mind Temple.[5] Zen Center gained national publicity with the publication of Suzuki Rōshi's book, *Zen Mind, Beginners' Mind*, and, shortly after, Edward Espe Brown's *The Tassajara Bread Book*.[6]

Suzuki Rōshi's health began to deteriorate in 1971; before his death he suggested the idea of a farm practice place. The following year his *dharma* heir Richard Baker took the lead in orchestrating Zen Center's purchase of Green Gulch Farm, which became Green Dragon Temple. George and Hope Wheelwright had owned the land for thirty years, long before the coast highway was built, when Muir Beach was a small village of Portuguese fishermen. George raised cattle there, including award-winning prize bulls. To improve pasturage for his cattle, he sprayed 2-4D herbicide on the hills to limit shrub growth. The creek was channeled to produce a series of reservoirs for water storage. The land still bears tracks of cattle trails; the creek passes through a concrete ditch for much of the stretch through the valley.[7]

Compared with the wooded side canyons of neighboring Franks Valley, Green Gulch was heavily cut over after the San Francisco 1906 earthquake. Many redwoods and Douglas firs were transported out of Big Lagoon dock at Muir Beach to help rebuild the city. To reforest the lower valley, Wheelwright planted lines of non-native eucalyptus along the entrance road. When Zen Center became the Green Gulch land steward, students undertook significant efforts to build a twenty-acre organic farm and a one-acre organic garden. To protect and restore the land, they planted windbreaks of Monterey cypress and Monterey pine between the agricultural fields. Since 1975 tree plantings have been carried out yearly and non-native invasive plants (acacia, broom, ivy) have been culled back. Field soils have been improved by large-scale compost-making and legume cover crops. The farm grows and markets certified organic lettuce, squash, pumpkins, potatoes, and kitchen greens. The garden supports a variety of perennial dahlias, Siberian iris, and roses, along with annuals such as sweet pea, anemone, larkspur, and Peruvian lilies. In the greenhouses flowers, vegetables, and native plants are propagated for community and private gardeners.[8]

Spirit Rock Meditation Center lies in San Geronimo Valley, a connecting link between the urban corridor of San Rafael, north of San Francisco, and the open space of Point Reyes National Seashore and Samuel B. Taylor State Park. The valley is relatively sparsely settled, remaining in rural ranchlands and dairy farms. Intensive development pressure has been held at bay due to the fiercely protective conservation and planning efforts of the San Geronimo Valley Planning Group. The center is named for a prominent outcrop of rock thought to be sacred to the local Miwok tribes. Rising up behind Spirit Rock lie rolling grassy foothills graced by scattered coast live oaks and bay laurels.

In the 1960s a number of Western students traveled to Southeast Asia to study *vipassanā*, or insight meditation practice. In the 1970s they returned home and began teaching at various retreat centers, including Naropa Institute in Boulder, Colorado. On the East Coast, in 1976 a group of senior students and teachers led by Jack Kornfeld and Joseph Goldstein purchased a Catholic seminary in Barre, Massachusetts, and established the Insight Meditation Society as a permanent retreat center. On the West Coast, interest in *vipassanā* practice grew with the national publication of the *Inquiring Mind* newsletter and an increasing number of retreats at various local centers (including Green Gulch). In 1983 a small group of Californians began meeting regularly to consider establishing a retreat center on the West Coast. Three years later, Jack Kornfeld found a four hundred-acre parcel in San Geronimo Valley for sale by The Nature Conservancy, which wanted to contribute the purchase money to Amazon rain forest preservation. The land seemed ideally suited to their purposes—classes, daylong retreats, staff housing. After extended negotiations with the landowners as well as representatives of the San Geronimo Valley Planning Group, the deal was closed.[9]

This land, in contrast to Green Gulch Farm, was undeveloped, with few previous buildings. Ongoing fundraising has generated enough support to build the necessary infrastructure for hosting regular retreats. Several temporary trailers were installed in 1990 to house a meditation hall and office. In 1995 a dining hall was built so meals could be served on the premises. Future design plans include four residence halls for eighty-four retreatants, a larger meditation hall to seat two hundred, staff housing for twenty resident staff, additional parking areas, a family program building, four family apartments, teacher housing, a Council House with meeting rooms, and an adjacent hermitage with eighteen private huts, a small meditation hall, and two teacher rooms. In early 1996 the plan received approval from Marin County Department of Public Works and all other necessary official agencies. The next building phase is expected to begin soon.[10]

A brief comparison of these two rural Buddhist centers shows a number of strong similarities and differences that are significant in the evolution of ecological culture and values at each place. Both sites are physically part of the larger landscape system surrounding Mount Tamalpais, a prominent local peak extending to 2,571 feet. The mountain is flanked by Douglas fir and redwood forest, coastal scrub, serpentine outcrops, and luxuriant moss-lined creeks. Green Gulch lies at the base of the southwest-facing slope; Spirit Rock lies below the northeast-facing flank. The distance between the two centers is a long day-hike of twenty-two miles over the edge of the mountain. The centers draw their primary vertical reference point from the mountain, a beloved landmark in northern San Francisco matched by the taller Mount Diablo east of Oakland.

Both sites lie in affluent Marin County, an area with a strong conservation history and a well-established plan to limit development to the highway corridor along the San Francisco Bay. The western two-thirds of the county has been protected as open space, due to the tireless efforts of the Marin Conservation Association and others over the last seventy years. This relatively pristine open space is a magnet for hikers, joggers, mountain bikers, and sightseers from not only the nine-county-wide Bay area but the entire United States as well. The connecting national and state parks offer over two hundred miles of hiking trails, mountain views, and majestic stretches of open beach. Before either Buddhist center was established, the land itself was a spiritual draw for thousands of people.

Visitors and students to Green Gulch and Spirit Rock frequently express their appreciation for the beauty of the rural country settings of these retreat centers. They come for the Buddhist teachings, but they also spend time walking in the garden, on the beach, or across the hills. The landscape itself is spiritually inspiring and is seen as part of the meditative experience. Teaching in both centers takes place outdoors as well as in the meditation hall—in particular, instruction in walking meditation. Through one or many visits to Green Gulch or Spirit Rock, practitioners come to associate their experience with the *dharma* as connected to these specific pieces of land.

The differences between the two centers are also significant. Though both are surrounded by large vistas of open space, almost all of the Green Gulch landscape is held in public trust, whereas neighboring land at Spirit Rock is private property. For Green Gulch, good neighbor relations require ongoing cooperation and negotiation with primarily public agencies; for Spirit Rock these are with private landholders. Though both centers are located in Marin County, the two microclimates are quite different. Green Gulch, on the coast, has somewhat milder winters and much foggier summers. Spirit Rock, in an inland valley, experiences more temperature extremes and is more subject to fire hazard in the fall dry season.

Green Gulch inherited its main buildings from the Wheelwrights and adapted them to retreat center use despite existing flaws. For example, the meditation hall, formerly the cattle and horse barn, has lovely high ceilings and a thick wood floor, but because it was built over the original creekbed, it retains a certain dampness through the winter. Spirit Rock has been able to design site-appropriate buildings from the start, drawing on state-of-the-art environmental design principles wherever possible. Green Gulch has committed twenty acres to organic farming and gardening, with all the related challenges of soil building, water management, marketing, and integration with other Zen Center activities. Spirit Rock has no organic farm or garden and no plans for anything on this scale other than minor landscape plantings.

Perhaps most significant of all, Green Gulch has been a residential center from the start. Those who live there perceive it to be their home; Sunday guests and retreatants are visitors with relatively little influence. Decision-making power for the land is in the hands of the staff, the board of directors, and, to some extent, the stewardship committee. Almost all the members of the two governing bodies and volunteer committee are or have been residents at one of the Zen Center sites. In contrast, Spirit Rock has never been residential, except for minimal caretaking, and will not be for several more years. Fundraising for even the land purchase depended on extensive lay involvement and volunteer activity beyond that of the very limited staff.[11] This difference in governance has shaped the way land relations have evolved in each center, according to the number and seniority of those responsible for land-management decisions.

Ethics of Ecological Living: Toward Reinhabitation

Frameworks for environmental ethics can be based on a number of different principles.[12] For example, Holmes Rolston III ennumerates human ways of valuing nature (economic, scientific, recreational, aesthetic, sacramental) in contrast to the intrinsic value of organisms, landforms, and so on—"for what it is in itself."[13] One could evaluate religious centers according to which values they promote and how these preferences are reflected in spiritual practices. Ecofeminist Valerie Plumwood frames human relations with nature in the context of social power relations and the perpetuation of oppressive dualisms.[14] One could evaluate religious centers as to the degree they reproduce cultural hierarchical attitudes toward nature. Conservation biologists Reed Noss and Edward Grumbine, among others, set forth ethical principles based on protecting and enhancing biodiversity.[15] Ecophilosopher David Abram suggests guidelines based in reciprocal sensory communication with the "more than human" world.[16] Each framework offers a radically different lens through which to consider cultural practices.

For the purposes of this assessment, I am interested in the transmission of ecological culture. I want to see how religous institutions use spiritual principles to support ecologically sustainable ways of life. In his classic essay, "The Land Ethic," Aldo Leopold, the wildlife biologist of the 1930s and 1940s, defines ethics as "a kind of community instinct in the making." For Leopold, all ethics rest upon the premise that "the individual is a member of a community of interdependent parts":

> The land ethic. . .enlarges the boundaries of the community to include soils, waters, plants, and animals, or collectively: the land. . . . In short, a land ethic changes the role of *Homo sapiens* from conqueror of the land community to plain member and citizen of it.[17]

As people live on the land over time, they become part of the land, the land comes to include them. They no longer live *on* the land but rather *with* the land and all its members. Here I explore the proposal that institutional practices (as opposed to individual isolated practices) reflect the evolution of a community instinct in the making.

Gary Snyder suggests that a useful orientation for an ecological community instinct would be "reinhabitation" as an ecosystem-based culture. He refers to biogeographer Ray Dasmann's distinction between *ecosystem cultures* whose "life and economics are centered in terms of natural regions and watershed" and *biosphere cultures* that are directed from urban centers and oriented to global use and plunder of natural resources.[18] Native and rural peoples are almost entirely ecosystem-based cultures, generally having less impact on the health of the surrounding system than biosphere cultures. *Reinhabitory peoples* are those who are committed to a life based in place, "making common cause" with the life-styles of the original inhabitory peoples.[19] This means a life identified with a specific place, understanding the local community of plants and animals as companions, neighbors, and supporters of human life. Over time, this sense of place deepens with familiarity, and place-based knowledge is passed on from generation to generation.

Snyder suggests three aspects that are the core of the practice of a reinhabitory ecological ethic: "feeling gratitude to it all; taking responsibility for your own acts; keeping contact with the sources of the energy that flow into your own life (namely dirt, water, flesh)."[20] On the surface this seems to be deceptively simple, yet the implications are very broad and particularly suited to a review of religious centers. As Snyder puts it, "the actual demands of a life committed to a place. . .are so physically and intellectually intense that it is a moral and spiritual choice as well."[21] He suggests that to survive as an ecosystem person, one must draw on moral and spiritual resources. These are strengthened through knowledge of place and, reciprocally, through knowledge of self as dependent on place.

The first of these three aspects, "feeling gratitude," generates humility and a sense of awareness of the wider self. Mixed in are awe, caution, fear, and common sense. Prayers of thanks are offered for the gift of life, for freedom, for the moment, from the death-dealing forces of nature. Reinhabitants remember that human lives are dependent on other lives, that nothing lasts forever, that no food, water, or shelter are ever guaranteed. The practice of gratitude in a Buddhist context carries understandings of no-self, impermanence, and interdependence.

The second aspect, "taking responsibility for your own acts," implies the exercise of restraint, recognizing the rippling effects of each action in the jeweled net of Indra.[22] The practice of acting responsibly means minimizing destructive human impact on the land and allowing room for the flourishing of nonhuman others. Contained in this practice are the Buddhist precepts for self-restraint, including no killing and no abusive relationships.[23]

The third aspect, "keeping contact with the sources of energy. . . flow," may be the most subtle and easily overlooked. Snyder is speaking of "wild mind," the original source energy, and the need always to be nourished directly by this primordial wisdom. This is the energy shared with other life-forms, the force of weather, place, and history commingled. An individual at a Buddhist center may contact this energy through walking meditation, gardening work practice, or mindful food preparation. But how does an institution maintain contact with wild mind in its structures and organizational culture? I suggest that in addressing this challenge Buddhist retreat centers begin to approach reinhabitation, allowing the land to influence local ecological practice significantly. The three elements of Snyder's ethic describe a method for transmission of ecological culture on American soil. This look at two Buddhist centers can provide a preliminary assessment of the degree to which these "new settlers" may be headed toward long-term reinhabitation.

Evaluation of Two Buddhist Centers

Green Gulch Zen Center

Looking first at Green Gulch Zen Center, I will begin by examining practices of *gratitude to the land*. These are usually mixed in with gratitude for the Buddha, *Dharma*, and *Saṅgha* (the Three Jewels) to various degrees, but certain practices specifically highlight relationship with and dependence on the land. On a daily basis, students recite the Zen meal chant:

Innumerable labors brought us this food
We should know how it comes to us
Receiving this offering let us consider whether our virtue or practice
    deserve it
Desiring the natural order of mind, let us be free from greed, hate,
    and delusion
We eat to support life and to practice the way of Buddha.[24]

For Zen students at Green Gulch, the innumerable labors are obvious: moving irrigation pipes, cropping salad greens, propagating greenhouse seedlings, turning compost. The meal chant is a regular reminder to offer gratitude for the food upon which they depend.

Across the course of the seasonal year, dedication *eko*s are offered at the four turning points of the year. At the spring equinox service outside on the east-facing side of the valley, gratitude is offered on behalf of the community for the rising sun of the new year. On the summer solstice, at mid-day, gratitude is offered for the bountiful garden and the produce of the fields. The autumn equinox dedication is offered at dusk, facing west, accepting the teachings of impermanence and death. And the winter solstice is marked at midnight under the dark sky, with gratitude for the vast wild mind of no-self.[25]

In addition to these natural points of the sun's shifting motion across the ridges, Green Gulch Zen Center also marks the bounty of the farm harvest at Thanksgiving. *Zendo* and dining-room altars are decorated with offerings of beets, pumpkins, lettuce, chard, herbs, and potatoes and the *Heart Sūtra* is chanted with gratitude for the riches of the land. On Buddha's birthday in April, children collect representative flowers of each of the wild species in the watershed and add them to the elephant flower cart for bathing the baby Buddha. The dedication chant at this ceremony lists all the flowers (over one hundred!) in a long, entertaining drone, occasionally marked by the further amusement of Latin names. In the repetition is the transmission of gratitude for the wild hills and diversity of flowers.[26]

The second aspect of Snyder's ecological ethic, *taking responsibility* for one's acts, is a complicated undertaking at a rural center such as Green Gulch. I will report on previous and current efforts, but certainly much more can be done to act fully responsible on this ecologically complex piece of land. I will describe institutional efforts to take responsibility in four arenas: land stewardship, community relations, ecological culture, and education.

Land stewardship activities focus primarily around two areas: land restoration efforts and the organic farm. Some of the restoration efforts take the form of doing nothing, allowing the wild mind of the coastal habitats to surface again. The hills are no longer sprayed with herbicides to control vegetation, and cattle no longer trample the soil. Along the creek, a thicket of shrubs has been left to grow into a healthy wildlife corridor, well populated by local songbirds. Of the more proactive restoration efforts, annual tree plantings on Arbor Day in February have been carried out since 1975. Windbreaks of Monterey cypress and Monterey pine are now easily fifty feet tall and play a significant role in deflecting the powerful ocean winds that ravage the coastal soils. Since 1991, in addition to plantings of redwood and Douglas fir, coast live oak acorns gathered from the neighboring valley have been planted on protected sites to replace those grazed down by the cattle.[27] Though somewhat controversial, staff and volunteers have also made an effort to remove non-native eucalyptus shoots, acacia, German ivy, and broom where they are choking back native vegetation. A preliminary landscape ecology report was drawn up in 1991 with detailed recommendations for further tree work and land restoration.[28] Forward motion is restricted by the lack of a staff person designated as Land Manager. Though positions exist for Head of Farm and Head Gardener, as well as Head Maintenance, no one staff person assumes responsibility for the overall health of the landscape ecosystem.

The twenty-six-acre organic farm is a model of good farming stewardship and is recognized throughout the state for its ecological practices. It is a certified member of the California Organic Farming Association, meeting the standards for soil free of pesticides and chemical fertilizers. Heavy machine use is moderate, primarily for plowing the fields and transplanting seedlings. Weeding and cropping are done by hand as part of mindfulness work for Zen students. The soil is built through careful application of compost made from kitchen scraps, green waste, and horse manure; a cover crop of fava beans is planted each winter and turned under as green manure in spring. Insect pests and diseases are managed through observation, crop rotation, and selected organic and mechanical pest

controls.[29] Because it is accountable to the standard-setting association for organic produce as well as to the community of organic
farmers in the wider Bay area, the Zen community at Green Gulch
has an incentive to maintain a high degree of institutional responsibility for its actions. Likewise, the one-and-a-half-acre perennial
garden is organic, with all cultivation in double-dug beds and all
cropping done by hand.

Community relations regarding the land require ongoing conversations with Muir Beach residents and staff of the Golden Gate
National Recreation Area (GGNRA). With each, being a good
neighbor means cooperating to share land and water resources,
acknowledging the institutional impact of Zen Center. Green Gulch
Creek empties into Redwood Creek near its mouth to the ocean at
Muir Beach. In dry summers, the farm has drawn on these combined
water supplies to irrigate the lower fields. Because water in coastal
California is limited, rates of water use have been a source of
conflict with the local community, other ranchers, and Muir Woods
(a part of GGNRA). To maintain navigable levels of water for
salmon in Redwood Creek through Muir Woods, and to share the
remaining water with neighbors, Green Gulch has reduced tillage
areas in dry years.[30]

Relations with the GGNRA are also an ongoing part of Green
Gulch institutional life. Staff have been asked to comment on plans
for bike routes through Green Gulch, control of escaped South
African capeweed, and restoration of Big Lagoon at Muir Beach.
Over the years a strong relationship has developed between the
garden staff at Green Gulch and the park rangers at Muir Woods,
as they have cooperated in plant propagation and volunteer planting
days together. GGNRA resource staff have been helpful in offering
advice for land-management decisions at Green Gulch which affect
the surrounding landscape.[31]

The farm and garden encourage community interaction through
outreach projects with other farms and gardens. Seedlings and plant
starts are often donated to other fledgling farms, such as the Hunter's
Point jail project and Schoolyard Garden in Berkeley. Volunteers
are encouraged to join farm staff for potato and pumpkin harvest
days. Farm and garden staff often consult with other farm projects
to offer advice on soil building, planting design, and propagation
techniques.

By *ecological culture* I mean everyday activities which promote sound environmental habits. At Green Gulch three arenas reflect a high degree of institutional responsibility: food practices, waste recycling, and water use. As a Buddhist center, Green Gulch has chosen a policy of not cooking or serving meat in the dining hall. Though vegetarianism is often associated with Buddhism, it is not strictly mandated by the teachings. However, since Zen Center *is* committed to vegetarian practice, it does not support the often inhumane institutional practices associated with factory animal farming and animal slaughter. Further, by adhering to vegetarianism, the institution is not contributing to the accelerated clearing of global rain forests for cattle pasture and beef imports. Food served at Green Gulch includes as much in-season produce as possible from the organic farm. Other produce is purchased from local dairy and vegetable farms to support neighboring farmers. Though these aspects of Green Gulch food contribute to ecological responsibility for the land and for the regional economy, some residents urge even stronger ecological practices, such as serving *only* organic food.

Food waste goes into large compost piles adjacent to the farm and garden. After several months of "cooking" with green clippings and manure from the neighboring horse farm, the compost is ready to spread on the fields. Green Gulch also recycles white paper, magazines, glass and plastic bottles, cans, cardboard, motor oil, and batteries. The farm reuses wood and cardboard produce crates from regular customers by picking them up on produce runs; the garden reuses gallon pots and seedling trays for propagation. Paper towels, napkins, and toilet paper as well as most office paper purchases are from recycled paper sources. Fallen trees become firewood; trash lumber is used for kindling or is burned. Relatively little waste is hauled away from Green Gulch besides the recyclables. These efforts to simplify food and waste flows to and from the center are motivated both by the high cost of trash removal and the Zen aesthetic of tidiness.

Water conservation is mandatory at Green Gulch as water supplies are limited to local springs and Green Gulch Creek. These supply all the water needs year-round for the thirty to forty residents, ten to fifteen guest students, two to three hundred Sunday visitors, and additional conferences and retreats. Located in the highly developed San Francisco Bay region, Green Gulch is unusual in being water self-sufficient. Its entire water system is self contained and locally maintained, drawing on five reservoirs, three storage tanks, and a well. The valley is not connected to the Marin County Water District for backup supplies of additional water, so water use is managed according to what is actually available. In summer and early fall, rates of flow drop significantly, bringing added pressure to conserve water. Low-flow toilets and showers are installed in the guest and residence areas; drip irrigation is used in some of the garden beds.[32] During meditation retreats, frugal use of water is practiced in formal Japanese *oryoki* meals, where each person washes his or her bowls with less than a cup of water per meal. Water conservation depends on continual reminders to the ever-changing population of guests and staff, particularly during dry months.

Education for environmental awareness is an ongoing effort at Green Gulch Zen Center, spearheaded almost entirely by the garden staff. Farm and garden classes are offered year-round on composting, perennials, vegetable gardening, and other topics. Children's classes and other groups receive tours of the farm and garden, meditation hall, and residential buildings. For several years Green Gulch has hosted a "Voice of the Watershed" series of walks and guest lectures on topics of local natural history. Each year before Arbor Day, senior staff lead a ridge circumambulation of the valley to place the center in a larger landscape context. The 1992 summer practice period focused specifically on "Environment and Meditation," drawing together texts, teachers, and daily practice engaging environmental issues. One result was an educational pamphlet on "Environmental Practice at Green Gulch," a summary of institutional efforts to be environmentally conscious and responsible.[33] Although various staff and students have carried these efforts forward, environmental concerns are not yet considered a top priority by those in leadership positions.

Through each of these four areas—land stewardship, community relations, ecological culture, and education—Green Gulch has made some effort to systematize an ecological ethic of taking institutional responsibility for the center's actions. This is no guarantee that each individual who passes through Green Gulch receives the spark of this ethic, but at least while they are visiting, they are expected to follow the established environmental practices of the local culture.

The third aspect of Gary Snyder's ecological ethic entails *keeping contact with the sources of energy* that flow into one's life, in this case, the life of Zen Center. This is perhaps the least easy to ennumerate of the three elements of the ethic, and yet it is most crucial to the vitality of Snyder's framework. Individual Zen students report gaining access to this energy flow through working in the garden, sitting among the redwoods, or walking by the ocean. But these receptive activities are seldom undertaken by Zen Center as a whole. Several practices at Green Gulch do, however, support the possibility of increased contact with this energy flow of the wild.

The first of these derives directly from the traditional Zen emphasis on *work as practice*. Many classic Zen stories find their context in sweeping, cleaning, farming, or chopping wood.[34] In Soto Zen, enlightenment often happens in the mundane activities of everyday life. Guest students work two or three mornings in the farm and/or garden, usually engaged in silent mindfulness practice. Staff, other than farm and garden staff, join in solidarity with the summer farm effort once a week before breakfast, planting, weeding, or cropping in silence. These efforts are both practical, in terms of getting the necessary work done, and spiritually unifying, for all community members experience together the energy of soil, fresh air, and landscape on a regular basis.

A second area, which I will call *sacralizing the landscape*, involves institutional commitment to outdoor ceremonies, walks, and commemorations which include the land. In a very traditional way, Zen Center engages the landscape for weddings and memorial sites. Ashes of Zen Center elders—Gregory Bateson, Alan Watts, and Alan Chadwick, among others—are buried on the hillside above the garden. Memorial trees or shrubs have been planted by the pond or in the garden for several dozen people, including Zen teachers Katagiri Rōshi and Maureen Stuart Rōshi. Silent ceremonial ridge walks, as distinct from natural history strolls or recreational hikes, are part of the Center's annual calendar on Arbor Day and New Year's. These place the Center in the larger landscape, meeting the nearby wild zone through the act of walking, receiving the land into the feet. In a similar way, walking meditation sacralizes the garden, bringing human attention to the cultivated space.

Green Gulch has also adopted specific ceremonies to acknowledge nonhuman members of the land- (and mind-) scape. On Earth

Day practice leaders offer ceremonies for animals and trees, acknowledging their presence in the community. In December 1995 a beloved coast live oak crashed to the ground after a severe windstorm; later, a Monterey pine near the meditation hall had to be taken down because of bark beetles. On each occasion an altar was set up near the tree, and people were encouraged to offer incense and to include the dead or soon-to-be-dead tree in their practice.[35] I interpret this as an invitation to practice with the wild energy flow of death and destruction.

Last, in considering this third element of Snyder's ecological ethic, I suggest that practices of *simplifying* the institutional schedule and life-style promote contact with the energy flow that sustains life. Many of the traditional Zen practice forms emphasize restraint and moderation. Sensory impact from mechanical noise and bright lights is minimized; *zendo* clothing is dark and unobtrusive. Guest students are expected to maintain silence from early evening through breakfast the next day. During one-day and seven-day retreats, students remain silent the entire time, and the voices of great-horned owls, ocean waves, and blowing wind define the soundscape. To conserve energy and also darkness, Green Gulch has restricted night lighting to what is necessary for minimal safety needs. This leaves the hills dark and unmarked by human light sources, the night animals undisturbed by human presence.

Taken together, these institutional practices in all three aspects of Snyder's ecological ethic generate tangible evidence of a Buddhist practice response to the land at Green Gulch. Offerings of gratitude, commitments of responsibility in several arenas, and regular contact with the energy flow of the wild in the "valley of the ancestors" load the odds for transmitting ecological culture and moving toward reinhabitation. Graced by the rolling hills to the east and west and by the wild ocean to the south, Green Gulch Zen Center is in a strong position to promote an ecological land ethic as an institution and emerging culture for those who come to visit. These practices can be kept vital and evolving with support from those in leadership positions and with ongoing community involvement in environmental issues.

Spirit Rock Meditation Center

Though Spirit Rock Meditation Center does not have the same length of history on the land as Green Gulch, its ecological practices draw on well-established traditions of one of the oldest Buddhist denominations of Southeast Asia. The relationship with the land at Spirit Rock, in its very newness, is still in a honeymoon stage, growing and flourishing as the center attracts more practitioners. Much of the fundraising for the land purchase was motivated by a spontaneous bonding with the land for those leading the effort.[36] With more and more students using the land for retreats, the "falling in love" process seems to be multiplying and self-reinforcing.

Looking first at the element of "feeling *gratitude* to all," two core practices at Spirit Rock appear to support this element of Gary Snyder's ecological ethic. One-, seven-, ten-day and three-month retreats emphasize attentiveness practice, as described in the *Satipatthana Sutta* (the Four Foundations of Mindfulness), and mindfulness of breathing (*Anapanasati Sutta*). Guided meditations support practitioners in cultivating subtle awareness of mental and emotional states as well as sensory alertness. Gratitude practice naturally arises in relationship to food as attention to flavor, preparation, and source are noted with each meal. Vietnamese Zen teacher Thich Nhat Hanh has led several day-long meditation retreats at Spirit Rock, each with an elaborate guided eating meditation. Tangerines or apples are distributed to crowds of up to one thousand who may take up to an hour to appreciate the many causes and conditions arising in a single piece of fruit.[37]

Another major practice at Spirit Rock is the loving kindness meditation (*Metta Sutta*). At the close of each retreat day or class, some form of loving kindness meditation is recited. Many of the Spirit Rock teachers have extended the traditional meditation verses to include the land, the animals and trees of the land, and the gifts of sun and rain. Expression of gratitude takes the form of wishing for the safety, physical and mental well-being, and peacefulness of all members of the land community.

The second element of Snyder's ethic, *taking responsibility* for one's actions, has been central to the land purchase from the start. The Spirit Rock property had long been a prized piece of real estate in the valley; a number of other uses had been proposed for the property earlier. However, the citizens' San Geronimo Valley Planning Group, in their watchdog role of protecting open space and scenic landscapes, managed to prevent unsightly development along the Sir Francis Drake corridor. Negotiations for the Spirit Rock sale and planning design included important agreements about building sites, scale of operation, and stewardship for the land.[38] For the center to be a welcomed member of the West Marin community, Spirit Rock leaders needed to assure local residents of their commitment to protecting the integrity of the land.

The first decisions involved traffic management, both to limit congestion on the two-lane highway and to limit the amount of paved parking on the land. Early on, parking on the dry grass caused some spark-induced brushfires, alarming planners and reinforcing the need for careful attention to car placement. A carpooling policy was implemented by charging parking fees. Parking areas were laid out in curving tree-lined patterns to slow visitors down as they arrived. Center staff made consistent efforts to take responsibility for the potential impact on neighbors from car noise, increased traffic, and grassland fires.

Much of the land stewardship effort thus far has been directed toward careful planning of building projects. The Spirit Rock Design Committee and several architects meet regularly to discuss the scope and scale of the development vision for the land. Factors under consideration are relative invisibility of the buildings from the road, stream bank allowances, and impact on the stately coast live oaks which shape the character of the land. Temporary buildings for the office and meditation hall have been in place since 1990; a dining hall, the first construction project, was completed in 1995 to serve guests on retreat days. Future buildings will be added with additional funds and ongoing monitoring of the cumulative impact on the land and water systems.

Monthly work days are now part of the Spirit Rock tradition of land stewardship. In the beginning, volunteers pulled invasive star thistle and removed old fence posts and barbed wire from the pasture. They cleared brush and cut fallen trees for firewood. As part of one day's meditation, the teacher asked forgiveness of the plants, insects, birds, and animals for the disturbances to their homes. Heavy-labor tasks included digging trenches and sand pits for power, water, and phone lines as well as irrigation lines and a septic system. Many native trees were planted in the parking area and along the entrance road. Volunteers built bluebird boxes and posted them around the land. In the summer of 1995 several small ponds were excavated and dams built to retain the water. An altar and ceremonial area in Oak Tree Canyon were completed and a trail along the creek was marked out. The ponds are meant both for human enjoyment and as a water source for frogs, birds, badgers, raccoons, fox, deer, bobcats, perhaps even mountain lions.[39]

In the arena of *community relations*, Spirit Rock caretakers have continued to establish relationships with local neighbors and members of the San Geronimo Valley Planning Group. Though much of the land on the other side of the western ridge is publicly protected open space (Mount Tamalpais State Park and Marin County Water District), all the land adjacent to Spirit Rock is in private hands. In other rural situations in the United States, Buddhist and Hindu retreat centers have sometimes been resented as strange outsiders, bringing a new and not necessarily welcomed culture to the region. Spirit Rock teachers and staff have been consistent in their efforts to fit in with the local community and be cordial neighbors. This has been accomplished through community meetings, public hearings, and regular local contact with residents in the immediate area and nearby towns. Because center members are not versed in land practices, this has meant making a special effort to learn from those who know the territory, bringing in caretakers who could help with the transition from ranch to retreat center.

As part of taking responsibility for institutional actions, Spirit Rock is in the process of developing an *ecological culture* on the land. Though there are few residential staff at the moment (in contrast with Green Gulch), the number of staff and residents will increase as new buildings are added. Spirit Rock, like Green Gulch, is commited to vegetarian meals, thereby limiting their contribution to global environmental destruction caused by beef, chicken, and hog production. Recycling and composting systems have been set up to accommodate retreatants as well as residents and day guests. Fire safety protection is an important drill during the dry summer and fall months when fire danger is high.

To increase awareness of the land and promote a culture of ecological responsibility, Spirit Rock offers a number of *education* programs for children and adults. Volunteer naturalists lead nature walks across the diverse habitats of the four hundred acres, pointing out wildflowers and birds. Monthly children's programs explore the *dharma* teachings of the creek and oak trees. For several years, Spirit Rock has hosted an alternative *"Inter*dependence Day" on the Fourth of July, a chance to appreciate quietly the web of life with members of the spiritual community.

The Spirit Rock Center vision statement explains that the center "is being created as a living mandala: a western dharma and retreat center dedicated to discovering and establishing the dharma in our lives."[40] Six *Dharma* paths are described: retreats, right relationship, study, hermitage, integration in daily life, and service to the community. A practitioner can develop concentration, understanding, morality, and compassion through any or all of these paths. Cultivating right relationship includes people and also the earth; the service path is based on care and respect for all beings. This statement provides an introductory education on the founding principles of the center, which include respect for the land.

The third of Snyder's guidelines, *keeping contact with the sources of energy* that flow into one's life, is attended to at Spirit Rock primarily through walking meditation. Slow, careful walking practice, noting each step and breath, is a predominant aspect of *vipassanā* practice. At Spirit Rock, long periods of group walking meditation are practiced outdoors, offering opportunities for the feet and mind to absorb the wild energy of the land. One community member leads longer walking pilgrimages across Mount Tamalpais from Spirit Rock to Green Gulch. He specifically seeks to encourage the embodying of landscape knowledge through extended pilgrimage in local wild areas (as opposed to pilgrimages in Nepal or India).[41] Pilgrimage is also a way to bring members of the community together to share the experience of making contact with the land.

One of the six *Dharma* paths of the Spirit Rock vision is hermitage, offering the opportunity "to experience the simplicity and dedication of the renunciate life."[42] Though hermitage cabins have not yet been built at Spirit Rock, teachers encourage students to incorporate hermitage principles in everyday life through simpli-

fying consumer habits, spending more time in silence, and highlighting *dharma* study. The hermitage path is perhaps the path of
minimum impact and maximum exposure to the other plants and
animals inhabiting the land. With this as part of the master plan,
the center has built into its practice expectations the possibility that
deeper, longer-term connection with the land will develop through
hermitage retreats by senior students.

Taken together, these institutional practices, reflecting the three
aspects of Snyder's ecological ethic, show evidence of an emerging
Buddhist ecological culture in response to the land. Offerings of
gratitude, commitments of responsibility to mindful stewardship and
community relations, and contact with the energy flow of the wild
are helping to establish this center as an environmental model for
Buddhist practice. Held by the forested ridges to the south and the
open grasslands to the north, Spirit Rock presents another strong
opportunity for deepening ecological relations in a practice setting.
With the efforts of both centers contributing to the culture of
northern California, it is possible that American Buddhism can have
a significant influence on environmental practice and reinhabitation
in this region. This process, however, is not without its points of
tension.

Points of Tension

Though both of these centers now include certain ecological
practices as part of their religious cultures, neither is specifically
committed to the goal of ecological sustainability or self-reliance.
This degree of reinhabitation would stretch the capacities of staff
and residents beyond their current loads. For both centers the top
priority is to transmit Buddhist teachings and provide a supportive
place to practice. It is simpler and more convenient to depend on
external sources for food, energy, supplies, and funding. The choice
to draw on diverse trade sources, however, often involves certain
advantages of class and cultural privilege. Can reinhabitation take
place if residents are primarily dependent on goods produced away
from the land?

If ecological sustainability were to become an institutional goal,
debates would arise over how to *use* the land: could the open space
areas remain protected given the need to grow more food? Much

of the current attractiveness of both places depends on the sense of spaciousness from undeveloped land. This provides a kind of literal "breathing room" from the urban pressures of noise, pollution, and population. However, this aesthetic use of the land might be threatened by the choice to move further toward reinhabitation.

Buddhist centers in the United States and elsewhere have the opportunity to apply Buddhist analysis and self-study to their own institutions. Green Gulch and Spirit Rock have already done this in examining governance and economic structures and student-teacher relations. To do the same depth of work around ecological matters would mean investigating institutional habits around the relationship between nature and culture. To what extent do American Buddhist centers reproduce the dominant cultural attitudes of culture as superior, nature as inferior; culture as control, nature as chaos; culture as male, nature as female?[43] At Green Gulch this is manifest in giving weight and value to *zazen* meditation over ecological work practice. Farm and garden workers are seen by some as inferior to those who spend more time in the *zendo*, even though this is not supported by the teachings.

Another area of tension is around the need for community. In indigenous cultures, inhabitation goes hand in hand with culture and community. Generation after generation inhabits the same land, passing on knowledge of place through culture and social inter-action. Religious centers such as Green Gulch and Spirit Rock are explicitly *not* permanent communities but rather learning or training centers where people stay for different lengths of time. Can ecological culture be transmitted by example, if not through successive generations? There is a built-in conflict here: the more a practitioner engages in environmental work or contact with the land, the more he or she participates in a sense of community with others sharing the same experience. This leads to the desire to become a more permanent resident on the land—a move toward reinhabitation. However, because of the land's limited carrying capacity, this can constrict others from having access to the place at the same level of commitment. How can these religious centers serve as transmitters of ecological culture and values without the generational element of residential community?

Perhaps one of the most difficult questions lies in governance: who carries the burden of landownership and ecological stewardship? Legally, it is the board of directors and the staff they hire who are responsible; spiritually, the leadership role falls to the abbot and practice leaders. In contrast to the single head-of-household owner who makes most decisions for an individual piece of private property, the governing bodies of Green Gulch and Spirit Rock handle land responsibilities in diffuse arenas with various people carrying pieces of the land's history, capability, and management needs. Ecological monitoring is uneven and primarily related to human needs (water, wood, garden spaces, farm produce). Long-term planning for restoration of degraded habitats and expanded human use has been discussed informally but not incorporated into master plans for the sites.

Challenges for the Future

This evaluation documents ecological practices at two of the larger Buddhist centers in the San Francisco Bay area. Though some steps have been taken toward reinhabitation, many areas of ecological stewardship still need attention. In the course of this study, I have noted some of the immediate needs as well as future institutional challenges which are unresolved at present.

Green Gulch Zen Center and Spirit Rock Meditation Center both face issues of carrying capacity as they become increasingly attractive to students of Buddhism. This will require a closer look at pressures on parking spaces (always full on Sundays at Green Gulch and on Monday evenings at Spirit Rock), considering whether to limit attendance or pave more land to accommodate cars. Pressures on sewage, water, and energy sources will also rise with increasing numbers of visitors. Green Gulch, for example, may need to hold fewer programs and conferences in the fall when water supplies are at their scarcest.

Land-management issues already plaguing other parts of Marin County may sooner or later become problems for these two properties. Among these are the spread of feral non-native pigs who gouge the land and root up acorns and seedlings. On some nature preserves they are systematically hunted to prevent encroachment. This problem will likely affect Spirit Rock sooner than Green Gulch, but with so much open space connecting the two, it may be only a matter of time before the pigs are on the coast as well. Fire management is also an issue since coastal scrub, grassland, and coastal forests have evolved with fire in the California landscape. Fire suppression around human habitations often only postpones the inevitable. Both centers, as environmental stewards, will need to consider controlled burns or other fire-management methods to reduce fuel load.

People at Green Gulch are already raising questions about extensive stands of non-native trees on the property. The acacias in particular are quite fire-prone and present some danger to the adjacent dining area.[44] In earlier rounds of tree planting, Monterey pines were chosen to hold the soil and generate fast-growing poles and firewood. Locals have criticized these trees as non-native to the northern coastal regions as well as subject to bark beetle infestation. The prominent Australian eucalyptus, appreciated by many for its hanging strips of bark, drips oils that poison the soil below, reducing the biodiversity under these trees. Which of these trees should come out? Which should remain? Taking responsibility in this case means asking difficult ethical and ecological questions.

Both centers have small creeks on the land, though Green Gulch Creek is the larger and more managed. Water quality and aquatic habitats will need to be monitored, especially where dams impound water and holding basins have become clogged with silt. Waterways are natural corridors for songbirds and small mammals and can easily be enhanced to serve their food and shelter needs by allowing understory plants and aquatic insects to flourish. As for larger scale challenges, some of these will require creative initiative from either residents or guest/lay members to encourage a developing environmental conscience. In her book, *Campus Ecology*, April Smith outlines key areas for academic institutions to evaluate their ecological practices.[45] Many of these are applicable to religious institutions such as Green Gulch Zen Center and Spirit Rock Center. In the arenas of waste and hazard management, these two centers can work toward reducing the volume of solid waste beyond what is composted or recycled. This means attention to precycling, or choosing products with little or no packaging. It also means providing adequate disposal of potentially hazardous substances, such as used batteries, old tools, paints and solvents, autoshop chemicals, and concentrated organic pesticides.

More work can be done in the area of resource flow and infrastructure. While water is closely monitored at Green Gulch, energy use is dispersed and responsibility for energy conservation is uneven. Electrical heaters are often left on when rooms are empty. Food flows are managed closely at both centers to save money and as part of a commitment to vegetarian meals. Although perhaps half of the produce eaten at Green Gulch is organic, the center could in the future commit to an entirely organic menu, supporting local farmers as much as possible. As hazards from chemical agriculture are documented, particularly as hormone disrupters and immune system depressers,[46] one of the greatest supports to practitioners at both centers might be safe and healthy food.

Smith advocates institutional procurement policies to streamline product use, especially for recycled paper products in restrooms and offices. Both centers could make the choice to buy unbleached paper where possible, to minimize chlorine and dioxin hazards to users. Both centers currently have reusable dishware, eliminating the waste of disposable cups and plates; residents at Green Gulch are debating the option of cloth napkins and personal cups. For picnics and outdoor celebrations, the centers could encourage people to bring their own flatware and dishes, rather than using paper or plastic products.

As each center grows, their budgets grow. Funds are banked in institutions or held in stocks and bonds. The boards of these two centers can promote and implement a policy of socially responsible investing, to carry institutional weight into the arena of greening financial management. Taking responsibility at these levels will require more committee work and more volunteers helping the institutional structures evolve in their ecological ethics. As this work is engaged, it will be important for the centers to publicize their efforts among their own members as well as visitors to generate support and solidarity for this ecological work.

Buddhist Centers as Ecological Role Models

This first piece of comparative research on two Buddhist centers raises many interesting questions which will require additional case study work with diverse centers. Future research may include reviews of ecological practice at some of the following institutions: Rochester Zen Center (New York), Mt. Tremper Zen Center (New York), Karme Chöling Tibetan Center (Vermont), Manzanita Village (California), Shambhala Center (Colorado), Mountains and Rivers Temple (California), and others.[47] At this point it is unclear whether ecological practices are primarily motivated by Buddhist tradition or by American environmentalism. Will ecological culture become a mark of American Buddhism? It is also unclear how ecological practice relates to meditation practice and other aspects of Buddhist training in the specific centers. In future work, I would like to find out which aspects of Buddhism, as taught or practiced at individual centers, actually *dis*courage the evolution and adoption of ecological culture.

If institutions such as Buddhist retreat centers are to become more ecological in practice and concerns, upon what elements does such an evolution depend? Some possible significant factors may be: 1) the role of center leadership in establishing ecological priorities; 2) the creativity and efforts of key staff people; 3) the degree of teaching emphasis on the role of the environment; 4) methods for preserving and transmitting religious and cultural traditions; 5) the practice place itself and its ecological history and management needs; 6) outside development pressures. Some of these may be operational for certain centers but not for others; each center will have a distinct and complex story of environmental involvement. By examining both rural and urban centers, centers from diverse Buddhist traditions, and centers of different scale and leadership patterns, I may then be able to discern some patterns of ecological practice.

From this preliminary review of these two centers, it seems clear that Green Gulch Zen Center and Spirit Rock Meditation Center are beginning to demonstrate what is institutionally possible in living an ecological ethic. Religious centers in the past have served as role models for the wider community; perhaps these Buddhist centers can show others in Marin County and the wider Bay area how people can live more simply and environmentally. By offering gratitude, taking care of the land effectively, and keeping access open to the wild energy flow of the land, these centers support the very foundations of *dharma* practice. Working together as Buddhist neighbors and institutional *kalyana mitta* (spiritual friends), they can encourage others to act in environmentally responsible ways for the health of humans and nonhumans on the land. Over time, the incorporation of ecological culture into the everyday life of these centers may inspire visitors to transfer these practices to other institutions and households. Thus, seeds of ecological culture based in spiritual practice can support the beginnings of reinhabitation, drawing on the energy flow that sustains all life.

Notes

1. I am grateful to Kenneth Kraft and Wendy Johnson for their comments on earlier drafts of this article.

2. Sidney Piburn, ed., *The Dalai Lama: A Policy of Kindness* (Ithaca, N.Y.: Snow Lion Publications, 1990); see also, among Thich Nhat Hanh's many books, *For a Future to Be Possible* (Berkeley: Parallax Press, 1993) and *Being Peace* (Berkeley: Parallax Press, 1987).

3. For example, see Joanna Macy's treatment of Buddhist philosophy in *Mutual Causality in Buddhism and General Systems Theory: The Dharma of Natural Systems* (Albany: State University of New York Press, 1991).

4. See, among others, Lester Brown, "Launching the Environmental Revolution," *State of the World 1992* (New York: W.W. Norton, 1992), 174–90; and Herman E. Daly and John B. Cobb, Jr., *For the Common Good* (Boston: Beacon Press, 1989).

5. Michael Wenger, "History of Zen Center," *Wind Bell* 27, no. 1 (spring 1993):15–17.

6. Shunryu Suzuki, *Zen Mind, Beginner's Mind* (New York: Weatherhill, 1970); Edward Espe Brown, *The Tassajara Bread Book* (Boulder, Colo.: Shambhala Books, 1970).

7. Wendy Johnson and Stephanie Kaza, "Landscape Ecology and Management Concerns at Green Gulch Zen Center: A Report to the Zen Center Board of Directors," 5 November 1991.

8. Interview and site visit with Wendy Johnson, Green Gulch garden staff, June 1992.

9. Meredith Moraine and Jerry Steward, "The Story So Far," *Spirit Rock Meditation Center Newsletter*, September-January 1995, 5.

10. *Spirit Rock Meditation Center Newsletter*, February-August 1996, 4.

11. "Sangha of 1000 Buddhas," *Spirit Rock Meditation Center Newsletter*, February-August 1996, 10–11.

12. These have been catalogued in various taxonomies; see, for example, Warwick Fox, *Toward a Transpersonal Ecology* (Boston: Shambhala Books, 1990); and Steven C. Rockefeller, "Principles of Environmental Conservation and Sustainable Development: Summary and Survey," prepared for the Earth Charter project, April 1996.

13. Holmes Rolston III, *Philosophy Gone Wild* (Buffalo, N.Y.: Prometheus Books, 1989), 111.

14. Valerie Plumwood, *Feminism and the Mastery of Nature* (New York: Routledge, 1993).

15. Reed F. Noss and Allen Y. Cooperrider, *Saving Nature's Legacy: Protecting and Restoring Biodiversity* (Washington, D.C.: Island Press, 1994); and R. Edward

Grumbine, *Ghost Bears: Exploring the Biodiversity Crisis* (Washington, D.C.: Island Press, 1992).

16. David Abram, *The Spell of the Sensuous* (New York: Pantheon Books, 1996).

17. Aldo Leopold, *A Sand County Almanac* (New York: Oxford University Press, 1949; reprint New York: Ballantine Press, 1970); all quotes here are from the paragraph on pp. 239–40.

18. Described in Ray Dasmann's paper, "National Parks, Nature Conservation, and 'Future Primitive,'" presented at the South Pacific Conference on National Parks, Wellington, New Zealand, February 1975.

19. Gary Snyder, "Reinhabitation," in *A Place in Space* (Washington, D.C.: Counterpoint Press, 1996); first published in *The Old Ways* (San Francisco: City Lights Books, 1977), 191.

20. Ibid., 188.

21. Ibid., 190–91.

22. See Macy, *Mutual Causality in Buddhism,* chapters 2, 3, 10, and 11; also Francis H. Cook, *Hua-Yen Buddhism: The Jewel Net of Indra* (University Park: Pennsylvania State University Press, 1977).

23. See Elizabeth Roberts and Elias Amidon, eds., *Earth Prayers* (San Francisco: Harper, 1991), 120–21, for one environmentally based version of the precepts prepared for Earth Day 1990 at Green Gulch Zen Center.

24. Daily service and meal chants provided by the office of the *Eno* (Head of *Zendo*) at Green Gulch Zen Center.

25. Stephanie Kaza, "A Community of Attention," *In Context* 29 (summer 1991):32–35.

26. Annual wildflower lists on file with and prepared by Wendy Johnson, garden staff, Green Gulch Zen Center.

27. Annual records and documentation of Arbor Days provided by Wendy Johnson.

28. Johnson and Kaza, "Landscape Ecology and Management Concerns at Green Gulch Zen Center," 3–5.

29. Site visit with Peter Rudnick, Head of Farm, June 1995.

30. Ibid.

31. Some of the principal contacts in these consultations have been Mia Munroe, Muir Woods National Monument park ranger; Yvonne Rand, Zen teacher; and Wendy Johnson, Green Gulch garden staff.

32. Green Gulch site visit, June 1992.

33. Prepared by Stephanie Kaza in consultation with Green Gulch staff; brochure available in Green Gulch office.

34. See collections such as *The Mumonkan* (various translations) and *Book of Serenity*, trans. Thomas Cleary (Hudson, N.Y.: Lindisfarne Press, 1990).

35. Wendy Johnson, "Sitting Together under a Dead Tree," *Wind Bell* 30, no. 2 (summer 1996):34–36.

36. Meredith Moraine and Jerry Steward, "The Story So Far," *Spirit Rock Meditation Center Newsletter*, September-January 1995, 5.

37. *Spirit Rock Meditation Center Newsletter*, February-August 1996, 2.

38. *Spirit Rock Meditation Center Newsletter*, September-January 1995, 5.

39. Ibid., 3, 9.

40. Vision Statement for Spirit Rock Meditation Center, 1995, 1.

41. Dharma Aloka, "Pilgrimage Here and Now," interview by Anna Douglas, *Spirit Rock Meditation Center Newsletter*, February-August 1996, 12–13, 16. Dharma Aloka describes the walking: "The ritual nature of formal pilgrimage sets it apart from everyday life. It's a kind of liturgical drama enacted in a sacred landscape."

42. Vision Statement for Spirit Rock Meditation Center, 1995, 3.

43. See Sherry Ortner, "Is Female to Male as Nature Is to Culture?" in Michelle Zimbalist Rosaldo and Louise Lamphere, eds., *Woman, Culture and Society* (Stanford, Calif.: Stanford University Press, 1974), and many subsequent feminist theory articles discussing her assertions.

44. Johnson and Kaza, "Landscape Ecology and Management Concerns at Green Gulch Zen Center."

45. April Smith and the Student Environmental Action Coalition, *Campus Ecology* (Los Angeles: Living Planet Press, 1993).

46. See new evidence gathered in Theo Colborn, Dianne Dumanoski, and John Peterson Myers, *Our Stolen Future* (New York: Dutton, 1996).

47. Also see Jeff Yamauchi's article on "The Greening of Zen Mountain Center: A Case Study," included in this volume.

The Greening of Zen Mountain Center: A Case Study

Jeff Yamauchi

Introduction

We have seen, during recent years, more American Zen centers making efforts to incorporate an environmental ethic into their communities. An environmental ethic appears, in theory, well suited to Zen Buddhism, as Zen advocates a sensitivity toward all life and encourages restraint, moderation, and simplicity. It is still worthwhile, however, to see how an American Zen center actually applies its practice in relation to environmental concerns.

I have chosen Zen Mountain Center of Mountain Center, California, as the case study site for several reasons. I have resided at the center and have served during the past two years as one of the principal participants in the center's environmental program. The natural setting of Zen Mountain Center provides an ideal location for promoting outdoor education and other related activities that foster an appreciation for the environment. In developing an environmental program, the center's head administration is particularly concerned with preserving the integrity of the property and is willing to take steps to protect its native beauty. A general attitude of low environmental impact has always been the approach taken in on-going development of the center. A stewardship approach, however, was, until very recently, more one of implication than one of operational policy. Zen Mountain Center is on the verge of implementing an environmental program that is, in my opinion, both unique and wide-ranging in application. Thus, Zen Mountain Center has the potential to become a significant advocate in developing and promoting an environmental ethic within American Zen. It is perhaps the best example of an American Zen center that is attempting to "green" its practice.

Zen Mountain Center also offers a fine representation of American Zen in general. The center was founded in 1979 by one of the early Japanese Zen masters (*rōshi*), Hakuyu Taizan Maezumi, who taught for over thirty years, until his death in 1995. Maezumi Rōshi was one of the important pioneers who initiated and helped to establish Zen in the United States, Europe, and Mexico. Zen Buddhism has been a part of the American religious landscape for a longer time and more extensively than other Buddhist traditions. The history of the integration of Zen in America, therefore, may lead to a better understanding of Buddhism's impact on Western cultures. As I hope to show in this case study, through its active role in integrating Zen Buddhism and environmentalism, Zen Mountain Center may serve as an indicator of a general trend in American Zen.

The primary purpose of this essay, however, is to focus on Zen Mountain Center itself, rather than to examine in depth any future directions of American Zen that may occur. I hope, through an environmental assessment of this particular Zen center and a concluding proposal, to demonstrate the viability of American Zen as one religious path that could be taken to address the environmental crisis.

An Overview of Zen Mountain Center

Zen Mountain Center is located at the head of Apple Canyon in the southwestern slopes of the San Jacinto Mountains of Southern California, at an elevation range of 5,440 to 6,800 feet. Apple Canyon is a tributary to the south fork of the San Jacinto River. The beginning of the watershed is less than a mile upstream from Zen Mountain Center at the Desert Divide, a prominent ridge which divides the western and eastern slopes of the San Jacinto range.

Three general types of soils have been described in Apple Canyon: "Wind River medium, sandy loam, 2–15%, well-drained alluvial fan soils from granitic bedrock; Lithic xerothent, rock outcrop complex, 50–100% slopes, depth to hard, unweather rock is less than 20 inches; and rock outcrop, 30–100% slopes, contiguous bare bedrock with less than 15% inclusions of soil capable of supporting plants."[1] The sandy loam is associated with the canyon bottom, while the less decomposed rocky soils are primarily on the slopes of the canyon.

The property of Zen Mountain Center—160 acres (or a quarter section)—contains a mosaic of habitats: riparian, rock outcrops, meadows, montane chaparral, oak woodlands, and mixed conifer forests. In addition, much of the property of the center is relatively undisturbed. In fact, a substantial portion of the adjacent land is federally designated wilderness. The variety and relatively intact nature of the landscape in and around Zen Mountain Center supports a rich diversity of flora and fauna. A detailed biological impact report of the center lists as present in the area 216 species of plants, 63 species of birds, 24 species of mammals, and 16 species of reptiles and amphibians.[2]

Besides a few private residential homes, Zen Mountain Center and Pine Springs Ranch (a large retreat and conference facility one mile south of Zen Mountain Center, operated by the Seventh Day Adventist Church) impose the most significant human impact in Apple Canyon. Nearby communities of Mountain Center, Garner Valley, Pine Cove, and Idyllwild comprise the majority of the population in the general vicinity. The center is thus in a secluded location, even though Los Angeles and San Diego are only about one hundred miles east and south, respectively.

The relative isolation of Zen Mountain Center contributes, in part, to the rich diversity of species. Moreover, a significant number of rare, endangered, or sensitive species have been observed or are known to be present on the center's property (see table 1). Many of the thirty rare species in the vicinity of Apple Canyon are sensitive to human disturbances. A biological survey has identified seven rare animals (the spotted bat, northern San Diego pocket mouse, California spotted owl, mountain quail, northern goshawk, southern sagebrush lizard, and the San Diego mountain kingsnake) and two rare plants (Johnston's rock cress and the California penstemon) within the center's property. Indicator species, such as the California spotted owl, are typical and reflect the condition of a mature conifer forest.

The biological impact report gives clear and substantial evidence that the property of Zen Mountain Center is located in a rich habitat

TABLE 1: RARE, ENDANGERED, OR SENSITIVE SPECIES OBSERVED OR
EXPECTED TO OCCUR NEAR ZEN MOUNTAIN CENTER (ZMC)

SPECIES	STATUS*	NEAREST KNOWN LOCATION
MAMMALS		
Peninsular bighorn sheep *Ovis canadensis cremnobates*	C2,CE,NF	<10 miles from ZMC
Spotted bat *Euderma maculatum*	**C2,NF**	**observed at ZMC**
Pacific western big-eared bat *Plecotus townsendii townsendii*	C2	<5 miles from ZMC
San Bernardino northern flying squirrel *Galucomys sabrinus californicus*	C2	<10 miles from ZMC
Northern San Diego pocket mouse *Perognathus fallax fallax*	**C2**	**observed at ZMC**
BIRDS		
California spotted owl *Strix occidentalis*	**C2**	**observed at ZMC**
Southern bald eagle *Haliaeetus leucocephalus*	FE,CE	<5 miles from ZMC
Northern goshawk *Accipiter gentilis*	**C2,NF**	**observed at ZMC**
Mountain quail *Oreotyx pictus*	**C2,NF**	**observed at ZMC**
REPTILES AND AMPHIBIANS		
San Diego horned lizard *Phrynosoma coronatum blainvillei*	C2	<5 miles from ZMC
Coastal western whiptail *Cnemidophours tigris umbratica*	C2	<5 miles from ZMC
Southern sagebrush lizard *Sceloporus graciosus* *vandenburgianus*	**C2**	**observed at ZMC**
San Diego mountain kingsnake *Lampropeltis zonata*	**C2,CR**	**observed at ZMC**

* FE = Federally endangered; C1 = Federal candidate 1; C2 = Federal candidate 2;
CE = California endangered; CT = California threatened species; CR = California
rare; CNPS = California Native Plant Society list 1B; NF = National forest sensitive
species.

San Diego ringneck snake *Diadophis punctatus similis*	C2	<5 miles from ZMC
Southern rubber boa *Charina bottae umbratica*	CT	<5 miles from ZMC
Large-blotched ensatina *Ensatina eschscholtzii klauberi*	C2	<10 miles from ZMC
Mountain yellow-legged frog *Rana muscosa*	C2,NF,CR	<10 miles from ZMC

PLANTS

California bedstraw *Galium californicum ssp. prinum*	CNPS,NF	<10 miles from ZMC
California penstemon ***Penstemon californicus***	**C1,CNPS,NF**	**observed at ZMC**
Hall's Monardella *Monardella macrantha var. hallii*	CNPS,NF	<5 miles from ZMC
Hidden Lake Blue Curls *Trichostemma austalmonatum ssp.* *compactum*	C1,CR,CNPS	<10 miles from ZMC
Johnston's rock cress ***Arabis johnstonii***	**C2,CNPS,NF**	**observed at ZMC**
Lemon lily *Lilium parryi var. parryi*	C2,CNPS,NF	<10 miles from ZMC
Munz's hedgehog *Echinocereus engelmannii* *var. munzii*	C2,CNPS,NF	<5 miles from ZMC
Parish's chaenactis *Chaenactis parishii*	C2,CNPS	<5 miles from ZMC
San Jacinto spiny phlox *Leptodactylon jaegeri*	CNPS,NF	<5 miles from ZMC
Shaggy-horned alum root *Heuchera hirsutissima*	CNPS	<5 miles from ZMC
Slender-horned spine flower *Dodechama leptoceras*	FE,CE,NF	<15 miles from ZMC
Tahquitz ivesai *Ivesia callida*	CR,NF	<5 miles from ZMC
Ziegler's tidy tips *Layia ziegleri*	C1,CNPS,NF	<1 mile from ZMC

Adapted from Michael Hamilton, *Biological Impact Report of Zen Mountain Center* (Idyllwild, Calif.: Michael P. Hamilton and Associates, 1994), 40–43.

with an abundance of species. Southern California is notorious for losing, through urbanization and all manner of developmental projects, a large percentage of its natural habitats. This loss alone makes it even more imperative that the center at least consider the fragile uniqueness of Apple Canyon.

History of Land Use at Apple Canyon

The Cahuilla Indians first occupied the San Jacinto Mountains and surrounding areas about twenty-five hundred to three thousand years ago.[3] Periodic visits into Apple Canyon by Cahuilla occurred primarily during the months of October and November when the acorns were ready to harvest. Cahuilla families would camp beside groves of oaks, spending several weeks gathering the ripening acorns—their most important food staple. The nutritional value of acorns compares favorably with grains such as wheat and barley: though acorns are somewhat lower in protein and carbohydrates, they are higher in fat and calories. The oaks generally provided a reliable yield of acorns—up to several hundred pounds from each mature tree. A large boulder and several grinding mortars that the Cahuilla used for grinding the acorns into meal are located just south of the center's property. A grove of mature California black oaks (*Quercus kelloggii*) near the mortars offers further evidence that Cahuilla came to Apple Canyon to collect and process acorns for the winter months.

During the late nineteenth century, Apple Canyon was originally part of Thomas, then Garner, Ranch, which at one time consisted of ninety-five hundred acres.[4] From the 1800s to about the 1960s, cattle grazing occurred on what would become the property of Zen Mountain Center and adjacent areas. Although there was a substantial timber industry in the area at one time, only selected harvest of trees occurred in the upper reaches of Apple Canyon due, in part, to difficult access. There are now, within the property, scattered old-growth stands of Coulter pine (*Pinus coulteri*), Jeffrey Pine (*P. Jeffreyi*), and incense cedar (*Calocedrus decurrens*). A core sample taken from an exceptionally mature Jeffrey pine at Zen Mountain Center indicates its age to be about five hundred years.

Since the purchase of the quarter section in Apple Canyon in 1979, Zen Mountain Center has gradually grown into an intensive

Zen training center. This process started in 1982 with the first three-month-long meditation retreat (*ango*). The center has generally concentrated its practice in the summer months; the rest of the year remains relatively quiet, with only a small number of staff maintaining the buildings and grounds. Little impact to the environment has occurred during most of the tenure of Zen Mountain Center, primarily due to the minimal development of the property. The early buildings, for example, were only a bathhouse, kitchen, small meditation hall, and several outhouses. A few small trailers were added to house residents and guests. The earlier building complex was confined to a small area. When time, money, and appropriate personnel became available, buildings were constructed over the next ten years. The facilities now include a larger meditation hall, five cabins, the abbots' quarters, a workshop, a small dormitory, and a two-story bathhouse. With the exception of three small cabins, the building complex is situated on only three acres at the southern end of the property. The restricted location of human use has thus significantly lessened the impact on Apple Canyon and directly contributed to the continued vigor and health of the local environment.

Although economic constraints have slowed the development of Zen Mountain Center, an ecological sensitivity is a factor in considering the appropriate way to approach building a Zen center in the mountains. An article by an early resident of the center reflects this environmental "awareness":

> The primary form of sacred space in the Buddhist tradition has been the temple or monastery. Because it was built by man it could be located in different places. Generally they were built either in the city near the source of political power or in the mountains near another source of sacred power. The combining of two forms of sacred space, that of the temple and the natural one of the mountains, made a powerful center for practice.[5]

Apple Canyon has been viewed as an extension of the center, which has naturally fostered, with care and attention, both the buildings and the grounds. Minimal disturbance to the environment is the direct consequence of treating the canyon as a sacred place. The biological impact study conducted has substantiated that there has been minimal environmental impact to the center's property:

[D]isturbances apparently are minimal because of low noise levels, limited lighting, no hunting or trapping, lifestyle characteristics which favor biological diversity, and limited human visitations. . . . As a result, the biological diversity of the property is unusually rich.[6]

Even after seventeen years of occupation by the Zen Mountain Center community, the habitat within the center's property and its rich diversity of species has been preserved. Although not explicitly stated, a land ethic has certainly played a part in the development of Zen Mountain Center during its early years.

Stewardship Practice at Zen Mountain Center

The Zen Mountain Center mission statement has recently been revised, in part because of the need to devise a comprehensive approach for managing the center's property. The center's goals, as stated, are:

> To provide a supportive environment for teaching, training, and practice in Zen Buddhism. To incorporate sound ecological principles in the development and function of Zen Mountain Center. To ensure continuation of the Buddha Dharma for future generations.

During the process of revising the mission statement, a general consensus was reached by the center's community and members, then formally approved by the board of directors on 14 April 1996. The revised mission statement set the stage for the development of a stewardship practice that would preserve the natural integrity of the property. The empirical data on the rich diversity of habitats and species included in the 1994 biological impact report and biological inventory provided the Zen Mountain Center administrators with a base line from which to evaluate appropriate ways of managing the center's natural resource.

A registered forester was also recently hired to assist in developing an appropriate stewardship plan to sustain the health and diversity of the biota residing on the center's property. The center's overall goal was to leave over 90 percent of its property undisturbed. A comprehensive plan submitted by the forester in 1995 gave specific instructions for improving the vigor of the different habitats, reducing the high-risk fire potential, and enhancing wildlife habitat diversity.[7] Carried out over a six-year period, the stewardship program requires appropriate forest resource management practices that include thinning excessive vegetation, removing diseased and bug-infested trees, planting desirable conifers and oaks, and retaining standing dead trees (snags) and downed decaying logs as animal nesting sites. A grant of two thousand dollars has been received through the Stewardship Incentive Program (SIP) for the first stage of thinning accumulated vegetation that poses a serious fire risk. Sponsored by state and federal agencies, the intention of SIP is to encourage and assist private woodlot owners in actively managing their land in a sustainable fashion for themselves and for future generations.

The recommendation given first priority in the stewardship plan is the reduction of dense undergrowth, especially near buildings. Fire suppression has been the most significant human impact in Apple Canyon and the rest of the San Jacinto Mountains. United States Forest Service fire history maps reveal that no major fire has occurred on the property of Zen Mountain Center for at least fifty years. However, four fires have recently threatened Apple Canyon. In 1982, a large fire that originated on the desert side of the San Jacinto Mountains headed toward Apple Canyon from the north. Only updraft winds prevented the fire from coming down the ridge. Another fire in 1993 was caused by a careless hunter who left a smoldering camp fire just outside the center property. It was quickly extinguished by water dropped from a fire-attack helicopter. The following year, lightning strike caused another fire near Lake Hemet, about five miles south of Zen Mountain Center. A volunteer evacuation was initiated for all residents of Apple Canyon and adjacent May Valley. The Lake Hemet fire was contained in three days. Finally, more than ten thousand acres south of Idyllwild and Pine Cove were burned in early July 1996 during a fire that also endangered the towns. No lives or homes were lost in that fire, but it was serious enough to require an evacuation of both communities. These episodes offer clear evidence of the real danger posed by fire.

In meeting the first stewardship goal, six acres around the center's building complex have been thinned to act as a fire break. Scheduled clearing of dense undergrowth and deadwood will

continue to be done around the facilities, going beyond the pre-
scribed thirty feet to at least twice that distance or longer whenever
appropriate. Controlled burns in selected areas—given the right
supervision and conditions—are also a possible way to lessen the
fire threat. Fire can occur too readily, either by natural causes or
through human actions, and spread too rapidly, given the hazards
of the narrow canyon which, acting as a wind tunnel, can contribute
to an incredible amount of destruction. The restoration of the Apple
Canyon forest and replacement of old-growth trees following a fire
would take centuries. Thus, a strong commitment to fire prevention
is the primary element in developing a stewardship management
program for the center's property.

Another stewardship practice is to limit the growth of development. Only a small percentage of the property is slated for
development, and this is largely confined to three acres. The vast
majority of the center's land will be left undisturbed. Limited access
to the upper canyon (California spotted owl habitat) will remain
restricted, with no new trails added, in order to inhibit direct
disturbance to wildlife and plants.

The potential increase in the number of residents, students, and
guests will also have a significant impact and must be considered
when devising appropriate measures to limit adverse growth.
Currently, there are thirteen full-time residents at Zen Mountain
Center, with periodic visitations ranging from one to one hundred
individuals lasting from one day to three months. During the
summer months the numbers of visitors are highest, with an average
of about ten and, periodically, as many as twenty or slightly more.
The projected increase is up to thirty-six full-time residents, with
the range of numbers of visitors and length of stay probably
remaining the same. The average number of visitors, however, is
projected to climb. Adequate housing to accommodate the influx of
people will have to be considered in the overall complex design.

The full-time residents themselves will have the most impact and
will most significantly affect the development and functioning of
Zen Mountain Center. Most importantly, the center's residents set
the tone for stewardship behavior and, to a large degree, dictate
environmental policy because they essentially implement it. A
resident, for instance, may wish to have a cat or dog which, if
allowed to roam free, will have an impact on the wildlife in the

general area. Though at this point in time there have been no real deviations from an attitude of stewardship, guidelines may be needed in the future to outline explicitly precautions necessary to minimize the impact on the environment of the center's property and its inhabitants.

Zen Mountain Center can be viewed, in many respects, as a nature preserve, with 98 percent of the property currently undeveloped and sustaining the biodiversity of the center's property. The center's meditative activities readily lend themselves to a stewardship approach of property management, because only a small area needs to be developed for Zen training and because of the overall Zen perspective of causing as little harm as possible. In other words, demands on the environment need only be minimal for the center to function properly. The natural beauty of Apple Canyon actually enhances Zen training of contemplation and meditation. It is therefore in the center's best interest to protect and properly manage the habitats that support a rich and diverse biota.

A Proposal for an Environmental Program at Zen Mountain Center

I am in the process of incorporating a nonprofit educational course, to be known as "Earth Witness Foundation," that will focus on the development of an environmental program. The name, "Earth Witness Foundation," is derived from the moment when the Buddha touched the earth as a sign of validating his enlightenment. The comprehensive nature of the environmental program warrants an organization that will pay particular attention to carrying out its objectives effectively and appropriately. The primary purposes of this public benefit corporation, as stated in the bylaws of Earth Witness Foundation, shall be: 1) to provide environmental educational retreats and workshops that are contemplative in approach; 2) to provide indigenous educational workshops; 3) to sponsor special events, such as presentations by guest speakers and symposiums, that foster environmental awareness and ecological consciousness; 4) to publish environmental information on a quarterly basis and occasional texts; 5) to provide an open forum on the integration of religion and ecology, particularly with

Buddhism; and 6) to implement outreach programs that address environmental issues.[9]

The purposes of the environmental program are deliberately broad in scope to accommodate future objectives. The program is currently in the planning stage, and most of the projects are still in the process of being developed. Depending on the amount of commitment among the Zen Mountain Center's *sangha*, the actual implementation of the environmental program may take as long as two years.

The Earth Witness Foundation initiative, however, coincides with new leadership at Zen Mountain Center; the time for beginning an environmental program appears to be ideal. The center's administration is particularly open to innovative ideas that extend the function of the center beyond strictly formal Zen practice. The current dynamic atmosphere at Zen Mountain Center makes the environmental program more acceptable and potentially viable to the community as a whole, especially considering the center has already developed and implemented a stewardship plan.

The next step in the center's stewardship plan, I believe, is to develop a sustainable life-style at Zen Mountain Center that utilizes sound ecological principles and minimizes material consumption. When the community of Zen Mountain Center seriously tries to live a more sustainable life-style, stewardship will become a more engaging process for all. Although Zen Mountain Center cannot be totally self-sufficient, producing or making everything on its own, the community can promote positive changes toward sustainability that limit the impact to the environment.

There are numerous ways of reducing both the impact and demand of consumption. The following two examples are initial steps toward sustainability at the center; they will serve to give some indication of what Zen Mountain Center is doing to foster the so-called greening of its Zen practice.

Using a more renewable form of energy is one way to reach the environmental program's objective of sustainability. The center already has in place a photovoltaic (PV) system in tandem with a six kilowatt propane generator that provides electricity during heavy periods of use and the shorter winter days. Currently, the inadequate number of solar panels prevents full-capacity use of the sun as the source for electrical energy at the center. One of Zen Mountain

Center's goals is to have a PV system that will utilize the sun as virtually the only source of electricity. The key to success, in a PV system, however, is conservation of electricity, such as using low watt bulbs, turning off unnecessary lights, and restricting the use of electric appliances.

A second way for the community of Zen Mountain Center to become more sustainable is by organically growing some of its food. Located on the southwest corner of the center's property is a small apple orchard (twenty trees) and a vegetable garden (thirty-three feet in diameter). This is a modest beginning, but there is potentially an acre at the center suitable for gardening. The garden's produce would only be a food supplement and would be limited to the summer and early autumn months. Still, in order to raise a variety of fruits and vegetables, knowledge of the general life cycles of plants and familiarity with the local climate and growing conditions are necessary. Gardening, in other words, could be considered a successful method of educating residents and visitors about the immediate environment.

Sustainability thus plays a key role in developing an environmental program at Zen Mountain Center. The center's conscious efforts to reduce consumption and find some sustainable alternatives provide an environmental awareness that underlies Zen practice. The environmental program is also enhanced by the fact that Zen Mountain Center takes an active role in actually living according to its stewardship ideals.

Environmental Workshops at Zen Mountain Center

Only one workshop has been held as part of the recently proposed Zen Mountain Center environmental program, but a few more have been scheduled to take place in the near future. The main purpose of the environmental workshops is to foster an appreciation for the environment. Given the natural beauty of Apple Canyon and the sensitive way in which the center manages its land, environmental workshops seem the next logical step in promoting ecological awareness.

Because the workshops are still in the initial stages of development, a comprehensive and organized presentation of them is difficult to make. My intention in describing the workshops is to indicate the general direction the center may take in developing programs other than formal Zen training. Zen Mountain Center, I believe, is moving beyond the traditional boundaries of what a Zen center normally does. More environmentally oriented workshops can provide the center with opportunities to explore other avenues of education besides Zen training. The workshops described here only hint at the possibilities that could be adapted for the Zen Mountain Center environmental program.

Some of these workshops are on indigenous American arts. Indigenous American traditional methods and symbols contain not only native cultural elements but also express an affinity with their local environments. The indigenous arts workshops fit into the natural setting of the center: participants are able to share in developing a better understanding of and relationship to their environment while exploring their creativity and gaining a heightened awareness of the crafts and traditions of American indigenous cultures.

In a recent pottery workshop, for instance, Juan Quezada spent five days demonstrating and teaching to students his technique—one which originates in the Casas Grandes ceramic tradition. The pottery reflects his culture and represents the surrounding deserts of his home in Mexico. Quezada is the founder of a thriving ceramic industry in Mata Ortiz, Mexico. His work has been exhibited in the Heard Museum, the Lowie Museum, and the Southwest Museum in Los Angeles. Pottery workshops with Juan Quezada will become a yearly event. Another indigenous arts workshop that has been scheduled is on Cahuilla basketmaking, which demonstrates the craft of this local native culture. Cahuilla basketmaking requires native plants that are abundant at Zen Mountain Center. Identification of the plants and knowledge of the techniques used to make Cahuilla baskets will foster a greater appreciation of Cahuilla culture and the local environment.

A wilderness retreat in the backcountry of the San Jacinto Mountains is another upcoming environmental workshop. A three-day backpacking trip will combine mindful hiking, morning and evening meditation, and extended solo periods. The underlying concept of the retreat is to facilitate nature as the "teacher," while a contemplative state of meditation and silence enhances the insights gained during the time on the trail and in the forest. Four such retreats are scheduled to coincide with the four seasons. The retreats may take place at other locations besides the San Jacinto Mountains: possibilities include the nearby Santa Rosa Mountains, the Anza-Borrego Desert, and Joshua Tree National Park.

Environmental workshops may become another way to apply mindful Zen practice while encouraging a better appreciation and understanding of the natural world. These workshops will also introduce Zen Mountain Center to people who might not otherwise come. Scheduled Zen meditation instruction will continue to be offered during the workshops, but it will generally be optional or incorporated whenever appropriate. There has been enough interest to support the further development of such environmental workshops. I believe there is a place for these kinds of workshops and retreats in an on-going environmental program: not only will they provide an appropriate means of financial support for the center, but they will also contribute to diversifying Zen Mountain Center by accommodating a broader educational perspective.

Summary and Conclusion

Since the purchase of the quarter section of property at Apple Canyon, Zen Mountain Center has continued to evolve as a "green" Zen center. The greening of the center implies the existence of a stewardship management ethic. The biological impact report conducted in 1994 offers substantial evidence that the center has acknowledged, to a significant degree, the value of the biota of Apple Canyon by consciously minimizing the center's impact on the surrounding environment. The report is also a scientific document that reveals the richness of the habitats and species found on the center's property.

Zen Mountain Center's environmental ethic, in turn, has been developing, in a more coherent and organized fashion, into a vital part of its general policy. The center's revised mission statement, which incorporates sound ecological principles, is a reflection of the seriousness of its commitment to following appropriate environmental guidelines. A six-year stewardship plan proposed by a registered forester was implemented and is well past its first year.

There is, overall, a higher degree of sophistication in the management of the natural resources on the center property. Conscious efforts toward a sustainable way of life are also being made as Zen Mountain Center develops more fully into a Zen center that deeply considers environmental consequences, both morally and spiritually.

A proposal for creating an environmental program at Zen Mountain Center is the next step in the greening of the center. Though still in its inception, the environmental program appears promising. Although it is too early to make a general assessment of the center's environmental program, some conclusions can be drawn. Zen Mountain Center has an opportunity to develop a unique environmental program, given the natural setting of the center, its progressive administration, and the environmental elements in Zen. There are literally hundreds of American Buddhist centers that offer a variety of meditative practices, but only a very few of them have well-developed environmental programs—if they have such programs at all. Zen Mountain Center has the potential to be a Buddhist center deeply involved in integrating environmental concerns with its Buddhist practice.

Zen Mountain Center also has the potential to contribute significantly in the exploration of the connections between religion, nature, and the environment. Clearly, all the major religious traditions must come to terms with the current situation of increasing environmental degradation and destruction. The center already has a good track record of preserving the natural integrity of the land, and it is now beginning to find ways of fostering that attitude in the larger community. By continuing its strong emphasis on stewardship and environmental education, Zen Mountain Center can serve as an example for other American Buddhist centers that hope to begin greening their practice.

Notes

1. Michael Hamilton, *Biological Impact Report of Zen Mountain Center* (Idyllwild, Calif.: Michael P. Hamilton and Associates, 1994), 11.

2. Ibid., 60–99.

3. John W. Robinson and Bruce D. Risher, *The San Jacintos* (Arcadia, Calif.: Big Santa Anita Historical Society, 1993), 23.

4. Ibid., 64.

5. Carl Mugai Gibson, "Mountain Center: A Special Place to Practice," *Ten Directions* (Los Angeles: Zen Center Publications) 6, no. 1 (1985):5.

6. Hamilton, *Biological Impact Report*, 48.

7. James F. Bridges, *Forest Stewardship Plan for Zen Mountain Center at Apple Canyon, Riverside County, California* (Redlands, Calif.: n.p., 1995), 48.

8. Ibid., 45.

9. Jeff Yamauchi, *Bylaws of Earth Witness Foundation, A California Public Benefit Corporation* (Mountain Center, Calif.: Zen Mountain Center, 1996).

Applications of Buddhist Ecological Worldviews

Nuclear Ecology and Engaged Buddhism

Kenneth Kraft

This is an interesting time to be a Buddhist thinker or a thinking Buddhist, inside or outside the academy. Wherever one looks within the Buddhist tradition, one can find doctrinal tenets and forms of practice ripe for reinterpretation. One stimulus of current innovations in Buddhist thought and practice is the worldwide environmental crisis. Scholars and practitioners alike are asking: Is it possible to address contemporary ecological issues from a Buddhist perspective? Is it possible to transform Buddhism authentically in light of today's ecological challenges? One way to approach such questions is to take up a specific environmental problem. A particular issue redirects one's attention in unfamiliar ways because one has to grapple with alien disciplines, immerse oneself in concrete details, and cultivate new groups of colleagues. Yet that very process can reflect light back toward Buddhism.

Nuclear waste is one such problem. Fifty-plus years into the nuclear age, the disposition of nuclear waste has stymied industrial societies scientifically, technically, socially, politically, and ethically. Radioactive waste repels most people even as a subject for consideration, in part because the present formulations of the problem are stale, blocking both insight and action. We lack a fresh conceptual framework that incorporates the relevant resources and engages our imaginations. So I have been experimenting lately with the concept of *nuclear ecology*.[1] At first, nuclear ecology sounds like an oxymoron: radioactive materials are so flagrantly unecological that nuclear *non*ecology might be more plausible. Yet some apparent contradictions eventually make sense (such as engaged Buddhism or dependent origination, which still sound oxymoronic to the uninitiated). Putting the words nuclear and ecology side by side may spur us to consider nuclear realities in a larger context that incorporates present and future effects on the biosphere—in a word, ecologically. Ideally, potential threats to beings and ecosystems would be a first thought rather than an afterthought.[2]

As a field, nuclear ecology might also serve to integrate the disparate disciplines and individual roles required for the long-term management of nuclear materials. Observers concede the inadequacy of today's overcompartmentalized approaches:

> The issues involved require a greater understanding of physics, engineering, medicine, epidemiology, geology, economics, systems analysis, psychology, management techniques, and so on, than any individual can muster. So, in a sense, there are no experts, no individuals who have special insights into all the technical areas, let alone the nontechnical ones.[3]

Under the ecology rubric alone there are several subfields that pertain to nuclear materials but have never been consolidated in the service of nuclear-waste management. These include radiation ecology (also called radiobiology), applied ecology, industrial ecology, restoration ecology, and deep ecology. In recognition of the rights of future generations, a unified nuclear ecology should embody some vision of stewardship or guardianship, derived from secular or religious sources. Buddhism, with its "cosmic ecology"[4] and a range of other resources, may indeed have something to contribute.

Buddhist Responses to Nuclear Issues

Buddhists have been sensitive to nuclear issues for several decades, especially in North America. Concern about radioactive waste was prefigured by varied expressions of opposition to nuclear weapons, including marches across the United States, sit-ins at the United Nations, demonstrations at the Nevada Nuclear Test Site, individual acts of civil disobedience, and participation in local watchdog groups. In 1974, poet Gary Snyder wrote in his Pulitzer Prize-winning book *Turtle Island*, "No more kidding the public about nuclear waste disposal: it's impossible to do it safely."[5] Vietnamese

Zen master Thich Nhat Hanh speaks of nuclear waste as "the most difficult kind of garbage" and a "bell of mindfulness."[6] The Dalai Lama's five-point peace plan for Tibet, first announced in 1987, has an explicit antinuclear plank: it calls for "the abandonment of China's use of Tibet for the production of nuclear weapons and dumping of nuclear waste."[7]

The most influential Buddhist thinker-activist in this area is Joanna Macy, author of the concept of nuclear guardianship. Macy's ideas and example have inspired many, including me. Rather than shrink in dread from nuclear waste, she argues, we must take responsibility for it. Macy cultivates an awareness of future beings, imagining that one of their urgent questions to us might be: "What have you done—or not done—to safeguard us from the toxic nuclear wastes you bequeathed to us?" She proposes the creation of guardian sites, former nuclear facilities where radioactive materials are monitored in a manner that reflects a widely shared moral commitment to the task. Such sites might also have religious dimensions, serving as places of pilgrimage, meditation, or rituals associated with stewardship. The Nuclear Guardianship Project, a group led by Macy, flourished from 1991 until 1994. In study groups and public workshops, participants experimented with futuristic ceremonies that expressed the vision of guardianship. Although the Nuclear Guardianship Project has not developed organizationally, some of its ideas have circulated as far as the Energy Department's Office of Environmental Management.[8]

Several American Buddhist communities have incorporated concern about nuclear issues into their religious practice. In 1995, the Green Gulch Zen Center, north of San Francisco—probably the most active in this regard—staged an evocative multimedia ceremony-and-performance to commemorate the fiftieth anniversary of the bombing of Hiroshima and Nagasaki. Members of a small Zen group in Oregon became so determined to do something about the "poison fire" of nuclear waste that they added a fifth vow to the traditional four vows of a *bodhisattva*:

> Sentient beings are numberless; I'll do the best I can to save them.
> Desires are inexhaustible; I'll do the best I can to put an end to them.
> The Dharmas are boundless; I'll do the best I can to master them.
> The Poison Fire lasts forever; I'll do the best I can to contain it.
> The Buddha way is unsurpassable; I'll do the best I can to attain it.[9]

The intersection of Buddhism and nuclear issues has sparked varied works of art, often with an activist thrust. Mayumi Oda creates colorful prints and banners of goddesses and bodhisattvas protecting Earth from nuclear contamination. Kazuaki Tanahashi leads participatory art performances at government nuclear sites, using colored cloths or huge calligraphy brushes to create circles as large as a hundred feet in diameter. Tanahashi's circles allude to two disparate sources: the time-honored Zen circle (Japanese, *ensō*) that symbolizes oneness, and recent Energy Department documents that pledge to "close the circle" on the splitting of the atom.[10] There is also an expanding corpus of Buddhist-related poetry using nuclear themes. One of the earliest of these poems is by Gary Snyder:

> L M F B R
>
> Death himself,
>   (Liquid Metal Fast Breeder Reactor)
>   stands grinning, beckoning.
> Plutonium tooth-glow.
> Eyebrows buzzing.
> Strip-mining scythe.
>
> Kālī dances on the dead stiff cock.
>
>   Aluminum beer cans, plastic spoons,
> plywood veneer, PVC pipe, vinyl seat covers,
>   don't exactly burn, don't quite rot,
>   flood over us,
>
>   robes and garbs
>   of the Kālī-yūgā
>
>   end of days.[11]

In April 1994, about fifty American Buddhists commemorated Buddha's birthday at the Nevada Nuclear Test Site. The outdoor ceremony, created collaboratively the previous evening, included offerings at an altar, recitation of *sūtra*s, and circumambulation. One by one, participants expressed their concerns and their aspirations. A woman spoke tenderly of her stepfather, who as a young soldier had been forced to witness aboveground atomic tests. A college student stoutly declared, "I dedicate my life to working for the Earth and all beings." Others placed handwritten messages on the altar, silently pinning scraps of paper under rocks. The possibility of nonviolent civil disobedience was an integral part of the event because the walking meditation led right up to a boundary guarded by men in uniform. Some of the walkers deliberately stepped over the line and were arrested. The ceremony concluded with the following invocation:

> All merit and virtue that may have arisen through our efforts here, we now respectfully turn over and dedicate to the healing of this beautiful sacred land and to all beings who have been injured or harmed by the weapons testing on this place, so that the children of this world may live in peace free from these profane weapons, and thus may have their chance to realize the Buddha's Way.[12]

Mindfulness in a Nuclear Age

If one were to select the term most often used to characterize Buddhist practice in the West today, it would be mindfulness. This is the case not only in Buddhist circles but also in the popular press. A recent article in *USA Today*, cleverly entitled "Buddhism: Religion of the Moment," called mindfulness "the heart of Buddhist meditation. . . . the ability to live completely in the present, deeply aware and appreciative of life."[13] Such definitions are unobjectionable. However, in a nuclear or environmental context a practitioner may be prompted to ask: What is the scope of my mindfulness?

If mindfulness is misinterpreted, it may actually move in a direction away from environmental awareness. Misapplied, mindfulness can be used to shut out unwanted thoughts, feelings, or perceptions. A practitioner who focuses too narrowly on "living completely in the present, deeply aware" may unwittingly disregard the larger impact of his actions or others' actions. What would we say, for example, of an atom-bomb designer at Los Alamos who attentively follows his breath as he drives to work, or an official of the Nuclear Regulatory Commission who is "mindful" in her daily life but tolerates safety lapses at nuclear power plants?

At times mindfulness involves complexities and challenges that cannot be reduced just to living in the present. Socially and environmentally concerned Buddhists recognize that they must attend to breadth as well as breath, and that their breath connects them to their breadth.[14] Authentic practice aims continuously to

broaden and deepen the scope of mindfulness. In this spirit, Thich
Nhat Hanh attempts to connect mindfulness with nuclear waste:

> The most difficult kind of garbage is nuclear waste. It doesn't need
> four hundred years to become a flower. It needs 250,000 years.
> Because we may soon make this Earth into an impossible place for
> our children to live, it is very important to become mindful in our
> daily lives.
>
>   Nuclear waste is a bell of mindfulness. Every time a nuclear
> bomb is made, nuclear waste is produced. There are vast amounts
> of this material, and it is growing every day. Many federal agencies
> and other governments are having great difficulty disposing of it.
> The storage and clean-up expense has become a great debt we are
> leaving to our children. More urgently, we are not informed about
> the extent of the problem—where the waste sites are and how
> dangerous it can be.[15]

Thich Nhat Hanh does not explain at greater length how mindfulness
in daily life might apply to nuclear-waste problems. In this case,
being mindful could entail research on local sources of energy and
possible alternatives, efforts to alter one's own life-style and the life-
styles of others, broader political activism, and so on. The society-
wide vigilance required to keep radioactive materials out of the
biosphere now and in the future can also be seen as a kind of
collective mindfulness.[16]

Karma Isn't What It Used To Be

The far-flung effects of modern technology, in space and time, go
beyond any previous human experience. In many cases we really
have no idea what the consequences of our actions will be. Technology dilutes, amplifies, or camouflages the effects of action in
such complicated ways that ethical evaluation of action becomes
commensurately complex. One way to assess nuclear waste from a
Buddhist perspective would be to analyze certain issues in karmic
terms. For example, what are the karmic implications of creating
long-lived radioactive materials that put perhaps thousands of
generations of descendants at risk? From a scholarly standpoint or
a religious one, traditional understandings of *karma* do not readily

accommodate problems of this nature. Are we talking about the karmic import for us, for our descendants, for the environment, or for all of those? At the very least, previous thinking about karma needs to be extended or adapted.

In Buddhism's long history, understandings of karma have varied considerably. A classic definition of karma equated it with intention (Sanskrit, *cetanā*); this emphasis on intention was an important Buddhist departure from the mechanistic Hindu and Jain notions of karma prevalent at the time. Although some Buddhists seem to have believed that as long as their intentions were sound they were karmically in the clear, others have acknowledged that intention is complicated and at times problematic. A person may not really know what his or her intentions are in a given situation. Most behavior reflects multiple intentions. The intended effects of action and the actual effects of action often differ. The very best intentions can be thwarted or can cause harm. And sometimes people do the right thing even when doing so goes against their deepest intentions. So intention cannot be the whole story.

Among contemporary Buddhists in the West, karma principally signifies the moral implications of action, with the sense that causation mysteriously operates in the realm of ethics as well as the realm of physics. A basic tenet of engaged Buddhism is that— whatever one's intentions—it is not possible to follow a spiritual path in a social or political or environmental vacuum. While practicing mindfulness in daily life, even while meditating in a meditation hall, one's actions and nonactions continue to have wider repercussions. Sometimes, to our dismay, we realize that we are reinforcing large systems based on privilege and ecological blindness.[17] There is no such thing as a karma-free zone.

It was not uncommon in Asia to use beliefs about karma to *evade* responsibility ("It's their karma to be poor—why should I try to help them?"). However, according to other interpretations, karma enjoins a radical degree of responsibility: even though we cannot possibly know all the causes and conditions that have led us to be who we are, we have to take responsibility for our past and our present anyway. Karma can be seen positively as a recognition of the interrelatedness of all beings and phenomena. The work of bodhisattvas and aspiring bodhisattvas takes place in this realm of relatedness.

In all Buddhist cultures the primary arena of karmically significant action has been the individual: one's present situation is supposed to be the fruit of one's past actions, and one's future will be similarly conditioned by one's current actions. Although Buddhism holds that the laws of moral causation operate over vast spans of time, in practice the significant effects of karma were usually thought to extend over a few lifetimes at most. The operation of karma was considered to be orderly and relatively comprehensible—otherwise, the karmic worldview would lose its persuasiveness. So bad things can happen to good people, and vice versa, without necessarily destroying one's faith in some kind of cosmic system of justice.

Current nuclear and environmental problems challenge these assumptions in several ways. Where Buddhism has focused on individual karma, now we also need better ethical analysis of collective behavior. How might notions of group karma be rendered in modern terms? Our understanding of institutional discrimination offers a parallel: even if individuals do not have the intention to discriminate, the institution as a whole may function prejudicially. In that sense, entire systems can have intentions. In the face of systemic problems, engaged Buddhists seek ways to act effectively in groups. Karma theory must also be able to account for the relation between individual and collective responsibility. One analogy compares the simultaneous presence of these two kinds of karma to a doubly exposed photograph. Another analogy is a newspaper photograph: a field of dots (individual dimension) reveals recognizable patterns (collective dimension) from a proper distance.

Where Buddhism has focused on the immediate future, now we also need ways to account for the effects of our actions over time spans of geologic proportions. (Plutonium remains toxic for 250,000 years, or about 100,000 generations.) And where Buddhism has focused on seemingly comprehensible laws of moral cause-and-effect, now we also need to confront the increased opacity of moral consequences in a nuclear, technological age.

For karma doctrine to demonstrate relevance in contemporary contexts, it will have to survive some difficult leaps: from Asian cultures to Western cultures, from premodern eras to modern and postmodern eras, from religious milieus to secular and pluralistic milieus, from low-tech societies to high-tech societies. Admittedly, that is a tall order. If new understandings of karma are in the offing, will they involve changes in the application of long-standing principles, or changes in basic principles themselves? If the latter, it would not be the first time that cardinal tenets of karma doctrine have shifted. The early Buddhist focus on intention, an innovation at the time, has been noted. Another epochal revision, crucial to the development of Mahāyāna Buddhism, was the notion that good karma could be transferred to others, a striking abrogation of the ancient Indian belief that karmic retribution was inescapable.

Eco-karma

To rethink karma it will probably be necessary to develop some new concepts and terms. For example, the ethical implications of high-tech actions may differ from the ethical implications of low-tech or no-tech actions. Buddhist exegesis of high-tech ethics may call for a new category like *techno-karma*. By the same token, to illuminate the ethical dimensions of actions that affect the environment, a concept such as *eco-karma* may prove useful.[18] Today, we have a growing appreciation of the ways in which our past behavior has affected the biosphere, and of the ways in which our present behavior will shape the environment of the future. If, for example, we pondered global warming in light of eco-karma, we might be better able to address the ethical dimensions of the problem. Einstein may have (inadvertantly) enunciated a first law of eco-karma when he said, "Humanity will get the fate it deserves."[19]

As new terms are auditioned and defined, one of the tests will be their compatibility with prior Buddhist tradition. Initially, an expansion of karma in an ecological direction does not seem to conform very closely to Buddhism's past. Although Buddhists valued nature highly at different points in various cultures, one hesitates to call premodern Buddhism ecological in the present-day sense of that word. Cardinal virtues such as nonviolence and compassion were applied to individual animals but not to species or ecosystems. At the same time, other features of Buddhism could be cited to justify the invention of eco-karma. Animals, for instance, have been regarded as subject to the laws of karma. In comparison with Western religious and intellectual history, that belief alone is a significant step away from anthropocentrism (human-centered thinking).

A concept such as eco-karma may facilitate reexamination of the assumed boundaries between humans, animals, and plants. The nature of plants has been debated within Buddhism for centuries. Are plants sentient? Do they suffer? Can they attain buddhahood? Buddhist scholar Lambert Schmithausen has noted:

> The question arises why the Buddhists, unlike Jainas and most Hindus, have not also included plants into the karmically-determined rebirth system. Provided that we do not already presuppose the later view that plants are not sentient beings but rather the earlier one that they are sentient and hence exposed to suffering through being cut, mutilated, or the like, there is no reason why one should not—as the Jainas and many Hindus actually do—regard them, too, as owing their state to former karma, and hence as another possible form of rebirth.[20]

New concepts allow new questions: What is the eco-karma of a plant? What is the eco-karma of an animal, or a species, or an ecosystem? Is it helpful to think about the eco-karma of Earth as a whole? What is my eco-karma? yours? ours?

When one attempts to bring some of these considerations to bear on the specific problem of nuclear waste, the complexities intensify. Assigning agency, for instance, is no easy matter. To say that "we" are creating nuclear waste is accurate enough from a far-future perspective—most of us take full advantage of the opportunity to live a developed-world life-style, thereby exporting some of the true costs of privilege to distant places or distant generations. We take it for granted that we have abundant electricity twenty-four hours a day. Yet "we" can also be used too loosely. Before one makes blanket assertions about the karma of flicking a light switch, specific situations must sometimes be taken into account. Analysis of a particular region may reveal, for example, that the energy sources there are nearly or fully sustainable (wind power, solar power, and so on). It may also be necessary, on ethical grounds, to draw a distinction between an executive in the nuclear-power industry and, say, a homemaker who uses electricity drawn partially from nuclear sources.

Even if we recognize that nuclear waste puts untold future generations at risk, ethical scrutiny of that legacy depends on a host of factors, including the scientific and social nature of the risks themselves. As ethicist Kristen Shrader-Frechette notes, the magnitude of a risk is only one of the pertinent variables:

> Numerous other factors, in addition to mere magnitude, determine the acceptability of a risk: whether it is assumed voluntarily or imposed involuntarily; whether the effects are immediate or delayed; whether there are or are not alternatives to accepting the risk; whether the degree of risk is known or uncertain; whether exposure to it is essential to one's well-being or merely a luxury; whether it is encountered occupationally or nonoccupationally; whether it is an ordinary hazard or (like cancer) a "dread" one; whether it affects everyone or only sensitive people; whether the factor causing the risk will be used as intended or is likely to be misused; and whether the risk and its effects are reversible or irreversible.[21]

In some cases we know that we are contributing to a problem, but we are dependent on systems that offer no viable alternatives. For a Los Angeles commuter, being able to drive to work is a necessity, even though she may understand that hundreds of excess deaths and thousands of excess illnesses are caused annually in Los Angeles from the effects of too many cars. There is even a further complication here, one that is characteristic of technological societies: any *single* driver does not make the smog perceptibly worse or increase the number of its victims. In such situations assessing moral or karmic accountability is difficult.

In applying a concept like karma to contemporary Western life, at what point is its (re)definition constrained by Buddhist doctrine and tradition? The degree to which karma can be decoupled from literal interpretations of rebirth will be crucial in this regard. Contemporary Buddhists will also have to determine the limits of what to consider karmically. Do organ donation and organ reception have meaningful karmic implications? Do investment strategies? They may. However, over-karmacization—weighing the karmic repercussions of sharpening a pencil—is likely to give rise to conceptual absurdities and functional paralysis.

Precisely when we need to take more responsibility for bigger and bigger things, our sense of responsibility is being eroded by powerful social forces. Public figures who try to broach the subject of accountability in moral terms, using available Western principles and language, are often accused of being too, well, moralistic. The Buddhist tradition offers another way and another language. If today's engaged Buddhists manage to refine and enrich karma doctrine to suit current conditions, karma won't be what it used to be, but it may serve constructive purposes in unforeseen arenas.

Challenges of Buddhist-Environmentalist Practice

Although Buddhist environmentalism is a recent development, several practical challenges can already be identified. Some of these are common to any environmental issue; others pertain especially to nuclear waste. Let's imagine a Generic American Buddhist Environmentalist and call him Gabe for short. Assume that Gabe has a deep-seated aspiration to come to enlightenment and an equally deep-seated aspiration to protect Earth. Such a person is likely to encounter a number of stumbling blocks on the Buddhist-environmentalist path, among them the following: discontinuities between traditional Buddhist teachings and contemporary realities; the need to clarify the priority of Dharma work or environmental work; difficulties that attend the creation of public-interest groups; and doubts about the efficacy of symbolic actions in response to ecological threats.

We have seen above that karma doctrine is one domain that reveals potential gaps between past teachings and present circumstances. For someone interested in nuclear-waste issues, the topic of waste offers another example of apparent discontinuity. In Zen, monks and other serious practitioners are not supposed to waste anything or treat anything as waste. The instructive stories are graphic: a novice is scolded for discarding a single chopstick; a monk runs alongside a mountain stream to retrieve a single piece of lettuce; Zen master Dōgen uses only half a dipper of water to wash his face. "No waste" usually has two linked meanings in these contexts: "do not waste" and "do not perceive anything as waste." A contemporary Zen master declared, "Roshi's words that originally there is no rubbish either in men or in things actually comprise the basic truth of Buddhism."[22]

These doctrines and practices are exemplary, and they seem applicable in a broad sense to nuclear waste. If we related to Earth and all living beings with the respect and oneness exhibited by a Dōgen, we would probably not produce any nuclear waste in the first place. Or, if constructive purposes (for example, medical uses) unavoidably generated a limited amount of radioactive waste, our descendants would cheer if we were able to safeguard that waste with the intensity of the monk who chased the lettuce leaf downstream.

Yet these same examples also raise some questions. No pre-modern forms of waste were toxic in the ways that nuclear waste is. How might Zen teachings apply to toxic waste? Dealing with the waste produced by a monastery is one thing; dealing with the tens of thousands of tons of atomic waste generated by nuclear reactors and weapon plants is a problem on a different scale. While a monk may be able to retrieve a stray lettuce leaf before the rice is cooked, plutonium cannot be handled safely until 250,000 years have elapsed. We also notice a discrepancy between the focus on individual action in the Zen examples and the highly complex collective action required for the production and prospective containment of nuclear waste. Attempting to reconcile such gaps, engaged Buddhists seek new approaches that are transformative not only for the toxic waste but also for those who deal with it.

There are only twenty-four hours in a day, and at times an ecologically aware Buddhist must choose (however reluctantly) between one activity and another. If an apparent conflict arises between Dharma work and environmental work, what are the priorities of a Buddhist environmentalist? Imagine that Gabe, on a given day, has mindfully fulfilled family, job, and civic duties and then realizes that he has some free time. "Ah," he thinks to himself, "should I use this hour for some uninterrupted meditation, or should I use it to write my Congressman to oppose the makeshift plans for a nuclear dump in our state?" You may alter the hypothetical conditions and substitute any inner-directed practice for meditation, but there will always be situations in which there is a choice between one course and another. Thich Nhat Hanh and Thai Buddhist activist Sulak Sivaraksa once contemplated a much weightier choice between peace and the survival of Buddhism. This is Sivaraksa's recollection of their conversation:

Before the end of the Vietnam War, I asked Ven. Thich Nhat Hanh whether he would rather have peace under the communist regime, which would mean the end of Buddhism, or rather the victory of the democratic Vietnam with the possibility of Buddhist revival, and his answer was to have peace at any price.[23]

Pressed to clarify his priorities, Thich Nhat Hanh placed peace over the survival of Buddhism. Would he answer similarly if asked to choose between, say, the survival of a globally significant ecosystem and Buddhism's survival?

If a practitioner is meditating peacefully in her room, and suddenly outside the window she hears the screech of brakes, a loud thump, and a frantic scream, the proper course of action is obvious. At that moment, running to the scene *is* Buddhism. But when problems are more protracted and complex—as most environmental problems are—it is less clear when a situation calls for one to remain on the mat and when to leave it. Joanna Macy, questioned about the apparent discrepancy between Dharma practice and nuclear-related activism, emphatically replied, "This nuclear work *is* the Dharma. One of the aims of practice is to be able to transform our own actions. For those who are involved in this work, the 'poison fire' is a Dharma teacher."[24]

Some further distinctions are advisable here. If one were to conceive of Dharma practice narrowly (nothing but meditation, chanting, and prostrations) and then argue that by perfecting those activities one is thereby working on behalf of the planet, we would probably object that such a conflation is oversimplified. Similarly, if one were to argue that engagement in environmental work is also by its very nature Dharma work, we would have to say: "Wait a moment. That might depend on some other factors, like the degree of a person's spiritual maturity, or the mindstate with which one approaches the environmental task." The most common practical pitfall for Gabe and his colleagues is that the dharmic dimension of activism can evaporate all too quickly. There may not be a satisfactory answer to the question of priority in its rigid either/or form. The answer has to be lived, until one reaches a point where most activity expresses Buddhist awareness *and* environmental awareness, simultaneously.

With regard to environmental problems in general and nuclear issues in particular, Buddhists have experienced difficulty translating the values and practices of Buddhism into meaningful public action, concrete policies, and enduring organizations. Of course, Buddhist environmentalists are not alone in this regard. Charlene Spretnak, an ecofeminist and a practitioner of *vipassanā* meditation, asks:

> How can we induce people and institutions to think in terms of the long-range future, and not just in terms of their short-range selfish interest? How can we encourage people to develop their own visions of the future and move more effectively toward them? How can we judge whether new technologies are socially useful—and use those judgments to shape our society?[25]

If Gabe wants to work publicly to motivate leaders and citizens to do the right thing about nuclear waste, one of his first impulses may be to credit Buddhism as a source of inspiration—after all, he is a *Buddhist* environmentalist. But on second thought, he may decide that if he wants to reach mainstream America, a Buddhist label might be counterproductive, tending to confuse or alienate potential supporters. Ironically, it may be most skillful in today's public arena to take the "Buddhism" out of Buddhist environmentalism.

The Buddha's birthday ceremony at the Nevada Test Site, noted above, exemplifies another challenge of Buddhist-environmentalist practice. In the face of real environmental threats—in this case the hazards posed by nuclear weapons and nuclear waste—what is the significance of symbolic/ritual activities? And if rituals *are* among the appropriate responses, what relevance do traditional ritual forms have in contemporary contexts? Several of the participants in the Test Site ceremony wondered aloud if their vows, prostrations, and other gestures have any real impact on the terrible dangers they seek to address. For some, the discrepancies of scale seemed unsurmountable. For others, rituals addressing nuclear concerns would have a better claim to relevance in a society that handled nuclear matters responsibly. A third group recognized the severity of our nuclear plight yet reaffirmed the efficacy of ceremonial acts, even if such behavior appears futile to skeptics.

As Buddhists and others struggle to come to terms with nuclear waste, they find that their fears and hopes call out for vehicles of expression that go beyond what can be fashioned individually. The leader of a citizen watchdog group in Amarillo, Texas, where plutonium cores from former atom bombs are literally being stacked in bunkers, recently told me of two strong emotions that she experiences in tandem: joy that the stacked cores signify the end of the Cold War and near despair at the prospect of safeguarding all the plutonium that is accumulating in her community. We had just left the office of a high-ranking Energy Department official, and my friend was crying quietly as she spoke. Many of today's de facto nuclear guardians would be receptive to new rites of remembrance, innovative rituals of forgiveness, and ceremonies that connect present generations to future generations. Scholars of contemporary environmentalism have suggested that radical environmentalists are engaged in "a kind of ritualized guerrilla warfare over sacred space in America," a contest that pits the desire for consecration against the danger of desecration.[26] In that sense, the Buddhist activists at the Nevada Test Site may have been taking the initial steps in the creation of spiritually evocative nuclear rituals.

Ecology Koans

Members of the Zen group in Oregon expressed their sense of accountability for nuclear waste by modifying the four bodhisattva vows of Mahāyāna Buddhism, as cited above. A more literal rendering of the first bodhisattva vow is: "All beings, without number, I vow to liberate." What does a vow to save all beings mean in a nuclear age? In what ways does it include those who have already been harmed by nuclear-weapon production and nuclear-power production, from Japanese atom-bomb casualties to Navaho uranium miners and Chernobyl children? In what ways does it include the countless future beings, human and nonhuman, who will suffer the prolonged effects of current military and energy policies?

In eighth-century India the Buddhist monk and poet Śāntideva proclaimed:

> For as long as space endures
> And for as long as living beings remain,
> Until then may I too abide
> To dispel the misery of the world.[27]

When the current Dalai Lama alludes to Śāntideva's stanza he says, "No matter how extensive space, or how extensive time, I will save all beings."[28] This pledge was challenging enough—to comprehend and to actualize—in a premodern age. Today, with an appreciation shaped by science of the immensity of space and time, we also have a newfound awareness of the extensiveness of beings and the extensiveness of the threats to those beings. Thus, to affirm the bodhisattva vow with nuclear realities in mind is to declare a willingness to accept responsiblity for the fate of all the beings who will be exposed in the next 250,000 years and beyond to the wastes we have created in just the past fifty years. Contemporary Zen teacher John Daido Loori, referring to the planetary ecological crisis, maintains that one must begin to "take responsibility for the whole catastrophe." He writes:

> And because someone in South America is doing it, that does not mean we are not responsible. We are as responsible as if we are the ones clubbing an infant seal or burning a hectare of tropical forest.[29]

Such radical assertions of responsibility have well-established antecedents in Buddhism. A ninth-century Ch'an (Zen) text, *The Platform Sūtra of the Sixth Patriarch*, taught: "When others are in the wrong, I am partly responsible. When I am in the wrong, I alone am to blame."[30]

Our rational minds tell us that saving all beings, or taking responsibility "for the whole catastrophe," is a preposterously grandiose notion. Yet those who undertake to fulfill such an aspiration assert that the very incomprehensibility of the task pushes the mind to deeper and deeper levels, until it becomes possible to transcend constraints of time and space, saving or not saving. Anyone familiar with Zen *kōans* will recognize that bodhisattva vows and comparable declarations have a koan-like quality. A koan, strictly speaking, is a distinctive type of Zen practice:

> A koan is a spiritual puzzle that cannot be solved by the intellect alone. Though conundrums and paradoxes are found in the secular and sacred literature of many cultures, only in Zen have such formulations developed into an intensive method of religious training. What gives most koans their bite, their intellect-baiting hook, is some detail that defies conventional logic.[31]

For example: What is your original face before your parents' birth? A koan differs from a riddle in that the person attempting to solve it *becomes* something in the process. In a similar way, a bodhisattva vow consumes the devotee to the point where she realizes that she is part of the vow.

The ecological crisis itself has koan-like aspects. Nuclear waste is a good example: we have difficulty grasping the problem conceptually, and we flounder when it comes to practical action. There are no certifiably safe ways to contain radioactive materials, yet we do not even have the sense to stop producing them. So nuclear waste appears to be a problem without a solution. Several other questions raised in this essay can also be treated as koans to some degree. The aim is not to be inventive but to see if any of the time-tested tools of Buddhist practice can be of service in dealing with these new and pressing issues. There may be beneficial ways to engage the following questions as *ecology koans* (or *eco-koans*, if we can stand another neologism):

What is waste?

What is the scope of my mindfulness?

What is my/our responsibility for our environmental legacy? for our nuclear legacy?

What is a spiritually motivated environmentalist's first priority, spiritual work or environmental work?

When do we know that we have done all that can be done?

These questions may lack an "intellect-baiting hook" in the style of classic Zen koans, yet they can nonetheless be probed in the sustained, penetrating way that one probes a koan. The questioning itself is often more valuable than any "answers" that are produced. We may achieve a dependable understanding of these and other ecology koans only after we live with them, allow them to question us, and restrain our impulse to accept merely conceptual solutions.

Some of the classic koans and related texts also invite fresh interpretations in light of contemporary conditions. The ninth-century Ch'an master Nan-ch'uan was once asked, "When one realizes that *there is*, where should one go from there?" Nan-ch'uan replied, "One should go down the hill to become a buffalo in the village below."[32] If asked today, maybe Nan-ch'uan would say, "One should go to Nevada to become a nuclear guardian at Yucca Mountain." Here is another example:

> The priest Hsiang-yen said, "It is as though you were up in a tree, hanging from a branch with your teeth. Your hands and feet can't touch any branch. Someone appears beneath the tree and asks, 'What is the meaning of Bodhidharma's coming from the West?' If you do not answer, you evade your responsibility. If you do answer, you lose your life. What do you do?"[33]

The question "What is the meaning of Bodhidharma's coming from the West?" has the thrust of "What is the essential truth of Zen?" So the hapless protagonist must somehow demonstrate his Zen insight without opening his mouth and falling to his death.

Humanity's current predicament in relation to Earth resembles the predicament of the person hanging from the branch: beyond a certain point, action and nonaction are equally ineffective. Culture historian Thomas Berry seems to be elucidating a planetary version of Hsiang-yen's koan when he writes:

> By entering in to the control of the planet through our sciences and our technologies in these past two centuries, we have assumed responsibilities beyond anything that we are capable of carrying out with any assured success. But now that we have inserted ourselves so extensively into the functioning of the ecosystems of the Earth, we cannot simply withdraw and leave the planet and all its life systems to themselves in coping with the poisoning and the other devastation that we have wrought.[34]

If, in the spirit of a bodhisattva vow, we truly embrace the larger responsibilities that we customarily push out of awareness, our lives will change dramatically. Indeed, we will lose our (former) lives. A man hanging from a tree by his teeth, humans inserted irrevocably into Earth's ecosystems. . . "If you do not answer, you evade your responsibility. If you do answer, you lose your life." *What do you do?*

Notes

* I am grateful to Sōgen Hori, Stephanie Kaza, and Alan Senauke for their thoughtful comments on a draft of this essay.

1. I have not found any previous uses of "nuclear ecology" in Western-language sources, but I have learned of an institute in Moscow called (in translation) the Center for Nuclear Ecology and Energy Policy.

2. Any expression has its drawbacks. Nuclear ecology sounds too benign if it is misinterpreted as casting dangerous nuclear realities only in a positive light. Admittedly, nuclear ecology would stretch the meaning(s) of ecology (i.e., nuclear materials are not recyclable the way other materials are). For practical purposes, limits must be defined; I would suggest, for example, that nuclear-disarmament and nonproliferation issues fall outside the scope of nuclear ecology.

3. Douglas MacLean, "Understanding the Nuclear Power Controversy," in H. Tristram Engelhardt, Jr., and Arthur L. Caplan, eds., *Scientific Controversies* (Cambridge: Cambridge University Press, 1987), 578–79.

4. Francis H. Cook, *Hua-yen Buddhism: The Jewel Net of Indra* (University Park: Pennsylvania State University Press, 1977), 2.

5. Gary Snyder, *Turtle Island* (New York: New Directions, 1974), 94.

6. Thich Nhat Hanh, "The Last Tree," in Allan Hunt Badiner, ed., *Dharma Gaia: A Harvest of Essays in Buddhism and Ecology* (Berkeley: Parallax Press, 1990), 220.

7. The Dalai Lama, "Five-Point Peace Plan for Tibet," in Petra K. Kelly, Gert Bastian, and Pat Aiello, eds., *The Anguish of Tibet* (Berkeley: Parallax Press, 1991), 288, 292. For connections between nuclear waste and Tibet's threatened Buddhist culture, see International Campaign for Tibet, *Nuclear Tibet: Nuclear Weapons and Nuclear Waste on the Tibetan Plateau* (Washington, D.C.: International Campaign for Tibet, 1993).

8. Joanna Macy, *World as Lover, World as Self* (Berkeley: Parallax Press, 1991), 220–37 and passim. See also Kenneth Kraft, "The Greening of Buddhist Practice," in Roger S. Gottlieb, ed., *This Sacred Earth: Religion, Nature, Environment* (New York: Routledge, 1996), 492–94.

9. "Buddhist Vows for Guardianship," in Nuclear Guardianship Project, *Nuclear Guardianship Forum* 1 (spring 1992):2.

10. Office of Environmental Management, U.S. Department of Energy, *Closing the Circle on the Splitting of the Atom: The Environmental Legacy of Nuclear Weapons Production in the United States and What the Department of Energy Is Doing about It* (Washington, D.C.: Department of Energy, 1995).

11. Snyder, *Turtle Island*, 67.

12. Tenshin Reb Anderson, "Dedication for Buddha's Birthday at the Gate of the Nevada Nuclear Test Site," 10 April 1994.

13. "Buddhism: Religion of the Moment," *USA Weekend*, 15–17 September 1995.

14. For engaged Buddhists, some of the traditional injunctions of breath-focused meditation practice still have meaning with the added "d," as in "follow your breadth," or "become one with your breadth."

15. Nhat Hanh, "The Last Tree," 220.

16. A study of the "revenge effects" of technology calls for a kind of vigilance that bears resemblance to Buddhist mindfulness; see Edward Tenner, *Why Things Bite Back: Technology and the Revenge of Unintended Consequences* (New York: Alfred A. Knopf, 1996), 277.

17. On average, one American consumes as much of the world's limited resources as fifty citizens of India. Even if a conscientious American Buddhist cuts the average rate of consumption in half, there is still a significant disparity.

18. Here and elsewhere I am guilty of what Ian Harris has called "terminological revisionism." Yet Harris also suggests that redefinitions are "part of a seamless reflexive process inherent to the Buddhist tradition itself" (Ian Harris, "Buddhist Environmental Ethics and Detraditionalization: The Case of EcoBuddhism," *Religion* 25, no. 3 [July 1995]:201–2).

19. Quoted in Peter Weiss, "And Now, Abolition," *Bulletin of the Atomic Scientists* 52, no. 5 (September/October 1996):43.

20. Lambert Schmithausen, *The Problem of the Sentience of Plants in Earliest Buddhism* (Tokyo: International Institute for Buddhist Studies, 1991), 101.

21. K. S. Shrader-Frechette, "Ethics and Energy," in Tom Regan, ed., *Earthbound: New Introductory Essays in Environmental Ethics* (Philadelphia: Temple University Press, 1984), 121.

22. Morinaga Sōkō, "My Struggle to Become a Zen Monk," in Kenneth Kraft, ed., *Zen: Tradition and Transition* (New York: Grove Press, 1988), 17.

23. Sulak Sivaraksa, "Buddhism in a World of Change: Politics Must Be Related to Religion," in Fred Eppsteiner, ed., *The Path of Compassion: Writings on Socially Engaged Buddhism*, rev. ed. (Berkeley: Parallax Press, 1988), 16.

24. Personal conversation with Joanna Macy, 19 August 1992.

25. Charlene Spretnak, "Ten Key Values of the American Green Movement," in Roger S. Gottlieb, ed., *This Sacred Earth: Religion, Nature, Environment* (New York: Routledge, 1996), 536.

26. David Chidester and Edward T. Linenthal, eds., *American Sacred Space* (Bloomington: Indiana University Press, 1995), 21.

27. Stephen Batchelor, trans., *A Guide to the Bodhisattva's Way of Life* (Dharamsala: Library of Tibetan Works and Archives, 1979), 193.

28. Martin Wassell, producer, *Heart of Tibet: An Intimate Portrait of the Dalai Lama* (New York: Mystic Fire Video, 1991).

29. John Daido Loori, "The Precepts and the Environment," included in this volume (179) and also published in *Mountain Record*, spring 1996, 13.

30. My translation, following the Tun-huang version of the text. See Philip B. Yampolsky, *The Platform Sutra of the Sixth Patriarch* (New York: Columbia University Press), 1967, 161, and p. 18 of the Chinese appendix.

31. Kenneth Kraft, *Eloquent Zen: Daitō and Early Japanese Zen* (Honolulu: University of Hawaii Press, 1992), 58.

32. John C. H. Wu, *The Golden Age of Zen* (New York: Doubleday, 1996), 96 (slightly edited).

33. Robert Aitken, trans., *The Gateless Barrier* (San Francisco: North Point Press, 1990), 38.

34. Brian Swimme and Thomas Berry, *The Universe Story: From the Primordial Flaring Forth to the Ecozoic Era* (San Francisco: Harper San Francisco, 1992), 252.

Buddhist Resources
for Issues of Population, Consumption,
and the Environment

Rita M. Gross

This chapter applying basic Buddhist teachings to questions regarding fertility control and resource utilization is written by a feminist academic scholar of religion, for whom Buddhism is the long-standing religion of choice. Therefore, I bring to this chapter the perspectives of both an insider trained in Buddhist thought and an outsider with allegiance to the cross-cultural comparative study of religion and broad knowledge of major religious traditions.

As is the case with all major traditions, conclusions relevant to the current situation cannot be quoted from the classic texts; rather, the *values* inherent in the tradition need to be applied to the current, unprecedented crises of overpopulation and excessive consumption that threaten to overwhelm the biosphere upon which we are dependent. This task of applying the traditional values of Buddhism to such issues in the contemporary context is not difficult, in my view, since classic Buddhist values suggest highly relevant ways of responding to the current situation. In this essay, I will work to some extent as a Buddhist "constructive theologian," interpreting the tradition in ways that bring the inherited tradition into conversation with contemporary issues and needs. Reflecting my own standpoint both as a Buddhist and as a scholar, I will include materials not only from early Buddhist thought, but also from the Mahāyāna and Vajrayāna perspectives within Buddhism. At the same time, I shall try to be as nonsectarian as possible.

Defining the Issues: Environment, Consumption, and Population

When we try to bring traditional Buddhist values into conversation with the current situation, it is important to have a clear understanding of that situation. The assignment of this chapter is to address the interlocked issues of the environment, resource utilization, and population growth, from a Buddhist point of view. Since Buddhism always suggests that we need to deal with things as they are, not with fantasies, it is appropriate to begin with some brief consideration of how the ecosystem, consumption, and population actually interact. When relating these three concerns to one another, one can imagine three alternatives: a sufficiently small population living well on a stabile, self-renewing resource base; an excessive population living in degraded conditions on an insufficient resource base; or the present pyramid of a few people living well and large numbers of people barely surviving. Obviously, only the first option contains merit. How people could value reproduction so much that they could prefer the second option to the first is incomprehensible, and the current pyramid of privilege is morally obscene. It should also be clear that population is the only negotiable element in this complex. In other words, when we look at the three factors under discussion—the environment, population, and consumption—there are two non-negotiables and one negotiable. Fundamentally, it is not negotiable that the human species must live within the boundaries and limits of the biosphere. However it is done, there is no other choice, because there is no life apart from the biosphere. Morally, it is not negotiable that there be an equitable (*equitable*, not *equal*) distribution of resources among the world's people. These two non-negotiables mean that population size is the negotiable factor in the equation. It is hard to question the proposition that a human population small enough so that everyone can enjoy a decent standard of living without ruining the environment is necessary and desirable. We cannot increase the size of the earth and can increase its productivity only to a limited extent, but we, as a species, can control population. All that it requires is the realization that many other pursuits are at least equally as sacred and as satisfying as reproduction.

Religions commonly criticize excessive consumption but commonly encourage excessive reproduction. Therefore, though I will note the Buddhist values that encourage moderate consumption, I will emphasize the Buddhist values that encourage moderation and responsibility regarding reproduction, which are considerable. I emphasize these elements in Buddhism precisely because there has been so little discussion of religious arguments that favor restraining human fertility. The example of a major, long-standing world religion whose adherents lead satisfying lives without an overwhelming emphasis on individual procreation certainly is worth investigating. Buddhism can in no way be construed or interpreted as pronatalist in its basic values and orientations. The two religious ideas that are commonly invoked by most religions to justify pronatalist practices are not part of basic Buddhism. Buddhism does not require its members to reproduce as a religious duty. Nor do most forms of Buddhism regard sexuality negatively, as an evil to be avoided unless linked with reproduction, though all forms of Buddhism include an implicit standard of sexual ethics. Therefore, fertility control through contraception as well as abstinence is completely acceptable. The practices regarding fertility and reproduction that would flow from fundamental Buddhist values favor reproduction as a mature and deliberate choice rather than as an accident or a duty. Because of the unique ways in which Buddhism values human life, only children who can be well cared for, physically, emotionally, and spiritually, should be conceived. Few Buddhists would disagree with the guideline that one should have few children, so that all of them can be well cared for without exhausting the emotional, material, and spiritual resources of their parents, their community, and their planet.

By contrast, pronatalism as an ideology seems to be rampant on the planet; those who mildly suggest that unlimited reproduction is not an individual right and could well be destructive are derided. Suggest that there is a causal relationship between excessive reproduction and poverty and watch the fallout. Pronatalist ideology includes at least three major ideas, all of which are subject to question. Pronatalists always regard a birth as a positive occasion, under any circumstances, even the most extreme. To suggest that reproduction under many circumstances is irresponsible, and merits censure rather than support, makes one unpopular with pronatalists.

Furthermore, pronatalists claim that it is necessary to reproduce to be an adequate human being; those who choose to remain childless are scorned and suffer many social and economic liabilities. Finally, pronatalists regard reproduction as a private right not subject to public policy, even though they usually insist that the results of their reproduction are a public, even a global, responsibility. The tragedy of pronatalism is that although excessive populations could be cut quite quickly by voluntary means, lacking those, they probably will be cut by involuntary means involving great suffering—diseases, violence, and starvation. Therefore, it is critical to counter the mindless and rampant pronatalist religious doctrines, socialization, peer pressure, tax policies, sentiments, and values, which senselessly assault one at every turn.

Before beginning to discuss Buddhist teachings as a resource for developing an ethic of moderation concerning both reproduction and consumption, it is important to pause to acknowledge two controversial issues. They cannot be debated in this context, even though my conclusions regarding them will be apparent in my discussion of Buddhist ethics, the environment, consumption, and reproduction.

Because the Buddhist concept of all-pervasive interdependence makes sense, I see no way that individual rights can extend to the point that an individual exercising his or her supposed rights may be allowed to threaten the supportive matrix of life—a point that has been reached in both consumption and reproduction. Whatever wealth or values a person may have that drive them to inappropriate levels of consumption or reproduction, it is hard to argue that they have individual rights to exercise those levels of consumption or reproduction without regard to their impact on the biosphere. The rhetoric of individual rights and freedoms certainly has cogency against an overly communal and authoritarian social system. But today, that rhetoric and stance threaten to overwhelm the need for restraint and moderation to protect and preserve communities and species.

Furthermore, especially in the need to counter pronatalist ideologies and policies, we have reached a point beyond relativism. In the human community, we have learned too late and too slowly the virtues of relativism whenever it is feasible. We have been too eager to condemn others for having a worldview different from our own. Relativism regarding worldview is virtuous because diversity of worldviews is a valuable resource. On the other hand, relativism regarding basic ethical standards leads to intolerable results. Are we really willing to say of a culture in which women are treated like property or children are exploited that "that's just their culture"? There would be no possibility of an international human rights movement if people really believed that ethical standards are completely relative and arbitrary. And both consumption and reproduction are ethical issues of the highest order, since their conduct gravely affects everyone's life. We can no longer afford to let individuals who believe that they should reproduce many children do so, just as we no longer condone slavery, the exploitation of children, or treating women as chattel. Certain long-standing and deeply held cultural and religious values are at stake in the claim that pronatalism is an intolerable and inappropriate ethical stance, given current conditions. Some religions need to adjust their recommendations regarding fertility to the realities brought about by modern medicine, which has greatly reduced the death rate but not the birth rate, resulting in a dangerous growth in populations, all of whom want to consume at higher standards than have ever been known previously.

Walking the Middle Path in an Interdependent World: Basic Buddhist Resources for Moderation

One of the most basic teachings of Buddhism concerns interdependence (*pratītya-samutpāda* in Sanskrit and *paticca-samuppāda* in Pali), which is said to be one of the discoveries made by the Buddha during his enlightenment experience. This teaching prepares the ground for all further comments on consumption and reproduction, since interdependence is the bottom line which cannot be defied. Rather than as isolated and independent entities, Buddhism sees all beings as interconnected with one another in a great web of interdependence. All-pervasive interdependence is part of the Buddhist understanding of the law of cause and effect, which governs all events in our world. Nothing happens apart from or contrary to cause and effect according to Buddhism, which does not allow for accidents or divine intervention into the operations of cause and effect. Furthermore, since Buddhism understands cause and effect as interdependence, actions unleashed by one being have effects and repercussions throughout the entire cosmos. Therefore, decisions regarding fertility or consumption are not merely private decisions irrelevant to the larger world. Any baby born anywhere on the planet affects the entire interdependent world, as does any consumption of resources. It cannot be argued that either private wealth or low standards of material consumption negate this baby's impact on the universal web of interdependence. Nor can it be argued that private desires for children outweigh the need to take into account the impact of such children on the interdependent cosmos, since the laws of cause and effect are not suspended in any case. Similarly, utilization of resources anywhere has repercussions throughout the entire planetary system. Often, consumption of luxuries in one part of the world is directly related to poverty and suffering in other parts of the world. Thus, the vision of universal and all-pervasive interdependence, which is so basic to Buddhism, requires moderation in all activities, especially reproduction and consumption, because of their impact on the rest of the universe. When the Buddhist understanding of interdependence is linked with the scientific understanding of the planet as a finite lifeboat, it becomes clear that Buddhism regards appropriate, humane, and fair fertility control as a requirement. It is equally clear that Buddhism would regard ecologically unsound practices regarding reproduction or consumption as selfish, privately motivated disregard for the finite, interdependent cosmos.

The vision of cosmic interdependence presents the big picture regarding reproduction and consumption. This vision becomes more detailed when we look more specifically at the human realm within the interdependent cosmos. On the one hand, Buddhism values tremendously the good fortune of human rebirth, and on the other hand, Buddhism sees all sentient beings as fundamentally similar in their basic urge to avoid pain and to experience well-being. Thus, birth as a human is both highly valued and seen as birth into that vast universal web of interdependence in which what relates beings to each other is much more fundamental than what divides them into species. So two phrases, "precious human body," and "mother sentient beings," need always to be kept together when discussing Buddhist views about the human place in the interdependent cosmos. The preciousness of human birth is in no way due to human rights over other forms of life, for a human being *was* and could again be other forms of life—though Buddhist practice is also thought to promote continued rebirth in the human realm. On the other hand, all beings are linked in the vast universal web of interdependence and emptiness, from which nothing is exempt. This web is so intimately a web of relationship and shared experience that the traditional exuberant metaphor declares that all beings have at some time been our mothers and we theirs. Therefore, rather than feeling superior or feeling that we humans have rights over other forms of life, it is said over and over that, because we know how much we do not want to be harmed or to suffer, and since all beings are our relatives, we should not harm them or cause them pain, as much as possible.

As is commonly known, traditional Buddhism does believe in rebirth and claims that rebirth is not necessarily always as a human being but depends upon merit and knowledge from previous lives. Among possible rebirths the human rebirth is considered by far the most fortunate and favorable, favored even over rebirth in the more pleasurable divine realms. That belief alone might seem to encourage unlimited reproduction. But when one understands *why* human birth is so highly regarded, it becomes clear that excessive human reproduction destroys the very conditions that make human rebirth so valued. Rebirth as a human being is valued because human beings, more than any other sentient beings, have the capacity for the spiritual development that eventually brings the fulfillment and perfection of enlightenment. Though all beings have the inherent innate potential for such realization, its achievement is fostered by certain causes and conditions and impeded by others. Therefore, the delight in human rebirth is due to the human capacity for cultural and spiritual creativity leading to enlightenment, a capacity more readily realized if sufficient resources are available. *Mere* birth in a human body is not the cause for rejoicing over "precious human birth," since human birth is a necessary, but not a sufficient, condition for the potential inherent in humanness to come to fruition. It is very helpful, even necessary, for that body to be in the proper environment, to have the proper nurturing, physically, emotionally, and spiritually. This is the fundamental reason why a situation of a few people well taken care of is preferable to many people struggling to survive.

The conditions that make human life desirable and worthwhile are summed up in one of the core Buddhist values—that of the Middle Way or the Middle Path. This Middle Path is also discussed as right effort, not too much, not too little, not too tight, not too loose. To make the most appropriate use of the opportunity represented by the "precious human birth," a person needs to walk the Middle Way, and to be able to walk the Middle Way. To avoid extremes in all matters is one of the core values of Buddhism, learned by the Buddha before his enlightenment experience and a necessary precondition to it. First he learned that a life of luxury is meaningless, but then he had to learn that a life of poverty also leads nowhere. The Buddha concluded that, in order to become fully human, one needs to live in moderation, avoiding the extremes of too much indulgence and too much poverty or self-denial.

The guideline of the Middle Way emphasizes that too much wealth or ease can be counterproductive spiritually, since it tends to promote complacency, satisfaction, and grasping for further wealth—all attitudes that are not helpful spiritually. Thus, the concept of the Middle Way provides a cogent criticism and corrective for the rampant consumerism and overconsumption that are so linked with overpopulation. However, the concept of the Middle Way also makes the fundamental point that there are minimum material and psychological standards necessary for meaningful human life. Buddhism has never idealized poverty and suffering, or regarded them as spiritual advantages. Those in dire poverty or grave danger and distress do not have the time or inclination to be able to nurture themselves into enlightenment, into actually benefiting fully from their human rebirth, which is quite unfortunate. Buddhism celebrates moderation, but it does not celebrate poverty, because it sees poverty as unlikely to motivate people to achieve enlightenment—or even to allow them enough breathing time to do so.

Therefore, Buddhists have long recognized that before Buddhist teachings can be effective, there must first be a foundation of material well-being and psychological security. Buddhism has always recognized that one cannot practice meditation or contemplation on an empty stomach, or create an uplifted and enlightening environment in the midst of degradation, deprivation, or fear. Buddhists have known for a long time that deep spiritual or contemplative practice—which is seen as leading to the greatest joy and fulfillment possible to humans—is usually taken up after rather than before achieving a certain basic level of Middle Way comfort. Before that, people really do think that once they have enough material things, they will not suffer. One has to reach a certain basic level of satisfaction of basic desires before one begins to realize that desire and its attendant sufferings are much more subtle. At a point after basic needs have been met, when people begin to experience that desire and suffering are not so easily quelled, the basic message of Buddhism begins to make sense.

This point dovetails quite nicely with the point made by many who advocate that curbing excessive population growth is much more possible if people have an adequate standard of living. It is by now a well-known generalization that one of the most effective ways to cut population growth is to improve peoples' economic lives, that people who have some material wealth can see the cogency of limiting their fertility, whereas people who are already in deeply degraded circumstances do not. Buddhist thought consistently advocates investigating cause and effect, since the entire interdependent world is governed by cause and effect. Overpopulation does not just happen; it is the result of causes, one of which seems to be too much poverty, not being able to walk the Middle Way between too much luxury and too much poverty.

However, it is equally clear that too much reproduction would overwhelm all attempts to curb poverty, because a finite earth has limited resources. Thus, we return to the need to recognize the interdependence of excessive consumption, overpopulation, and poverty. If one of these key elements is left out, as is done by religious and cultural systems that have no guidelines limiting human reproduction, then an interdependent cosmos will be severely stressed. Again, it is important to point out that all religions and most cultures do have ethical guidelines limiting consumption. They are often not kept, but the guidelines do exist. Few religions, however, advocate limiting human fertility. Most encourage or require their members to reproduce, without providing any guidance about limits and without any recognition that there could be too many people. Therefore, examples of religious systems that can be invoked to provide religious reasons to limit fertility are critically important.

The vision of interdependence combined with the advice to walk the Middle Way in all pursuits certainly provides such guidance.

Taken together, these concepts of interdependence, of the value of human birth into appropriate circumstances, and of the Middle Way provide some sensible and obvious guidelines regarding fertility control and consumption. Regarding consumption, it is critical to see that the call for the Middle Path points in two directions. Clearly, excessive consumption violates the Middle Path. But so does too much denial. The advice to walk the Middle Path is *not* advice to pull in our belts another notch and make room for more people because reproductive rights are inviolable. It is advice to limit both fertility and consumption, which are interdependent, so as to make possible a life-style conducive to enlightenment for all beings. Certainly, too much fertility for the earth to sustain its offspring, and for communities to provide adequate physical and emotional nurturing, would be a contradiction of the Middle Way. It is crucial that human population not grow beyond the capacity of a family, a community, or the earth to provide a life within the Middle Way to all its members.

Simply providing sheer survival is not enough, and arguments that the earth could support many more people are not cogent because *quality* of life is far more significant than mere *quantity of bodies*. In addition to minimally adequate nutrition, sufficient space to avoid the overcrowding that leads to aggression and violence is important. Availability of the technological, cultural, and spiritual treasures that make life truly human is also basic. Therefore, globally, communally, and individually, it is important to limit fertility, so that all children actually born can have adequate material and psychological care. Not to do so would be wanton disregard for the spiritual well-being of those born into a human body. Neither the poverty nor the emotional exhaustion that results from trying to raise too many children is helpful to anyone—least of all to the children resulting from unlimited or excessive fertility. In Buddhist terms, this basic fact far outweighs private wishes for "as many children as I want" or pronatalist societal and religious norms and pressures.

These guidelines strike me as impeccable advice on how to negotiate problems of population pressure and resource utilization, though, clearly, reasonable and kind people could agree on the guideline and disagree on its implementation. Obviously, that Middle Way does not mean the mindless consumption of the first world, but neither does it mean the mindless pronatalism of much of the rest of the world, including large segments of the first world. And it does, in my view, include some technological basics that really enhance the quality of life—flower gardens, pets, computers, good stereo systems, international travel, electricity, refrigeration, cultural diversity, and humanistic education—things that cannot be provided to unlimited populations without extreme environmental degradation. Since many things in life are more sacred and more satisfying than reproduction, it would seem ludicrous to give up such cultural treasures in order to have large populations that lack those treasures.

Transmitting the "Enlightened Gene": The Mahāyāna *Bodhisattva* Path and Motivations to Reproduce

Many religions, including major Asian traditions with which Buddhism has coexisted, command perpetuation of one's family lineage as a religious obligation. For a Buddhist to have any children at all is not a religious requirement. In the Buddhist vision, one does not need to reproduce biologically to fulfill the acme of one's responsibilities to the interdependent web of mother sentient beings, or to realize the most exalted possibilities of human life. In fact, though the arguments, in their traditional form, elevate celibacy over the householder life-style, rather than childlessness over biological reproduction, a great deal of Buddhist tradition suggests that biological reproduction may interfere with helping the world or realizing one's highest potential. Since Buddhists are like other human beings, it is important and interesting to explore what inspires them to embrace religious ideas that do not require reproduction and also to investigate Buddhist discussions of appropriate reproduction.

The command to perpetuate family lineage is quite strong in some traditions and fuels pronatalist behaviors. Usually this command coexists with a complex of ideas and practices, including the judgment that one is unfilial and seriously remiss in one's religious obligations if one does not have a male heir, that everyone must marry and reproduce, and that women have few or no options or vocations beyond maternity. Traditions that insist that one must reproduce biologically to fulfill one's obligations rarely, if ever, also include the corollary command not to reproduce excessively, which could bring the preferred behaviors back from an extreme into some variant of the Middle Way. In fact, often such traditions discourage any attempts to limit fertility and people who want to do so are made to feel unworthy if they limit reproduction, even if they already have produced an heir to the family lineage. Buddhism, which sees such absolute concern with perpetuating the family lineage as merely an extension of ego, of the self-centeredness that causes all suffering, has never enjoined its adherents to do so. And Buddhism has come in for major criticisms from Asian neighbors for not requiring biological reproduction of its members.

The Asian criticism of Buddhists for being selfish in not requiring reproduction strikes Buddhist sensibilities as very odd. The Buddhist reply would be twofold. First, to contribute that which is most valuable to the interdependent web of mother sentient beings is in no way dependent on biological reproduction. Furthermore, biological reproduction is often driven by very self-centered and selfish motivations. Let us examine both of these ideas closely, because I think they are both important resources in countering the self-righteous moralism of much pronatalist thinking.

These conclusions regarding reproduction are not negative limits demanded of unwilling subjects. Rather, from the Buddhist point of view, they are rooted in deep knowledge of what people ultimately want, of what satisfies our deepest longings. Buddhists would say that the simultaneous pursuit of wisdom and compassion, to the point of enlightenment and even beyond, is what satisfies our deepest longings because it speaks to our fundamental human nature. Buddhists, contrary to much popular thinking, both Asian and Western, do not live their preferred life-style of moderation, meditation, and contemplation out of a self-centered motivation seeking to avoid pain. Buddhists do not reject family lineage as an ultimate value to seek individual fulfillment instead. Buddhists claim that we can never find fulfillment through reproducing or, equally important, through economic production and consumption, no matter how popular these pursuits may be or how rigorously religious or social traditions may demand them. Instead, we need to realize our spiritual potential. Finding life's purpose in either consumption or reproduction simply strengthens what Buddhists call "ego," the deeply rooted human tendency to be self-centered in ways that ultimately cause all our suffering.

Rather, Buddhists see perpetuating family lineage as trivial compared with cultivating and perpetuating our universal human heritage and birthright—the tranquility and joy of enlightenment. Rather than seek self-perpetuation through biological reproduction, Buddhists are encouraged to arouse *bodhicitta*, the basic warmth and compassion inherent to all beings. Then, to use a traditional Buddhist metaphor, having recognized that we are pregnant with Buddha-nature (*tathāgata-garbha*), we vow to develop on the *bodhisattva*'s path of compassion pursuing universal liberation. Rather than regarding this choice as a personal loss, it is regarded as joyfully finding one's identity and purpose in a maze of purposeless wandering and self-perpetuation. "Today my life has become worthwhile," reads the liturgy for taking the *bodhisattva* vow, the vow that is so central to Mahāyāna Buddhism. Upon taking this vow, one is congratulated for having entered the family and lineage of enlightenment.

Given that *bodhicitta* is regarded as the basic inheritance and potential of all sentient beings, including all humans, rousing and nurturing *bodhicitta* in oneself and encouraging its development in sentient beings is fostering family lineage in its most profound sense, beyond the narrow boundaries of genetic family, tribe, nation, or even species. The way in which such values actually foster perpetuation of our most valuable traits is pointed out by an idiosyncratic modern translation of the term *bodhicitta*. Usually translated "awakened heart-mind," my teacher sometimes translated *bodhicitta* as "enlightened gene," a translation that emphasizes *bodhicitta* both as one's inherited most basic trait and as one's heritage to the mother sentient beings. Who could worry about transmitting family genes when one can awaken, foster, and transmit the gene of enlightenment?

By contrast, the motivations to biological reproduction are often quite narrow and unenlightened. Many religious traditions have criticized material consumption as spiritually counterproductive. Few traditions have seen that biological reproduction can be equally self-centered and ultimately unsatisfactory, or that excessive reproduction stems from the same psychological and spiritual poverty as does excessive consumption. Buddhism, however, can easily demonstrate that biological reproduction is often driven by self-centered motivations, particularly by a desire for self-perpetuation or for the expansion of one's group. And self-centered desire always results in suffering, according to the most basic teachings of Buddhism. To expose the negative underbelly of emotionality and greed motivating much reproduction, to name it accurately, and to stop perpetuating false idealizations of the drive to biological reproduction is more than overdue. Such idealization is part of the pronatalist stance that drives many people, for whom parenthood is not a viable vocation, into reproduction. Regarding all reproduction as beneficial was always illusory, even in times of stable and ecologically viable population density; to continue to encourage or require everyone to reproduce their family lineage under current conditions is irresponsible.

Driven by a desire for self-perpetuation, parents often try to produce carbon copies of themselves, rather than children who are allowed to find their own unique lifeways in the world. The suffering caused by such motivation to reproduction is frequently unnoticed and perpetuates itself from generation to generation. As someone reared by parents who wanted a child who would replace them and reproduce their values and life-style, which I have not done, I am quite well acquainted with the emotional violence done to children who are conceived out of their parents' attachment, to fulfill their parents' agendas. Buddhist literature is filled with such stories. Frequently, personal neediness is the emotion fueling the desire to reproduce. Certainly, the mental state of some people who want to reproduce is far from the calmness and tranquility recommended by Buddhism. I am deeply suspicious of people who need and long to reproduce biologically, of their psychological balance, and of the purity of their motives. In my experience, most of my yuppie friends think population control is a vital issue—for some other segment of the population—but that their drive to reproduce as much as they want to is unassailable. The level of hostility and defensiveness that wells up upon the suggestion that maybe they are motivated by desire for self-perpetuation, rather than by *bodhisattva* practice, convinces me that, indeed, my suspicions are correct. My suspicions are deepened even further when such people endure extreme expense and go to extreme measures to conceive their biological child, instead of adopting one of the many needy children already present in the world. Finally, many people simply are overwhelmed by religious, family, or tribal pressures to reproduce and do not even make a personal decision regarding reproduction. Instead, they are driven by collective ego, which is not essentially different from individual ego. Like all forms of ego, collective ego also results in suffering.

Implicit in this call to recognize the negative underbelly of motivations to reproduction is the call to value and validate alternative nonreproductive life-styles, including gay and lesbian life-styles. One of the most powerful psychological weapons of pronatalism is intolerance of diversity in life-style and denigration of those who are unconventional. People who are childless should be valued as people who can contribute immensely to the perpetuation of the lineage of enlightenment, rather than ostracized and criticized. As a woman who always realized that, in order to contribute my talents to the mother sentient beings, I would probably need to remain childless, I am certainly familiar with the prejudice against women who are childless by choice. It begins with badgering from parents or in-laws about how much they want their family lineages perpetuated and how cheated they feel. It continues with continual feedback that one is self-indulgent to pursue one's vocation and will come to regret that supposed self-centeredness eventually. Then there is the loneliness, the outcasting, that results from friends who are too busy with their nuclear families to be proper friends. And, finally, most especially, there are the self-centered, self-indulgent middle-aged men whose goal in relationships, approved by many, is to have second families with young women. Patriarchal pronatalism is deeply prone to such prejudices.

Needless to say, of course, reproduction can be an appropriate agenda in Buddhist practice and much contemporary Buddhist feminist thought is exploring the parameters of reproduction as a Buddhist issue and practice. In my view, for reproduction to be a valid Buddhist choice and alternative life-style, it must be motivated by Buddhist principles of egolessness, detachment, compassion, and *bodhisattva* practice, not by social and religious demands, conventional norms and habits, compulsive desires, biological clocks, or an ego-based desire to perpetuate oneself. I also believe that such detached and compassionate motivations for parenthood are fully possible, though not anywhere nearly as common as is parenthood. In my own work as a Buddhist feminist theologian, I have also consistently stressed the need to limit both biological reproduction and economic production, as well as to share those burdens and responsibilities equitably between men and women so that meaningful lay Buddhist practice can occur.

The life-style that promotes the attainment of detachment, the Middle Way, wisdom, compassion, and the development of *bodhicitta* is encouraged and valued by Buddhists. Therefore, in many Buddhist countries, celibate monasticism is preferred over reproductive life-styles. Though the Buddhist record is far from perfect, in many, but not all, Buddhist societies this option is also available to women, who are no more regarded as fulfilled through childbearing than men are regarded as fulfilled through impregnating. In much of the contemporary Buddhist world, lifelong monasticism is less popular and less viable, but the movement toward serious lay Buddhist meditation practice is growing dramatically, not only among Western Buddhists but also in Asia, not only among laymen but also among laywomen. Serious Buddhist meditation practice is difficult and time-consuming. When laypeople become engaged in such practices, they must limit both their economic and their reproductive activities appropriately. Thus, both excessive consumption and overpopulation, the twin destructive agents rampant in the world, can be curbed at the same time by coming to value the human potential for enlightened wisdom and compassion and striving to realize them.

Enlightened wisdom sees the interdependence of all beings and forgoes the fiction of private choices that do not impinge on the rest of the matrix of life. Enlightened compassion cherishes all beings, not merely one's family, tribe, nation, or species, as worthy of one's care and concern. The great mass of suffering in the world would be dramatically decreased if the detached pursuit of the Middle Way more commonly guided the choices people make regarding both consumption and reproduction. According to the Buddhist vision of *bodhicitta* as inalienable enlightened gene, both inheritance from and heritage to the mother sentient beings, that which makes life fulfilling is developing compassion and being useful—not self-perpetuation, whether through individual egotism or biological perpetuation of family, tribe, or nation. In case it is not completely clear, this compassion is not regarded as something one has a duty to develop but, rather, as one's inheritance, the discovery of which makes life worthwhile and joyful. Pronatalism as religious requirement or obligation can have nothing to do with this membership in the lineage of enlightenment. Freed of pronatalist prejudice and valued for their contributions to the lineage of enlightenment, not their biological reproduction, human beings who have sufficient talent and detachment to become parents could do so freely, out of motivation more pure than compulsion, duty, or self-perpetuation— and those who make other, equally important contributions to the mother sentient beings would also be celebrated and valued equally.

Sexuality and Communication: A Few Comments on Vajrayāna Buddhism

A commandment to perpetuate the family lineage, combined with criticism of people who limit or forgo biological reproduction, is only one of the major religious sources of pronatalism. The other is at least equally insidious. Antisexual religious rhetoric is quite common in religion, including some layers of Buddhism. Frequently, sexual activity is claimed to be somehow problematic, evil, or detrimental to one's spirituality. Such guilt, fear, or mistrust surrounding sexual activity and sexual experience, grounded in religious rhetoric or rules, leads to several equations or symbolic linkages, all of which foster the agenda of pronatalism, among other negative effects. Regarding sexual experience as forbidden fruit in no way fosters mindful and responsible sexuality.

The first of the major equations that grows out of religious fear of sexuality is the identity between sexuality and reproduction that is so strong in some religious traditions. Some religions espouse the view that the major, if not the only valid, purpose of sexuality is reproduction. Sexual activity not open to reproduction is said to produce negative moral and spiritual consequences for people who engage in them. Therefore, the potential link between sexual activity and reproduction cannot and should not be questioned or blocked. Nonreproductive sexual activities, such as masturbation, homoerotic activity, or heterosexual practices that could not result in pregnancy, are discouraged or condemned. The effect of such views, however,

often aids the pronatalist agenda. Encouraging people to feel
negatively about their sexuality does not seem to curb sexual activity
significantly. But because people have been trained to link sexual
activity with reproduction, or because they have even been forbidden
to take steps to disassociate them, their sexual activity results in a
high rate of fertility, which, combined with the current lower death
rates, contributes greatly to excessive population growth.

Breaking the moral equation between sexual activity and reproduction is a most crucial task, for as long as nonreproductive
sexuality is discouraged or condemned, high birth rates are likely
to continue. That equation is easily broken by re-asking the
fundamental question of the function of sexuality in human society.
It seems quite clear, when we compare human patterns of sexual
behavior with those of most other animal species, that the primary
purpose of sexuality in human society is communication and
bonding. Unlike most other species, sexual activity between humans
can, and frequently does, occur when pregnancy could not result
because a woman, though sexually active, is not fertile. These
nonreproductive sexual experiences are actually crucial to bonding
and communication between human couples and thus to human
society. In addition, sexuality, properly understood and experienced,
is one of the most powerful methods of human communication.
Reproduction is, in fact, far less crucial and far less frequently the
outcome of sexual activity. Thus, it is quite inappropriate to rule
that sexual contact must be potentially open to pregnancy if sexual
activity is not to involve moral and spiritual defilement. Instead,
mindful sexuality, involving the use of birth control unless appropriate and responsible pregnancy is intended, should be the sexual
morality encouraged by all religions.

The view that sexuality should by inextricably linked with
reproduction is closely tied with several other equations that are
equally pronatalist in their implications. When sex cannot be
dissociated from fertility, and when females have no other valid and
valued identity or cultural role than motherhood, most women will
become mothers. Therefore, a symbolic and literal identity between
femaleness and motherhood is taken for granted. Not many years
ago, everyone assumed that a female deity would inevitably be a
"Mother-goddess." I remember well that such platitudes were
commonplace when I began my graduate study in the history of

religions. However, the assumption that even all divine females would be mothers proves to be incredibly naïve and culture-bound. When mythology and symbolism of the divine feminine is investigated free of prevailing cultural stereotypes about the purpose of females, it is discovered that divine females are many things in addition to, sometimes instead of, mothers. They are consorts, protectors, teachers, bringers of culture, patrons of the arts, sponsors of wealth. . . . Nor in mythology is their involvement in other cultural activities dependent on their being nonsexual. In mythology, one meets many divine females who are quite active sexually but who are not mothers or whose fertility is not stressed. Clearly, such religious symbolism and mythology of sexually active, but nonreproductive, females would not promote pronatalism. Therefore, a final caution is necessary. Great care must be taken in symbolic reconstructions of motherhood in contemporary feminist theology lest the symbols again reinforce the stereotype that to be a woman is to be a mother, literally.

The third equation links nurturing with motherhood, an exceedingly popular stereotype in both traditional religion and popular culture and psychology. The negative and limiting effects of this equation are various, not the least of which is the way in which this equation plays into the pronatalist agenda. If nurturing is so narrowly defined, then those who want to nurture will see no other option than to become parents. The equation between nurturing and motherhood also fosters the prejudice against nonreproducers already discussed, since it is easy to claim that they are selfish and non-nurturing. However, the most serious implication of this equation is its implicit limitation on the understanding of nurturing. If nurturing is associated so closely with motherhood, then other forms of caretaking are not recognized as nurturing and are not greatly encouraged, especially in men. The assumption that nurturing is the specialization, even the monopoly, of mothers, and therefore confined to women, is one of the most dangerous legacies of patriarchal stereotyping. Because of the strength of this stereotype, it is often assumed that feminist women, who will not submit to patriarchal stereotypes, would not be nurturing. But obviously, the feminist critique is not a critique of nurturing; it is a critique of the ways in which men are excused from nurturing and women are restricted to, and then punished for, nurturing within the prison of patriarchal gender roles. Feminism is not about restricting nurturing even further or discouraging it, but about recognizing the diversity of its forms and expecting it of all members of society. Since nurturing is valuable and essential to human survival, it is critical that our ideas about what it means to nurture extend beyond the image of physical motherhood to activities such as teaching, healing, caring for the earth, engaging in social action. . . . It is equally important that all humans, including all men, be defined as nurturers and taught nurturing skills, rather than confining this activity to physical mothers.

Because some of the grounds for fear, mistrust, and guilt surrounding sexuality are in religion, a religious, rather than merely secular or psychological alternative, view of sexuality would be significant to this discussion of religious ethics, population, consumption, and the environment. A religious evaluation of sexuality as sacred symbol and experience, helpful rather than detrimental to spiritual development, would certainly inject relevant considerations into this forum. Vajrayāna Buddhism—the last form of Indian Buddhism to develop, which is today significant in Tibet and becoming more significant in the West—includes just such a resource. Needless to say, it is crucial that such discussions of Vajrayāna Buddhism be disassociated from the titillating accounts of "tantric sex" that actually stem from fear and guilt about sexuality.

Symbolism and practice of sacred sexuality, such as that found in Tibetan Vajrayāna Buddhism, is radically unfamiliar to many religious traditions, including those most familiar to Western audiences. In Vajrayāna Buddhism, the familiar paired virtues, wisdom and compassion, are personified as female and male. Not only are they personified; they are painted and sculpted in sexual embrace, usually called the "yab-yum" icon. This icon is then used as the basis for contemplative and meditative practices, including visualizing oneself as the pair joined in embrace. After many years of working with this icon personally, I am quite intensely captivated by the liberating power and joy of this symbol. Rather than being a private and somewhat embarrassing, perhaps guiltridden indulgence, sexuality is openly portrayed as a symbol of the most profound religious truths and as contemplative exercise for developing one's innate enlightenment.

One of the most profound implications of the yab-yum icon and its centrality is the fact that the primary human relationship used to symbolize reality is that of equal consorts, of male and female as joyous, fully cooperative partners. This contrasts sharply with the tendency to limit religious symbolism to parent-child relationships, whether of Father and Son or of Madonna and Child, that are so common in other traditions. It also contrasts strongly with the abhorrence of divine sexuality that has been such a problem in those same traditions. One cannot help but speculate that this open celebration of sexuality as a sacred and profoundly communicative and transformative experience between divine partners would significantly defuse pronatalism based on a belief that sex without the possiblity of procreation is wrong.

In the realm of human relations rather than religious symbols—insofar as the two can be separated—this symbolism has led, in Vajrayāna Buddhism, to the possibility of spiritual and dharmic consortship between women and men. (The question of whether nonheterosexual relationships were also possible is more difficult to answer.) Such relationships are not conventional domestic arrangements or romantic projections and longings, but are about collegiality and mutual support on the path of spiritual discipline. Sexuality seems to be an element within, but not the basis of, such relationships. Though relatively esoteric, such relationships were, and still are, recognized and valued in late North Indian Vajrayāna Buddhism, as well as in Tibetan Buddhism. Western Buddhists are just beginning to discover or recover this resource, this possibility of consortship as collegial relationship between fellow seekers of the way and as mode of understanding and communicating with the profound "otherness" of the phenomenal world. To value, valorize, and celebrate such relationships would profoundly undercut pronatalist biases regarding the place of sexuality in human life, as well as contribute greatly to the creation of sane, caring, egalitarian models of relationship between women and men.

Buddhism, Global Ethics, and the Earth Charter

Steven C. Rockefeller

As the peoples and nations of the world prepare to enter the twenty-first century during a time of dramatic social change and increasing global interdependence, considerable attention is being given to the task of developing a new global ethics. An effort is now underway to create an Earth Charter that will give concise expression to those core ethical principles and practical guidelines necessary to ensure that Earth remains a secure home for humanity and the larger community of life. Those supporting this initiative hope that the Earth Charter will eventually be adopted by the United Nations General Assembly and that it will do for the protection and restoration of the environment and the cause of sustainable living what the Universal Declaration of Human Rights has done for the promotion of human rights and fundamental freedoms. It is the purpose of this essay to ask what distinctive contributions the Buddhist tradition might make to the development of the Earth Charter.

The need for ethical values that are shared worldwide and for what the Dalai Lama has called a sense of universal responsibility is fundamental and urgent.[1] Economic and technological forces are creating a new global community, and the process of globalization cannot be stopped. In addition, the industrial and technological revolutions sweeping the planet are causing severe worldwide problems that can only be resolved with global solutions and cooperation involving all sectors of society. Much can be done to address these problems through the development of new technologies, regulatory systems, and market mechanisms, but a change in attitudes and values on the part of individuals is also essential. It is the task of global ethics to chart the value changes needed and to guide the forces shaping the emerging global community in creative directions that promote planetary well-being.

If a new world ethics is to capture the minds and hearts of people throughout the world, it must have roots in their diverse traditions and emerge out of these many traditions. The new ethics will require transformations in the way people think and act, but it should not feel externally imposed. It should be constructed as the necessary extension or further development of basic values and principles that people respect and honor. In this regard, the contribution and leadership of the world's religions is of great importance.

Religions can help to further the growth of humanity's ethical consciousness in an age of global interdependence by applying the wisdom contained in their different traditions to the major problems of the time and by entering into interfaith dialogue in an endeavor to identify common concerns and values. As Hans Küng has put it:

> No survival without a world ethic.
> No world peace without religious peace.
> No religious peace without religious dialogue.[2]

In their relations with each other, the world's religions are called to model the kind of community that the diverse peoples of the world should be striving to realize. This means working out an agreement on the ethics of living together in a multicultural world that is interconnected ecologically, economically, and socially. The Earth Charter Project provides a unique opportunity for interfaith dialogue and for collaboration between religions and secular society on the ethical visions that inspire people's noblest undertakings.

International support for an Earth Charter has been slowly but steadily building since 1987, when the United Nations World Commission on Environment and Development called for the creation of a new charter that would "prescribe new norms for state and interstate behavior needed to maintain livelihood and life on our shared planet."[3] During the Rio Earth Summit, the 1992 United Nations Conference on Environment and Development (UNCED), significant efforts were made to develop an Earth Charter, but the time was not right. The Rio Declaration articulates a number of fundamental principles, but it did not meet the criteria that were set for the Earth Charter. In 1994 Maurice Strong, the former Secretary General of UNCED and the Chairman of the Earth Council, and Mikhail Gorbachev, in his capacity as Chairman of Green Cross International, together launched a new Earth Charter initiative. An international Earth Charter workshop was held at the Peace Palace in The Hague in the spring of 1995. The following year, a worldwide Earth Charter consultation process was organized as part of the Rio + 5 independent review directed by the Earth Council in coordination with the UN Rio + 5 review that will culminate with a special session of the UN General Assembly in June 1997. An international Earth Charter Commission will oversee the drafting of the Earth Charter in 1997.

The Earth Charter will be prepared as a relatively brief "soft law" instrument written in clear, inspiring language.[4] It will build on earlier international declarations, charters, and treaties, including some that have been drafted by a variety of nongovernmental organizations. Over the past twenty-five years, beginning with the Stockholm Declaration generated by the UN Conference on the Human Environment in 1972, substantial progress has been made in developing international law regarding the environment and sustainable development. A very significant international consensus is emerging around forty or fifty principles relevant to the Earth Charter.[5] These principles reflect a rational and pragmatic approach to the world's problems. They have been heavily influenced by the findings of the new science of ecology and by certain fundamental ethical concerns regarding human rights, social justice, economic equity, future generations, respect for nature, and environmental protection. The creation of an Earth Charter will involve further refining and developing the principles that form the emerging international consensus. This can be achieved through a global dialogue that draws on the insights of science, the practical experience of men and women who are living sustainably, and the extensive world literature on the ethics of environment and development as well as the wisdom of the world's religions.

The remainder of this essay focuses specifically on how the Buddhist tradition might respond and contribute to the development of an Earth Charter. How can Buddhists help to create a document that speaks to people in all cultures and all walks of life and that provides a foundation of shared ethical wisdom for healing the planet and creating peace, justice, democratic participation, sustainable development, and ecological well-being? What principles from a Buddhist perspective constitute the core of the emerging world ethic? How should these principles be formulated? What language should be used? Responses to these questions may, of course, involve a variety of answers reflecting different Buddhist perspectives. In addition, each different religion may want to formulate its own version of global ethics using its own distinctive language and worldview. However, the urgent question with regard to the Earth Charter concerns how far humanity can go in formulating its global ethics using a common language. Some further observations and questions regarding global ethics and Buddhism may help to stimulate some productive reflection on these issues.

First, at the recent centenary of the Parliament of the World's Religions held in Chicago in 1993, an interfaith committee headed by Hans Küng drafted a statement on global ethics that was adopted by the whole parliament. This statement identified the Golden Rule as the most fundamental moral principle shared by the world's religions.[6] In other words, the Golden Rule may be taken as the simplest formulation of the general meaning of the moral imperative to do good and to avoid evil. This is a position Buddhism can support. A variation on the theme of the Golden Rule is found, for example, in *The Precious Garland* (*Ratnāvalī*) by Nāgārjuna:

> Just as you love to think
> What could be done to help yourself,
> So should you love to think
> What could be done to help others.[7]

Further, the Dalai Lama has stated that all the teachings of the Buddha contained in both the Hīnayāna and Mahāyāna can be summarized in two ethical imperatives: "You must help others. If not, you should not harm others."[8] The imperative to help others and to avoid harming others corresponds to the general meaning of the positive and negative formulations of the Golden Rule. This teaching, asserts the Dalai Lama, is "the basis of all ethics." Here, then, is a shared first principle upon which people from very different cultures and religions can found their efforts to develop a global ethics. Awakening to the moral ideal of helping and not harming also, of course, introduces a person into the complexities and tragedy of the human situation, for we exist in a world where we cannot live without causing some harm and where at times the effort to help some beings inevitably involves harming others. This dilemma by itself is sufficient to keep ethical philosophers in business.

In a recent essay, "Toward the Possibility of a Global Community," the Confucian philosopher Tu Weiming has stressed the fundamental importance of the Golden Rule:

> The first step in creating a new world order is to articulate a universal intent for the formation of a global community. This requires, at a minimum, the replacement of the principle of self-interest, no matter how broadly defined, with a new golden rule: "Do not do unto others what you would not want others to do unto you."[9]

Professor Tu then adds that this negative formulation of the Golden Rule must also be augmented by a positive principle that reflects the reality of ecological and social interdependence: "In order to establish myself I must help others to establish themselves; in order to enlarge myself, I have to help them to enlarge themselves." This proposal seems very much in line with the Dalai Lama's succinct summary of the teaching of the Buddha, even though a Buddhist may not wish to use the language of establishing and enlarging the self. Buddhism also insists, of course, that the principle of helping and not harming should be extended to embrace the relations of people to all sentient beings, not just to human beings.

This discussion raises the question of whether the principle of the Golden Rule should be included in the Earth Charter as humanity's most fundamental shared ethical ideal, and, if so, how it should be worded. It has already been cited in the report of the World Commission on Environment and Development, *Our Common Future* (1987), in a section dealing with proposed new international law principles designed to prevent transboundary environmental harm. The report recommends that states, in their relations with other states, should adopt the principle, "Do not do to others what you would not do to your own citizens."[10] In addition, the most fundamental principle of environmental protection is widely recognized today to be a variation on the theme of *ahiṃsā*, or no harming. For example, the Draft International Covenant on Environment and Development prepared by the IUCN Commission on Environmental Law states in a section on "Fundamental Principles": "Protection of the environment is best achieved by preventing environmental harm rather than by attempting to remedy or compensate for such harm" (Article 6).[11] This principle is viewed as especially important in situations where there is the chance of irreversible environmental harm, as would occur, for example, if an endangered species were to be eliminated.

Second, contemporary international law is increasingly using the worldview emerging from the new physics, evolutionary biology, and ecology to justify many of the principles and guidelines being developed regarding the environment and sustainable development. Emphasis is put on the unity of the biosphere, the interdependence of humanity and nature, the interconnectedness of all members of the larger community of life, and the importance of biodiversity as well as cultural diversity. Does Buddhism support this new ecological worldview emerging from scientific inquiry? In recent years, a number of scholars have pointed out that there seems to be a significant convergence of Buddhist philosophy and contemporary physics, ecology, and environmental ethics.

Third, perhaps the single most important contribution Buddhism could make to the Earth Charter involves securing from the international community through the Earth Charter a stated intent to cooperate in providing all sentient beings with protection from cruel human treatment and unnecessary suffering. The Four Noble Truths focus attention on suffering as the fundamental problem from which sentient beings seek liberation, and Buddhist ethics regards compassion for the suffering of all sentient beings as the supreme ethical virtue. The emerging world ethic as expressed in international law does address many causes of human suffering, such as poverty, war, inequity, ignorance, and environmental degradation. However, existing international law does not identify the suffering of nonhuman beings as a moral issue. It is concerned with species and the protection of biodiversity, but it is not concerned with individual nonhuman sentient beings unless they are representatives of an endangered species. If protection is mandated for a member of a nonhuman species and its habitat, it is out of concern for the preservation of species, not prevention of suffering. International law does help to reduce some suffering of nonhuman creatures by calling for the protection of ecosystems, but the concern here is with ecosystem health and the integrity of biotic communities in line with Aldo Leopold's land ethic, and not with the suffering of individual creatures.

From a Buddhist point of view, a strong argument can be made that the Earth Charter should identify as a moral issue the suffering of nonhuman sentient beings caused by humans and it should include a principle that addresses this problem. The national law in many states does address the issue of cruelty to and abuse of animals. The question is whether international law should enter this field, joining the environmentalists' concern regarding ecosystems and species with compassion for all individual sentient beings. The IUCN guideline on the treatment of nonhuman creatures set forth in *Caring for the Earth* (1991) offers an example of what might be included in the Earth Charter. It states: "People should treat all creatures decently, and protect them from cruelty, avoidable suffering, and unnecessary killing."[12] Given its ethical traditions, Buddhism is in a unique position to influence the outcome of the debate over this issue. If its voice is not heard on this subject, it will be hard to win acceptance for such a principle. The animal liberation movement is not strong enough to carry the day on this matter without help from the world's religions.[13]

Fourth, there is a related issue which concerns the use of language. Reflecting the discussion in much contemporary environmental philosophy, international environmental law argues that all life-forms, that is, species, warrant respect and protection, because they are of intrinsic value quite apart from whatever value they might have for human beings. This claim is made, for example, in the World Charter for Nature (1982) and the Convention on Biological Diversity (1992). Here again, the assertion is that species, not individual creatures, are of intrinsic value. The importance of the appearance in international legal documents of the assertion that all species have intrinsic value cannot be overstated. It is a major breakthrough—a move beyond the traditional anthropocentric worldview that has dominated Western culture and much of the rest of the world in recent centuries. It establishes a basis for extending the community with which humans identify and for which they are morally responsible to include all life-forms. It means that nonhuman species deserve respect and care regardless of their instrumental value to humans. They are to be treated, in Kantian language, as ends-in-themselves and never as a means only. In other words, nonhuman species are to be regarded as subjects with moral standing and not merely as objects to be possessed and used.

However, even though Buddhism teaches compassion for all sentient beings, some Buddhist philosophers have expressed reservations about the notion of a being or species having intrinsic value. The issue seems to be that the concept of intrinsic value suggests the existence of some fixed essence or permanent self in things, which is contrary to the Buddhist doctrines of dependent co-arising, impermanence, emptiness, and no-self. In the light of these concerns, the question arises as to whether Buddhist philosophy can or should affirm the concept of the intrinsic value of all species and sentient beings. In considering this question, it should be kept in mind that if Buddhists were to reject the language of intrinsic value employed in international law, it could result in a major setback for the contemporary international effort to extend humanity's moral concern to include all sentient beings.

In seeking a solution to this problem, it is important to emphasize that Buddhist philosophy and ethics do not have a quarrel with the practical meaning or bearing of the concept of intrinsic value. The relevant point is that any being with intrinsic value is worthy of respect and care. Could there, then, be a Buddhist definition of the nature of intrinsic value consistent with the Buddhist doctrines of dependent co-arising and impermanence? It is true that some Western philosophers and theologians may try to explain the idea of intrinsic value with reference to the existence of a soul, an eternal self, or some permanent essential nature, but this is not the only way to explain that other beings are subjects worthy of respect and are not mere objects or means to be used and exploited. Could the intrinsic value of all beings be explained, for example, with reference to Buddha-nature?

Fifth, regarding a related question, some earth covenants and charters of nongovernmental organizations employ the language of the sacred.[14] For example, they affirm the sacredness of life or of all life-forms. This is another way of speaking about the intrinsic value of other life-forms and creatures. Buddhism affirms a reverence for life, especially sentient life. Would Buddhists support reference in the Earth Charter to the sacredness of life or to the sacredness of the community of life or the intricate web of life?

Sixth, this discussion of the idea of intrinsic value and the sacredness of life leads to questions about another important concept—the idea of rights. In the course of the past three hundred years, respect for human rights has come to be viewed as fundamental to the meaning of social justice. The Golden Rule has come to mean first and foremost: respect the basic dignity and the rights and fundamental freedoms of all persons. All people, it is agreed, have equal rights, because they are beings who are of intrinsic value. In the course of the last century, human rights law has been adopted throughout the world by nations with very different political orientations. Since World War II, the idea of human rights has been extensively developed in international law, and three binding global treaties on human rights have been adopted. Respect for human rights is one of the major elements of the emerging new global ethics. Many Buddhists strongly support human rights law. Should rights language be used with regard to nonhuman species? From a Buddhist perspective, people do have moral responsibilities in relation to nonhuman beings. Should the Earth Charter speak about the rights of nature or the rights of nonhuman sentient beings?

The use of rights language with reference to nature is very controversial, and it has not yet been used in any international document. It is not as fundamental an issue as establishing that nonhuman species and beings possess intrinsic value. The idea of intrinsic value establishes the essential foundation for affirming humanity's moral responsibility to respect and care for nature. Whether one uses rights language or not in the Earth Charter to clarify humanity's responsibilities in relation to nonhuman species is a secondary issue. In other words, the Charter could articulate a very strong ethic of respect and care for the community of life without employing rights language in relation to nonhuman species. The advantage of rights language is that it has a widely understood, clear moral and legal meaning, and its use would facilitate the legal protection of nonhuman species.

Finally, are there fundamental attitudes toward life and the world at large that from a Buddhist point of view might be given expression in the Earth Charter? In this regard, the leaders of the Earth Charter project would like the Charter to have spiritual depth. It could, for example, mention such attitudes toward life and the world as wonder, awe, reverence, humility, repentance, gratitude, compassion, and universal responsibility. This question recognizes the possibility that we may be at a point in the evolution of human consciousness and civilization where human beings from all cultures can affirm the value of a number of basic attitudes toward life as well as agree on a set of ethical principles that are consistent with and give expression to these attitudes.

The Earth Charter consultation process will continue throughout 1997 and beyond. When the Earth Charter has been drafted in final form, it will initially be circulated as a "peoples' treaty" for signature by individuals and adoption by religious organizations, nongovernmental organizations, and other groups throughout the world. It is hoped that, with a strong show of popular support, the Earth Charter will receive the approval of the United Nations by the year 2000.

The Buddhist community can make an important contribution to the ongoing Earth Charter consultation process. This essay has identified only a few of the many issues that must be addressed in drafting the Charter. Groups interested in a further introduction to the issues and principles that must be considered regarding the Charter may contact the Earth Council and take advantage of the resources that have been prepared by the Earth Council in support of the consultation process.[15]

The larger significance of the Earth Charter project is that it focuses the debate on global ethics in a very specific fashion and sets the stage for a very productive interfaith, cross-cultural dialogue. If it is carefully constructed, the Charter will provide ethical and practical guidance to individuals, schools, businesses, governments, religious congregations, nongovernmental organizations, and international assemblies. It can serve as an inspiring ethical compass for all humanity. It presents a challenge that is worthy of the best efforts of religious communities and thoughtful men and women everywhere.

Notes

1. The Dalai Lama, *A Policy of Kindness* (Ithaca, N.Y.: Snow Lion Publications, 1990), 17–19. The Dalai Lama explains that each and every person must assume responsibility for addressing the interrelated problems that face the larger world today, taking such action as is appropriate for the individual involved.

2. Hans Küng, *Global Responsibility: In Search of a New Global Ethic* (New York: Crossroad Publishing Company, 1991), 1, 71, 107, 138.

3. World Commission on Environment and Development, *Our Common Future* (New York: Oxford University Press, 1987), 332–33.

4. "Soft law" documents in the field of international law are considered to be statements that express the intention and aspirations of the states involved, but they are not viewed as binding treaties. However, some soft law instruments— the Universal Declaration of Human Rights, for example—eventually become hard law.

5. For clarification on these principles, see Steven C. Rockefeller, "Global Ethics, International Law, and the Earth Charter," in *Earth Ethics* 7, no. 3-4 (spring-summer 1996), and Steven C. Rockefeller, *Principles of Environmental Conservation and Sustainable Development: Summary and Survey* (1996). The latter document was prepared for the Earth Council in support of the Earth Charter Project, and it provides a summary overview of international law principles relevant to the Earth Charter. Copies may be ordered from Steven C. Rockefeller, P.O. Box 648, Middlebury, Vermont 05753, U.S.A.

6. Hans Küng and Karl-Josef Kuschel, eds., *A Global Ethic: The Declaration of the Parliament of the World's Religions* (New York: The Continuum Publishing Company, 1993), 23–24.

7. Nāgārjuna and the Seventh Dalai Lama, *The Precious Garland and the Song of the Four Mindfulnesses*, trans. Jeffrey Hopkins and Lati Rimpoche with Anne Klein (London: George Allen and Unwin Ltd., 1975), 55.

8. The Dalai Lama, *A Policy of Kindness*, 88, 96.

9. Tu Weiming, "Toward the Possibility of a Global Community," in Lawrence S. Hamilton, ed., *Ethics, Religion, and Biodiversity* (Cambridge: White Horse Press, 1993), 72.

10. The World Commission on Environment and Development, *Our Common Future*, 350.

11. IUCN is the International Union for the Conservation of Nature and Natural Resources, also known as the World Conservation Union. Its headquarters are in Gland, Switzerland, and its members include over eighty state governments and over four hundred nongovernmental organizations.

12. World Conservation Union (IUCN), United Nations Environment Programme (UNEP), and World Wide Fund for Nature (WWF), *Caring for the Earth* (Gland: Switzerland, 1991), 14.

13. Professor Jay McDaniel of Hendrix College, Arkansas, has drafted "An Open Letter to Authors of the Earth Charter" that addresses this issue and proposes that the Earth Charter include an animal protection principle using language very similar to that found in *Caring for the Earth*. The Open Letter substitutes the word "animals" for "creatures." One could also use "sentient beings." The Open Letter is being circulated by the Humane Society of the United States to other groups, including religious organizations, in the hopes that they will endorse it.

14. See, for example, "The Earth Covenant," which has been prepared and circulated by Global Education Associates, 475 Riverside Drive, Suite 1848, New York, New York 10115, U.S.A., and "The Earth Charter," designed and circulated by the International Coordinating Committee on Religion and the Earth, P.O. Box 67, Greenwich, Connecticut 06831-0767, U.S.A. Both of these documents have been reprinted in Joel Beversluis, ed., *A SourceBook for Earth's Community of Religions* (Grand Rapids, Mich.: CoNexus Press-SourceBook Project; New York: Global Education Associates, 1995), 201, 214–15.

15. The Earth Council has established an Earth Charter page on its internet website. Portions of the document, described above in note 5, on *Principles of Environmental Conservation and Sustainable Development: Summary and Survey* are included in this Earth Charter internet site. The Earth Council web page is located at http://www.ecouncil.ac.cr. In addition, a special Earth Charter double issue of the journal *Earth Ethics* 7, no. 3-4 (spring-summer 1996), has been published by the Center for Respect of Life and Environment, 2100 L Street, NW, Washington, D.C. 20037, U.S.A.

Theoretical and Methodological Issues
in Buddhism and Ecology

Is There a Buddhist Philosophy of Nature?*

Malcolm David Eckel

One of the most common and enduring stereotypes in environmental literature is the idea that Eastern religions promote a sense of harmony between human beings and nature. On the other side of the stereotype stand the religions of the West, promoting the separation of human beings and nature and encouraging acts of domination, exploitation, and control. Roderick Nash gave classic expression to this contrast when he said: "Ancient Eastern cultures are the source of respect for and religious veneration of the natural world" and "In the Far East the man-nature relationship was marked by respect, bordering on love, absent in the West."[1] Y. Murota drew a similar contrast between Japanese attitudes toward nature and the attitudes he felt are operative in the West: "the Japanese view of nature is quite different from that of Westerners. . . . For the Japanese nature is an all-pervasive force. . . . Nature is at once a blessing and a friend to the Japanese people. . . . People in Western cultures, on the other hand, view nature as an object and, often, as an entity set in opposition to mankind."[2]

This contrast between the East and the West owes much of its influence in recent environmental literature to the seminal article by Lynn White, Jr., "The Historical Roots of Our Ecologic Crisis."[3] White depicted the Judeo-Christian tradition as anthropocentric and argued that Judeo-Christian anthropocentrism stripped nature of its sacred status and exposed it to human exploitation and control. While he did not comment at great length about the Eastern traditions, he clearly understood them as the opposite of the traditions of the West.

The beatniks and hippies, who are the basic revolutionaries of our time, show a sound instinct in their affinity for Zen Buddhism and

Hinduism, which conceive of the man-nature relationship as very nearly the mirror image of the Christian view.

White's image of the contrast between East and West was taken up in the same journal seven years later by the Japanese historian Masao Watanabe.[4] Watanabe associated the Japanese people with "a refined appreciation of the beauty of nature" and said that "the art of living in harmony with nature was considered their wisdom of life." White's image continues to be reflected by some of the best-known contemporary writers in the environmental movement. In a recent collection of essays, Gary Snyder, the venerable and respected survivor of Lynn White's generation of "beatniks and hippies," drew a series of graceful connections between Henry David Thoreau's concept of the "wild," the Taoist concept of the *Tao*, and the Buddhist concept of *Dharma*:

> Most of the senses in this second set of definitions [of the wild] come very close to being how the Chinese define the term *Dao*, the *way* of Great Nature: eluding analysis, beyond categories, self-organizing, self-informing, playful, surprising, impermanent, insubstantial, independent, complete, orderly, unmediated, freely manifesting, self-authenticating, self-willed, complex, quite simple. Both empty and real at the same time. In some cases we might call it sacred. It is not far from the Buddhist term *Dharma* with its original sense of forming and firming.[5]

This image of an affirmative Eastern attitude toward nature must have lurked in the minds of the environmental activists and friends of the environment who gathered at Middlebury College in the fall of 1990 to hear the fourteenth Dalai Lama speak on the topic of "Spirit and Nature." Tibet, like traditional Japan, has been the focus of a certain Western yearning for the East as a place to discover not only a unique sense of wisdom (what one observer called "an intimate and creative relationship with the vast and profound secrets of the human soul") but a wisdom that can insure "the future survival of Earth itself."[6] There was a hush in the Middlebury field house as the Dalai Lama seated himself on the stage and began to speak.[7] It must have been a surprise when he began by saying that he had nothing to offer to those who came expecting to hear about ecology or the environment, and even more surprising when he interpreted the word "nature" as a reference to "the fundamental nature of all reality" and entered into a discourse on the Buddhist concept of Emptiness. To explain the connection between nature and Emptiness, he said: "When talking about the fundamental nature of reality, one could sum up the entire understanding of that nature in a simple verse: 'Form is emptiness and emptiness is form' (*The Heart Sūtra*). This simple line sums up the Buddhist understanding of the fundamental nature of reality."[8] And he went on to explain how Tibetan philosophers use logical analysis to develop their view of Emptiness and to pursue what he said was the "expressed aim of Buddhism," namely, the purification and development of the mind.

The Dalai Lama's words were surprising not because he seemed unfriendly toward the "natural" world in the prevailing sense of the word (that is, toward ecosystems of plants, animals, the atmosphere, the ocean, rivers, mountains, and so on), but because he so gently and easily shifted attention away from the natural world toward the development of human nature and the purification of the mind. The sense of surprise only became more acute when he began to develop the concept of Emptiness and indicated that it involved a denial of the reality of what he took to be "nature" itself. To say that "Form is Emptiness and Emptiness is Form," in the language of Mahāyāna philosophy, is to say that all things are "empty" of any inherent "nature" or identity.[9] The purification of the mind, which the Dalai Lama called the "expressed aim of Buddhism," comes from stripping away false concepts of the "nature" of things and resting content with their Emptiness. In other words, "nature" (in one possible meaning of the word) may very well be a barrier to overcome in a quest for human development.

What should we make of the gap between the Dalai Lama's words and the conventional image of the Buddhist attitude toward nature? Does the Dalai Lama see something in the Buddhist tradition that others do not? Is the image of Buddhism as an ecologically friendly tradition simply an artifact of the Western imagination? Or is it possible that the Buddhist tradition is a complex combination of ideas and aspirations, some of which are positively disposed toward the environment and some of which are not? If so, is it possible to reconcile the Dalai Lama's approach to the concept of nature with the image of a tradition that seeks to establish harmony between human beings and the natural world? The purpose of this essay is to explore the incongruity in the Dalai Lama's words, to ask where the incongruity comes from, and to ask whether it is possible to identify a "Buddhist philosophy of nature," a philosophy that is genuinely affirmative of what we have come to think of as the "natural" world and, at the same time, true to the complex impulses that shape the Buddhist quest for the purification and development of the mind.

To start with, where do we get the stereotype of Buddhist reverence for the natural world? Masao Watanabe began his account of the Japanese attitude toward nature by telling a story about the nineteenth-century art historian Lafcadio Hearn and the genesis of Western perceptions of Japan. Watanabe said that he read Lafcadio Hearn's account of his first visit to Japan to a group of American students. (It was the trip that led to Hearn's fascination with Japan and to his decision to make Japan his permanent residence.) Watanabe asked his students what they first noticed about Hearn's account of his visit. The answer was Hearn's image of the Japanese love of nature, symbolized in Hearn's story of a Japanese warrior who arranged vases of chrysanthemums to welcome his brother home from a journey. The students' answer then elicited Watanabe's own comments about the sense of natural beauty in Japanese land-scape design, flower arrangement, the tea ceremony, poetry, and cuisine.

Watanabe is right to suggest that Western people first approach Japanese views of nature through an aesthetic medium. When Japan opened to the West in the early 1850s, Japanese art flooded into Western markets and had a significant effect on the stylistic vision of Western artists as different as James McNeill Whistler and Vincent Van Gogh.[10] There are few more powerful and suggestive icons of the Japanese vision of nature than the gnarled rocks and empty spaces of a Zen garden like the one at Ryoan-ji in Kyoto, and few poets of the natural world can match the grace and intensity that is so evident in the works of the Japanese poet Bashō. It is sometimes said that to grasp the significance of Bashō's poem,

> Old pond—
> Frog jumps in-
> Sound of water!

is to grasp the whole meaning of Buddhism.[11] Certainly, the "meaning" of this poem must have something to do with the condensed appreciation of a single moment in the flow of the natural world, a moment in which the minds of the poet and the reader become absorbed in the natural event itself.

Bashō's poetic appreciation of nature has strong antecedents in Chinese literature, as in the work of the shadowy T'ang dynasty poet whose identity is known simply by the name Cold Mountain. In the lines of the Cold Mountain poet, the Buddhist "way" takes on a distinctly naturalistic flavor.

> As for me, I delight in the everyday Way,
> Among mist-wrapped vines and rocky caves.
> Here in the wilderness I am completely free,
> With my friends, the white clouds, idling forever.
> There are roads, but they do not reach the world;
> Since I am mindless, who can rouse my thoughts?
> On a bed of stone I sit, alone in the night,
> While the round moon climbs up Cold Mountain.[12]

This verse displays a distinctive sensitivity to the rough, unhewn aspects of nature, to mists, rocks, and trees—all the aspects of nature that Gary Snyder associated with Henry David Thoreau's concept of the "wild."[13] But it also expresses important Buddhist values. The lines reflect the traditional theme of the Middle Way, leading from the experience of suffering and ignorance in the world of ordinary people to the wisdom of a solitary and enlightened sage, and they map the contrast between these two realms of experience in a series of standard images. The ordinary world is one of entanglement, obscurity, and darkness, with "mist-wrapped vines," and "idling clouds." The world of enlightenment is one of detachment, coolness, and clarity, where the round moon that symbolizes the Buddha's awareness climbs up Cold Mountain. "Cold Mountain" is not merely the setting for the poems and a reference to the poet's own identity; it also expresses the path a sage has to tread to reach enlightenment and symbolizes enlightenment itself. To combine all of these meanings in a single, concrete image is to suggest that enlightenment involves a sense of fusion between the self and the natural world.

William R. LaFleur has shown that Bashō's poetry is the result of a long process of doctrinal reflection in East Asia about the religious significance of nature.[14] When one of Bashō's predecessors, the poet Saigyō (twelfth century), for example, depicts movement along the road to enlightenment as involving "just a brief stop" to linger in the shade of a willow, he raises a question about the nature of the way itself. Is it better to walk the road like a diligent pilgrim with your eyes fixed firmly on a distant goal or to step off the road and allow your consciousness to merge with some part of the natural world?

> "Just a brief stop"
> I said when stepping off the road
> into a willow's shade
> Where a bubbling stream flows by. . .
> As has time since my "brief stop" began.

Here it is the shade of the willow rather than the pilgrim's road that stops consciousness of the passage of time, and this "stopping" reflects the "cessation" of the Buddha's *nirvāṇa*. But why associate *nirvāṇa* with a willow rather than some other element of the natural world? LaFleur has shown that these lines reflect a complex doctrinal discussion about whether plants in particular can have "Buddha-nature," in other words, whether they can embody the state of enlightenment that the pilgrim is seeking. In China this question was first raised as part of the general discussion of the relationship between Emptiness and ordinary reality. The question then became focused as a specific question about vegetation. Did plants have Buddha-nature? Some Buddhist thinkers found an affirmative answer to this question in the chapter on "Plants" in *The Lotus Sūtra*, where it is said that the rain of the Buddha's teaching falls equally on all forms of vegetation, and each plant grows up and is nourished according to its own capacity.[15] In Japan this view evolved into the position represented by Saigyō's "brief stop." The natural world was treated as having special significance as a setting for the experience of enlightenment—enough significance to invite the poet to turn off the path and disappear in the shade of the willow.

Saigyō was not the only one, and his was not the only way, to explore the relationship between the natural world and the experience of enlightenment. Allan G. Grapard has shown that the concept of enlightenment can be mapped onto the physical landscape in even more complex ways.[16] The volcano Futagoyama on the Kunisaki peninsula, for example, was treated as a physical manifestation of the text of *The Lotus Sūtra*: its twenty-eight valleys were treated as equivalent to the twenty-eight chapters of *The Lotus Sūtra*; and its paths were lined with more than sixty thousand statues representing the total number of ideograms in the text. Here the landscape itself is the text, and the text is the *Dharma*. To walk the paths on the mountain and read its valleys as visual representations of the *Dharma* is to experience the relationship between nature and text, path and goal, and cessation and movement in a way that goes far beyond the simplicity of Saigyō's lines.

More examples could be cited of the relationship between Buddhist values and Japanese appreciation of the natural world, but these should be sufficient to show that Watanabe certainly had reason to say that reverence for nature plays a special part in Japanese culture, including its Buddhist dimension. There also are good reasons to think, however, that this is not the whole picture. In a remarkable article entitled "Concepts of Nature East and West," Stephen R. Kellert has given clear statistical shape to the suspicion that Eastern cultures are just as capable of showing disrespect for nature as their Western counterparts.[17] "In contrast," Kellert says, "to the foregoing descriptions of highly positive Eastern attitudes toward nature, modern Japan and China have been cited for their poor conservation record—including widespread temperate and tropical deforestation, excessive exploitation of wildlife products, indiscriminate and damaging fishing practices, and widespread pollution."[18] Kellert prepared a questionnaire to investigate and compare Japanese and American attitudes toward the natural world. He found that the most common approach to wildlife in both cultures, Japanese and American, was the one that Kellert called "humanistic": both cultures showed "primary interest and strong affection for individual animals such as pets or large wild animals with strong anthropomorphic associations." The percentage of people who held this opinion was 37 percent for Japan and 38 percent for the United States. The second most common attitude in the United States was the "moralistic": 27.5 percent of the American respondents showed what Kellert called a "primary concern for the right and wrong treatment of animals and strong opposition to overexploitation and cruelty toward animals." The second most common attitude in Japan, with 31 percent, was the attitude that Kellert called "negativistic": a "primary orientation [toward] an active avoidance of animals due to dislike or fear." The third most common Japanese attitude was one that he called "dominionistic" (28 percent): involving "primary interest in the mastery and control of animals." In other words, more than 50 percent of Kellert's Japanese respondents feared or disliked animals or were primarily concerned with their mastery or control. Kellert's findings received statistical confirmation from a 1989 survey by the United Nations Environmental Program: the survey found that Japan rated "lowest in environmental concern and awareness" of the fourteen countries surveyed.

Kellert pursued his investigation with a series of detailed interviews to elicit explanations of the Japanese attitudes. Many of the people interviewed "indicated that the Japanese tend to place greatest emphasis on the experience and enjoyment of nature in highly structured circumstances." The reasons for this emphasis were diverse but quite revealing. One person referred to "a Japanese love of 'seminature,' somewhat domesticated and tame." Another said that the Japanese "isolate favored environmental features and 'freeze or put walls around them.'" For all of Kellert's informants, the stress fell on the cultural transformation of nature, in which natural elements were refined and abstracted in such a way that they could serve as symbols of harmony, order, and balance. The stress on the cultural transformation of nature rather than nature in its pure, unrefined state has also been noted by Donald Ritchie who has said that "the Japanese attitude toward nature is essentially possessive. . . . Nature is not natural. . .until the hand of man. . .has properly shaped it."[19]

How can we explain the contradiction between Kellert's findings and the stereotype of the nature-loving Buddhist? How can the Japanese tradition appear to show such deep reverence for nature and yet tolerate, perhaps even encourage, such pervasive attitudes of cultural domination? One possible explanation is that the Japanese have so thoroughly absorbed a Western preference for the domination and exploitation of nature that the indigenous tradition has simply been overwhelmed in a rush for Western-style economic development. Kellert points out, however, that it is too simplistic to attribute this contradiction merely to the influence of the West. As W. Montgomery Watt noted in his account of alleged external influences on the formation of early Islam, it is difficult for one culture to "influence" another in a deep or significant way unless there are already tendencies or predispositions in the receiving culture that make such influence possible.[20] Could there be predispositions within the Buddhist traditions of Japan that tend to favor this "cultural transformation" of nature? Could Buddhism itself have contributed to such an attitude? The way to explore these questions at the most basic level is to move back to India, the homeland of the Buddhist tradition, and interrogate the tradition in its original setting.

How does the religious literature of India picture the natural world? India is a complex civilization, of course, and it is as complex in its approach to nature as any of the traditions of East Asia or the West, but it does not seem an oversimplification to say that there is a deep and abiding preoccupation in Indian civilization with the distinction between the "human" and the "natural." One of the best sources to use in reflecting on this distinction is the text of the *Bhagavad Gītā*, one of the best known and most beloved of Hindu scriptural texts. The text consists of a dialogue between two figures, the warrior Arjuna who is on the verge of a climactic battle, and his charioteer Kṛṣṇa, who reveals himself to be a manifestation of God. At the beginning of the story, Arjuna shrinks in grief as he contemplates the destruction to be wrought by the battle. Kṛṣṇa counsels him to pick himself up and do his duty as a warrior without feeling fear or grief about the consequences of his actions. The reasoning behind Kṛṣṇa's counsel reflects a fundamental feature of Hindu attitudes toward nature. Kṛṣṇa counsels Arjuna to distinguish between his "soul" (*puruṣa*), which is eternal and cannot die, and his "body," which is mortal, changeable, and destined eventually to be discarded as the soul makes its passage into another life.

> These bodies are said to end, but the embodied self is eternal, indestructible, and immeasurable; therefore, you should fight, O Bhārata. (2.18)

If Arjuna knows that his true identity is equated with the "soul" and not the "body," he does not need to be affected by grief or fear.

As the text develops the distinction between "soul" and "body," we find that the "body" is spoken of as *prakṛti*, a concept that is commonly translated as "nature." The distinction between soul and body is a reflection, in the microcosm of the personality, of the distinction in the cosmos at large between the principle of "spirit" and the principle of "nature." What does it mean to say that *prakṛti* is "nature"? The semantic range of the word *prakṛti* might seem at first to be considerably wider than the one that normally is mapped by the English word "nature." *Prakṛti* includes not only the material aspects of the cosmos but also the aspects of the personality called "mind" (*manas*) and "intellect" (*buddhi*). The basic distinction is not between body on one side and mind or spirit on the other but is, rather, between the complex of changeable elements in the personality (including body, mind, and intellect) and the eternal, unchangeable soul. The distinction between *puruṣa* and *prakṛti* comes close, however, to the distinction marked by the title of the symposium in which the Dalai Lama gave his Middlebury address: *puruṣa* is "spirit" and *prakṛti* "nature," in the sense that *puruṣa* is conscious, transcendent, and attainable through discipline (*yoga*) or reason while *prakṛti* merely reflects or obscures the consciousness of *puruṣa* and is subject to change and decay. The challenge for human beings in Arjuna's position, caught in the web of confusion spun by the strands of *prakṛti*, is to recognize their true identities as immortal souls and escape the bonds of nature.

In a technical sense, the distinction between *puruṣa* and *prakṛti* belongs to only two of the classic Hindu philosophical traditions, the Sāṃkhya and the Yoga, and these two traditions do not by any means serve as the dominant framework for the interpretation of reality in the Indian tradition. But the distinction has wide influence in Indian culture. When visitors make a journey, for example, to the great ruined temple of Elephanta in Bombay harbor, they travel across the waters of the harbor to a small island, climb a long line of stairs up to the rocky outcropping in the center of the island, then enter a cave where the central shrine has been cut out of the living rock. The journey across the water is a symbolic expression of a journey through the changeable, distracting world of "nature," and entry into the darkness and quiet of the temple represents an approach to the immovable center of "the soul." The religious drama of the journey depends on a basic cultural image of contrast between the world of *prakṛti* and the world of the soul. Even in nondualistic traditions, such as Advaita Vedānta, where the goal is to dissolve the distinction between self and world, the journey of enlightenment is still based on an initial insistence on the "distinction" (*viveka*) between the eternal self and all that is not-self.[21] One can argue with considerable force that the Hindu tradition is driven, even in its nondualistic dimension, by a conviction that eternal things have ultimate value and changeable things do not. "Nature" encompasses the things that change and pass away.

Buddhists do not share the Hindu conviction about the permanence of the individual soul, but they also are suspicious of the difficulties and dangers of the "natural" world. Lambert Schmithausen has noted that, in classical Buddhist sources, Buddhist peasants, townspeople, and even monks preferred the tamed and civilized world of the village and city to the virgin forest or the jungle.[22] The jungle and forest were symbols of death and rebirth (as was the ocean that the worshipper had to cross to reach the temple at Elephanta), and *nirvāṇa*, the cessation of death and rebirth, was represented as a city.[23] Images of Buddhist paradises, when they appear in Indian sources, are generally landscapes in which the "wild" aspects of nature have been thoroughly tamed. With trees laid out in symmetrical grids, rectangular ponds, golden lines, and shiny blue-black surface, the paradise of Sukhāvatī in the Indian *Sukhāvatīvyūha* is more reminiscent of a parking lot than it is of an untamed wilderness.[24] Schmithausen notes quite correctly that a significant number of Buddhist monks chose not to live in cities or towns. In the "hermit strand" of monastic life, one visualized the forest as useful for the practice of meditation. In the forest a monk can avoid the distractions of society and contemplate the impermanence of reality by observing the passage of the seasons. But even here the focus is on the natural world as a locus and a guide for the spiritual transformation of the monk himself, as it was in Grapard's account of the mapping of *The Lotus Sūtra* onto the ridges and valleys of Futagoyama.

It is important to be clear that this early strand in the Buddhist tradition is not hostile to nature as such: one does not attempt to dominate or destroy nature (in the form of either animals or plants) in order to seek a human good. But neither is the wild and untamed aspect of nature to be encouraged or cultivated. The natural world functions as a locus and an example of the impermanence and unsatisfactoriness of death and rebirth. The goal to be cultivated is not wildness in its own right but a state of awareness in which a practitioner can let go of the "natural"—of all that is impermanent and unsatisfactory—and achieve the sense of peace and freedom that is represented by the state of *nirvāṇa*. One might say that nature is not to be dominated but to be relinquished in order to become free.

In this context the significance of the Dalai Lama's approach to the topic of "spirit and nature" becomes clearer. He was not hostile to nature, but he had other important topics in mind, not the least of which was the purification of the mind itself. When he took up the question of nature in the philosophical style that was appropriate to his own tradition (linking it to the concept of Emptiness), his first step was not unlike Kṛṣṇa's first step in the *Bhagavad Gītā*: he distinguished between the realm of appearance or "nature" and the realm of ultimate reality or Emptiness. He said: "When talking about the fundamental nature of reality, one could sum up the entire understanding of that nature in a simple verse: 'Form is emptiness and emptiness is form' (*The Heart Sūtra*)."[25] The concept of "fundamental nature" might seem to function differently than the concept of *prakṛti* in the *Bhagavad Gītā*, and in a sense it does. It refers to the imagined "essence" or "identity" that a person imposes on reality (the reality of Emptiness) rather than to the distracting and alluring play of "material nature," but it performs the same discriminative function when it comes to the purification of the mind. In the Madhyamaka tradition, out of which the Dalai Lama speaks, the idea of "fundamental nature" (whether it is understood as the Tibetan *ngo bo nyid* and Sanskrit *svabhāva* or as the Tibetan *rang bzhin* and Sanskrit *prakṛti*) has to be stripped away in order to develop a purified awareness of Emptiness. The term "Emptiness" itself can refer either to the absence of such a "fundamental nature" in all things or to the purified awareness that perceives all things as empty in this way. For the Dalai Lama, the concept of "nature" elicits an image of Emptiness and suggests a practice of purification in which the illusions of "nature" are left behind.

Against this intellectual background, it is not surprising to find that Indian Buddhist literature contains very little of the reverence for the wild and "natural" world that one associates with the tradition of East Asia. Indian poetic accounts of insight or enlightenment often reflect a rhetorical distinction in which the teaching of the Buddha is "greater than" or "in contrast to" the possibilities of the natural world, as in the philosopher Dharmakīrti's exploration of the poetic relationship between the Buddha's teaching and the cooling rays of the moon.

> Were not a drop from the Moon of Sages,
> better than a flood of cooling moonlight,
> mixed within the vessel of its thought,
> how would this heart find happiness
> and, though it stood within a cold Himalayan cave,
> how would it endure the unendurable
> fire of separation from its love?[26]

When one puts Dharmakīrti's image of the superiority of the "Moon of Sages" (the Buddha) next to a comparable passage by the Chinese Buddhist poet Li Po (701–762)—

> Moonlight in front of my bed—
> I took it for frost on the ground!
> I lift my eyes to watch the mountain moon,
> lower them and dream of home.

—one clearly sees the aesthetic and ideological transformation that took place when the wine of the Indian Buddhist tradition was poured into its Chinese bottles. "Nature" in the Indian tradition was a world to be transcended, while in East Asia it took on the capacity to symbolize transcendence itself.

How then should we read the affirmative images of the natural world in the poetry of Saigyō and Bashō? Has the Japanese tradition been so thoroughly infused by Chinese attitudes toward the natural world that it has taken leave entirely from the Indian tradition? Certainly there is a striking contrast between the two traditions, but it is possible to see Indian Buddhism (including the Tibetan tradition of the Dalai Lama) in a way that gives us new eyes for the *Buddhist* dimension of the Japanese poetic tradition. Bashō's "Old pond / frog jumps in / sound of water" can be read as an expression of immersion in the flow of natural processes: a frog jumps into a pond, and the mind fuses with the event in a moment of intense perception. But the poem is not, strictly speaking, an expression of the frog or the water in themselves; it is an expression of a moment of perception. The force of the poem lies in the mind of the observer, not to the exclusion of nature, but in the mind's awareness of nature.

When Stephen R. Kellert probed the stereotype of Japanese attitudes toward nature, he found what one of his informants called an "emphasis on the experience and enjoyment of nature in highly structured circumstances." The stress fell less on nature in its raw

form than on the cultural transformation of nature: natural elements
were refined and abstracted so that they could serve as symbols of
harmony, order, or balance. Allan G. Grapard captured the ambiguity
and complexity of the same point when he suggested that

> what has been termed "the Japanese love of nature" is actually the
> "Japanese love of cultural transformations and purification of a
> world which, if left alone, simply decays." So that the love of culture
> takes in Japan the form of a love of nature.[27]

Nature may not need to be transformed in an overt, physical fashion
to be significant, although the design of a "natural" garden is
certainly a refined cultural act, but the significance of the natural
setting for the human observer lies principally in the act of
perception, and it may be appropriate or even necessary for nature
to be fashioned and controlled to make this "natural" mode of
perception clear.

   Then is there a "Buddhist" philosophy of nature? If the intention
of the question is to identify a simple, unified vision of the sanctity
of the natural world, the answer must be no. If anything, there is
the opposite. Beneath the evident differences between the Indian and
East Asian traditions lies a commitment to the view that human
beings work out their fates through the development and purification
of their own minds. Riccardo Venturini had something like this in
mind, no doubt, when he said that the Buddhist tradition develops
its attitude toward nature in the context of an "ecology of the mind"
and aims at a "purified" world with man as its steward.[28] Could it
be that the Buddhist tradition, which has seemed so promising as a
model to escape the destructive consequences of the Western
anthropocentric vision of nature, is as much compromised by the
flaws of anthropocentrism as its Western counterpart? The question
is crucial for understanding the possibility of a Buddhist response
to the ecological crisis, and much depends on the meaning of the
word "anthropocentrism."

   In the summer of 1981, the Dalai Lama gave a series of lectures
on Buddhist philosophy in Emerson Hall at Harvard University.[29]
At the beginning of the lectures a member of the Harvard community welcomed the Dalai Lama to Emerson Hall by referring to
an inscription over the portal of the building: "What is man that thou
art mindful of him?" He gave a Tibetan translation of the inscription

that related it to one of the key issues of Buddhist philosophy ("What are you referring to when you use the word 'man'?") and said that it seemed particularly appropriate to hear the Dalai Lama's words in a setting where the very issue of human identity had such a rich and controversial history. "What is man that thou art mindful of him?" In the Tibetan and Sanskrit traditions, the word "man" recalls a long controversy about the status of the *pudgala* (commonly translated as "person," but literally "man"). An ancient Indian Buddhist school known as the Personalists (*pudgalavādin*) took the position that a person's identity consisted in a *pudgala* that continued from one moment to the next.[30] This *pudgala* was related ambiguously to the momentary psycho-physical constituents (*skandha*) of the mind and body. The constituents changed at every moment while the *pudgala* continued, and the *pudgala* was neither identical to nor different from the constituents. It seems that the *pudgala* was considered to be something like the "shape" or "configuration" of the personality, so that one could say that a person retained the same "shape" even when all the individual constituents of the personality had changed, perhaps like a car in which all the individual parts have been replaced but which still retains the "shape" of the original car.

The Personalists have long since gone out of existence as an identifiable school, and the controversy about the *pudgala* could be relegated to the status of an obscure historical curiosity if it had not become a symbol for Buddhists of the classic mistake to be avoided when thinking about the nature of the self. One of the most basic themes in Buddhist philosophy is the claim that there is no "self," and by "no-self" is meant at least that there is no continuous *pudgala* that ties together the stream of the personality from one moment to the next. The *pudgalavāda*, the doctrine of the "man" or "person," is, as it were, the fundamental Buddhist heresy from which the tradition now chooses to distinguish itself. To ask, "What is man that thou art mindful of him?" or "What are you referring to when you use the word 'man'?" is to probe the foundations of the Buddhist view of the self at its most sensitive point: What is the most basic error that has to be avoided if one is to make progress toward the goal of enlightenment?

Herein lies the paradox of Buddhist "anthropocentrism." The tradition is genuinely concerned with the human achievement of human goals. At a deep historical and conceptual level, the tradition defends an ideal of self-reliance, as in the oft-quoted verse from the *Dhammapada*: "One is one's own Lord (or God or Protector). What other Lord can there be?" But the achievement of self-interest is tied in an equally fundamental way to the decentering of the self. On the intellectual level, the quest for *nirvāṇa* is tied up with a quest for an understanding of "no-self" as both a doctrine and a mode of awareness. On a more practical level, Buddhist discipline is built up of choices, both large and small, that challenge the naïve patterns of self-centeredness from which the fabric of ordinary life is woven. In traditional Buddhist societies in Southeast Asia, Buddhist monks go out each morning to beg their food from laypeople, meditating as they go on their "friendliness" or concern for all beings. Laypeople prepare the food and enact a model in which their own spiritual benefit is tied to a gesture of renunciation, of giving away the food that sustains the life of a monk. Moral precepts, particularly the prohibition against killing (which is extended not just to human beings but, in theory, to all sentient beings), cultivate a fundamental respect for life in all its forms. These ideals are realized with greater and lesser degrees of consistency in the Buddhist communities of Southeast Asia, but the theoretical connection between the self-interested decentering of the self and respect for life lies deep in the culture. One could paraphrase Grapard's claim that in Japan the love of culture takes the form of a love of nature by saying that in Buddhist culture at large the cultivation of the self takes the form of a decentering of the self and a concern for a wider network of life.

Steven C. Rockefeller has commented about the way anthropocentric and utilitarian approaches to environmental ethics take on a more biocentric character when they are combined with a scientific appreciation of ecological interdependence.[31] This conceptual development has its counterpart in the Buddhist tradition as well. To say that one's self-interest is served by realizing and enacting an ideal of no-self is to say that one's own self-interest is best understood by realizing one's location in a network of interdependence or "interdependent co-origination" (*pratītya-samutpāda*). The formulas that express the understanding of no-self in different versions of the Buddhist tradition often equate no-self (or its Mahāyāna counterpart, the doctrine of Emptiness) with interdependent co-origination. A famous verse in Nāgārjuna's root verses of the Madhyamaka school says: "We call interdependent co-origination Emptiness; this is a metaphorical designation, and it is the Middle Path." Other textual sources simply equate interdependent co-origination with the *Dharma* or with the Buddha himself, as in the common scriptural phrase, "He who sees interdependent co-origination sees the Buddha."

Whether one can interpret the concept of interdependent co-origination as genuinely "biocentric," however, is open to question. If a biocentric approach means recognizing "the intrinsic value of animals, plants, rivers, mountains, and ecosystems rather than simply. . .their utilitarian value or benefit to humans,"[32] then the word "intrinsic" presents a barrier. It seems to suggest precisely the substantial, permanent identity that the ideas of no-self and interdependent co-origination are meant to undermine. But the practical force of an "other-centered" position emerges quite clearly in different kinds of Buddhist meditative traditions. When the Dalai Lama teaches about Buddhist practice, he emphasizes the importance of compassion, as is customary in the tradition of the Mahāyāna, and one of his favorite sources for a meditation on compassion is the teaching about the "exchange of self and other" in the eighth chapter of Śāntideva's "Introduction to the Practice of Enlightenment."[33] Imagine, Śāntideva says, that on one side of a divide stands your own needy self and on the other side stand fifty or a hundred needy beings. Whose advantage is best to seek? Should you care just for yourself and cater just to your own limitations and fears? Or should you seek the benefit of the larger group? And what if the larger group is not just fifty or a hundred living beings but all the livings beings in the cosmos? Śāntideva says that the answer should be clear. The self's greatest benefit comes from seeking the widest possible benefit for the network of all living beings. Śāntideva's point can be construed as a practical centering of one's concern on others (on the network of *bios* or "life") in order to decenter the self (in the self's own interest).

Here lies another reason why the Dalai Lama was hesitant to address directly the themes and expectations of the Middlebury conference on "Spirit and Nature" and why he shifted attention so gracefully away from the natural world toward the purification of the mind. He was not insensitive to the claims of the natural world, but he felt that there was more important conceptual work to be done before its claims could be made clear. He had to begin with his own understanding of no-self (as expressed in the doctrine of Emptiness) before he could sketch the outline of an ethical response to the natural world, and the response continued to move in the orbit of "interdependence" and "compassion." One moves naturally, as it were, in a series of ever-widening, concentric circles, beginning with the impulse to purify the mind and cultivate one's own sense of self, through the sense of the self's interdependence with a network of all other beings, to a sense of affection and love for all existence. As the circles widen, the center comes under pressure, and the network of existence takes on the appearance of a circle whose center is everywhere and whose circumference is nowhere.

Some of the most forceful and perspicacious Buddhist writing about the environment explores the implication of this basic Buddhist conceptual movement from no-self to interdependence to compassion. In his reconsideration of E. F. Schumacher's famous concept of "Buddhist Economics," Stephen Batchelor points out that Buddhist economics has to start from a standpoint of nonduality and Emptiness, and from this point of view the concept of an ethical "center" comes increasingly into question. "In the West we are still caught in a struggle between theocentric and anthropocentric visions, which some Greens now seek to resolve through a notion of biocentrism. Such thoughts are alien to the Buddhist experience of reality, which, if anything, has tended to be 'acentric.'"[34] Joanna Macy has charted the same movement from the point of view of the Theravāda tradition, beginning with a sense of "the pathogenic character of the reification of the self," moving on to the concept of interdependence (*paṭicca samuppāda*), and then developing a sense of what might best be called universal "self-interest," in which the world is visualized as one's own body.[35] With the words of Arne Naess and the concept of "deep ecology" in mind, she turns the ethical argument about altruistic motives from one of "duties" rendered by the self to another into an argument about one's own "being." One protects nature in order to protect one's own self, and the circle of self encompasses the totality of the natural order.

Certainly, the sense of interdependence that is such a crucial part of Buddhist ethical theory gives good reason to be skeptical of any form of "centrism," whether it begins in the *theos*, the *anthropos*, or even more benignly in the *bios*. But do images of the "center" need to be entirely abandoned? Buddhist environmental literature abounds with metaphors of interconnection, from the jeweled "Net of Indra," in which every individual jewel is pictured as reflecting every other, to images of the "web" of existence. But there is another, relatively unexplored body of metaphor that has to do with a sense of "place" or "home." Buddhist sources speak from a very early period about a tradition of pilgrimage in which people visited sites that had been important in the life of the Buddha, "saw" them, and were moved by them. The sites of the Buddha's birth, his enlightenment, his first sermon, and his death were held in special reverence, and traditional sources speak of the throne of the Buddha's enlightenment, under the Bodhi Tree in Bodh Gayā, as the center of the cosmos. Some Mahāyāna texts pass on a tradition that every Buddha, of every era, is enlightened at exactly the same site, and beneath the spot where the Buddhas are enlightened sits a throne that is anchored at the center of the cosmos.[36] If there were a "center" in Buddhist ethical thinking about the environment, perhaps this is where it should be located, at the site where Buddhas attain their enlightenment.

But where is this site? Northern India is one possibility. The tradition, however, has a distinct aversion to literal conceptions of the Buddha. Embedded in Buddhist tradition is the idea that one finds the Buddha not in his physical form but by understanding the *Dharma*. ("What is there, O Vakkali, in seeing this vile body? He who sees the Dharma sees the Buddha. He who sees the Buddha sees the Dharma.") Where, then, is the "throne of enlightenment," the place where one understands the *Dharma*? One possible answer would be the mind itself. It is in the mind that one understands the nature of Emptiness. But the mind is located in a particular body, and the body is located in a particular place. While Emptiness, in a sense, is everywhere, it is realized only in *this* moment, *this* place, and *this* body. In a fine meditation on "Zen Practice and a Sense of Place," Doug Cochida quotes a reference by the Zen master Dōgen to the earth as the "true human body":

The meaning of "true" in "the entire Earth is the true human body" is the actual body. You should know that the entire Earth is not our temporary appearance, but our genuine human body.[37]

The earth is not, as it were, a mere illusion. It is the body of an enlightened sage, and it is as worthy of reverence as the throne of the Buddha.

In his essays on "The Practice of the Wild," Gary Snyder said: "In some cases we might call [nature] sacred."[38] To say only "in some cases" shows an appropriate Buddhist reticence toward attributing sacrality to nature in and of itself. But it is not completely implausible to use the language of "holiness" in speaking of the natural order. The natural world can function as a teacher when one meditates about impermanence. In some strands of the Buddhist tradition it can be thought of as possessing Buddha-nature. But most importantly, it is the place made holy by the quest for enlightenment. Enlightenment is made present in this body and this earth. To speak of the earth as the throne of enlightenment is a metaphor, of course, and it is not by any means a common metaphor in Buddhist writings. But it is one that resonates deeply with the theistic language of Erazim Kohák, the man to whom this essay is dedicated. Kohák's great meditation on the moral sense of nature, *The Embers and the Stars*, is alive with a sense of the holy or, as Kohák himself says, "the presence of God in the very fact of the world."[39] The Buddhist tradition has problems with the language of classic theism, but a sense of the presence of the holy is hardly unknown in Buddhist experience or imagination. It does not come, however, from the outside, nor is it ready-made. It has to be fashioned and developed by the application of human discipline, imagination, compassion, and awareness. This I take to be the force of the Dalai Lama's Middlebury address, as it is of the tradition more generally. Human beings have to take responsibility themselves for the harmony, the health, and the well-being of the setting in which the quest for enlightenment takes place.

Notes

* This essay was first presented in "Philosophies of Nature," a Boston University Symposium in Honor of Erazim Kohák, 13–17 November 1995.

1. Roderick Frazier Nash, *Wilderness and the American Mind* (New Haven: Yale University Press, 1967), 20–21, 192–93.

2. Y. Murota, "Culture and Environment in Japan," *Environmental Management* 9 (1986):105–12.

3. Lynn White, Jr., "The Historical Roots of Our Ecologic Crisis," *Science* 155 (1967):1203–7. Reprinted in *Machina Ex Deo: Essays on the Dynamism of Western Culture* (Cambridge, Mass.: MIT Press, 1968), 75–94.

4. Masao Watanabe, "The Conception of Nature in Japanese Culture," *Science* 183 (1974):279–82.

5. Gary Snyder, *The Practice of the Wild* (San Francisco: North Point Press, 1990), 10.

6. The words belong to Richard Gere, the Founding Chairman of Tibet House in New York, and appear in Marylin M. Rhie and Robert A. F. Thurman, eds., *Wisdom and Compassion: The Sacred Art of Tibet* (New York: Harry N. Abrams, 1991), 8. The image of Tibet as the "lifeboat of civilization" has been widely remarked upon in Asian studies, notably by Peter Bishop in *The Myth of Shangri-La: Tibet, Travel Writing, and the Western Creation of the Sacred Landscape* (Berkeley: University of California Press, 1989).

7. The Dalai Lama's speech appears in Steven C. Rockefeller and John C. Elder, eds., *Spirit and Nature: Why the Environment Is a Religious Issue* (Boston: Beacon Press, 1992).

8. Rockefeller and Elder, *Spirit and Nature*, 114.

9. This formula for the expression of Emptiness comes from the Madhyamaka school of Mahāyāna philosophy, the school within which the Dalai Lama himself speaks. For a more extensive account of this concept and for references to further literature, see Malcolm David Eckel, *To See the Buddha: A Philosopher's Quest for the Meaning of Emptiness* (San Francisco: HarperCollins, 1992; reprint ed., Princeton: Princeton University Press, 1994).

10. See, for example, Gabriel P. Weisberg et al., eds., *Japonisme: Japanese Influence on French Art, 1854–1910* (London: Robert G. Sawyers, 1975).

11. Robert S. Ellwood and Richard Pilgrim, *Japanese Religion: A Cultural Perspective* (Englewood Cliffs, N.J.: Prentice-Hall, 1985), 55.

12. Burton Watson, trans., *Cold Mountain: 100 Poems by the T'ang Poet Han-shan* (New York: Columbia University Press, 1970), 67.

13. Gary Snyder has produced some of the most powerful translations of the Cold Mountain poems. See his *Rip Rap and Other Poems* (San Francisco: Grey Fox Press, 1982).

14. William R. LaFleur, "Saigyō and the Buddhist Value of Nature," in *Nature in Asian Traditions of Thought*, ed. J. Baird Callicott and Roger T. Ames (Albany: State University of New York Press, 1989), 183–209.

15. *The Lotus Sutra*, trans. Burton Watson (New York: Columbia University Press, 1993), chapter 5.

16. Allan G. Grapard, "Nature and Culture in Japan," in *Deep Ecology*, ed. Michael Tobias (San Diego: Avant Books, 1985), 240–55.

17. Stephen R. Kellert, "Japanese Perceptions of Wildlife," *Conservation Biology* 5 (1991):297–308; "Concepts of Nature East and West," in *Reinventing Nature? Responses to Postmodern Deconstruction*, ed. Michael E. Soulé and Gary Lease (Washington, D.C.: Island Press, 1995), 103–21. See also Yi-Fu Tuan, "Discrepancies between Environmental Attitude and Behaviour: Examples from Europe and China," *Canadian Geographer* 12, no. 3 (1968):175–91.

18. Kellert, "Concepts of Nature East and West," 107.

19. D. Ritchie, *The Island Sea* (Tokyo: Weatherhill, 1971), 13; quoted in Kellert, "Concepts of Nature East and West," 115.

20. W. Montgomery Watt, *Muhammad: Prophet and Statesman* (Oxford: Oxford University Press, 1961).

21. "Distinction" (*viveka*) is one of the four "qualifications" for the knowledge of Brahman. See Eliot Deutsch, *Advaita Vedānta: A Philosophical Reconstruction* (Honolulu: University of Hawaii Press, 1969), 105.

22. Lambert Schmithausen, "Buddhism and Nature," in *Studia Philologica Buddhica: Occasional Paper Series*, 7 (Tokyo: International Institute for Buddhist Studies, 1991).

23. For sources see Schmithausen, "Buddhism and Nature," 15. The references to the "city of *nirvāṇa*" come from texts that are somewhat late. An interesting echo of the metaphor in an early source is a reference in *Suttanipata* 3.109 to *nirvāṇa* as a level piece of land (*samo bhūmibhāgo*).

24. *Buddhist Mahāyāna Texts*, trans. Max Müller, Sacred Books of the East, 49 (Oxford: Clarendon Press, 1894; reprint ed., New York: Dover Publications, 1969).

25. Rockefeller and Elder, *Spirit and Nature*, 114.

26. Daniel H. H. Ingalls, *An Anthology of Sanskrit Court Poetry* (Cambridge, Mass.: Harvard University Press, 1965), 438.

27. Grapard, "Nature and Culture in Japan," 243.

28. Riccardo Venturini, "A Buddhist View on Ecological Balance," *Dharma World* 17 (March-April 1990):19–23; quoted in Schmithausen, "Buddhism and Nature," 17.

29. The lectures have been published in His Holiness the Dalai Lama of Tibet, Tenzin Gyatso, *The Dalai Lama at Harvard*, trans. and ed. Jeffrey Hopkins (Ithaca: Snow Lion, 1988).

30. The classic account of the theory of *pudgalavāda* is found in the *Abhidharmakośa*, trans. L. de La Vallée Poussin, *Mélanges chinois et bouddhiques* 16 (1971). A useful English translation of the section from the *Abhidharmakośa* that deals with this theory can be found in Edward Conze, *Buddhist Scriptures* (Baltimore: Penguin Books, 1959), 192–97.

31. Steven C. Rockefeller, "Faith and Community in an Ecological Age," in Rockefeller and Elder, *Spirit and Nature*, 139–71. For further commentary on the issues of "anthropocentrism," see J. Baird Callicott, "Non-Anthropocentric Value Theory and Environmental Ethics," *American Philosophical Quarterly* 21 (1984).

32. Rockefeller, "Faith and Community in an Ecological Age," 143.

33. See chapters 8 and 9 of *The Dalai Lama at Harvard*. Śāntideva's own text is available in a number of translations, notably Stephen Batchelor, *A Guide to the Bodhisattva's Way of Life* (Dharamsala: Library of Tibetan Works and Archives, 1979).

34. Stephen Batchelor, "Buddhist Economics Reconsidered," in *Dharma Gaia: A Harvest of Essays in Buddhism and Ecology*, ed. Allan Hunt Badiner (Berkeley: Parallax Press, 1990), 178–82.

35. Joanna Macy, "The Greening of the Self," in Badiner *Dharma Gaia*, 53–63.

36. Etienne Lamotte summarizes Mahāyāna traditions about the throne of enlightenment (*bodhimaṇḍa*) in *The Teaching of Vimalakīrti (Virmalakīrtinirdeśa)*, trans. Sara Boin (London: Pali Text Society, 1976), 94–99.

37. Doug Cochida, "Zen Practice and a Sense of Place," in Badiner, *Dharma Gaia*, 106–11.

38. Snyder, *The Practice of the Wild*, 10.

39. Erazim Kohák, *The Embers and the Stars: A Philosophical Inquiry into the Moral Sense of Nature* (Chicago: University of Chicago Press, 1984), 188.

Green Buddhism
and the Hierarchy of Compassion

Alan Sponberg

Buddhist perspectives on nature and the environment have a long
and complex history, and it is thus not surprising that one finds
within this rich and varied tradition much that resonates with
contemporary concerns regarding nature and the place of humanity
within it.[1] While Buddhists of the past had little reason to formulate
an environmental ethic per se, there is much within traditional
Buddhist ethics that does indeed speak to the ethical aspects of the
environmental crisis confronting us today, a fact that has been well
noted and at least partially explored both by non-Buddhist environmental ethicists and by a growing number of contemporary
Buddhists themselves, advocates of what is frequently referred to
as "Green Buddhism."[2] My approach in the present article seeks to
bridge these two camps, and I shall thus be writing here both as a
practicing Buddhist and as an environmental ethicist, one with
academic training in philosophy and in the history of Buddhism. I
shall undertake a critique of certain features of Green Buddhism in
this article, and it is important for the reader to realize that I do
so from within the circle of this vital movement of contemporary Buddhism, seeking to identify the "near enemy" (*āsanna-
paccathika*) within, which, as Buddhaghosa commented in the fifth
century, is often more dangerous than the "distant enemy" (*dūra-
paccathika*) that remains more obviously (and safely) outside the
fold.

The "near enemy" I have in mind in this case is the view that
Green Buddhism is fundamentally incompatible with, and hence
necessarily opposed to, hierarchy in any and all forms. There are

good reasons why such a view appears quite plausible and attractive at first, though we must recognize that these reasons stem more from our own cultural history than from anything within Buddhism itself. While it is certainly true that Buddhism advocated, in its early forms at least, a radically decentralized institutional structure, this should not be misconstrued in the light of our current Western concerns to mean that the spiritual ideal in Buddhism was seen as nonhierarchical and egalitarian. The Buddha was indeed radical in that he recognized that all beings—not just human beings—have access to the liberation he proclaimed, but this does not mean that he felt that all beings were equal in the sense that there is no significant difference between species or individuals. To the extent that we fail to acknowledge this important sense in which Buddhism is non-egalitarian, we not only seriously misrepresent the tradition, we also risk disavowing an aspect of the *Dharma* that is sorely lacking in contemporary Western thought. Thus, in this article I shall seek to show, first, that the rejection of all forms of hierarchy is fundamentally *un*-Buddhist and, further, that such a view threatens, however unintentionally, to obscure and even reject a fundamental feature of Buddhism that may turn out to be crucial to the agenda of Green Buddhism.

To understand my argument we must reflect on the history of our current Western aversion to hierarchy in any form, and we must also clarify what place hierarchical structures do have in traditional Buddhism. If we find that hierarchy in some sense does have a place in Buddhism, then we shall have to ask whether it is the same kind of hierarchy that we are so anxious to banish from our own cultural history. I realize that discussion of "hierarchy" in any form will arouse very strong feelings among many Western Buddhists and environmentalists, yet I have intentionally chosen to use this provocative "h-word" for reasons that will become clear below. It is to those who find this word inherently objectionable that this article is respectfully dedicated. I truly share your concerns, and I ask only that you hear me out, bracketing for the moment whatever affront my thesis may initially elicit. Much of what Buddhism has to offer the West may, I fear, be lost, if we fail to see the quite specific sense in which Buddhism is, and must be, "hierarchical." By considering this apparently discordant assertion, we will, I submit, learn something quite important about Buddhism and also something about the cultural roots of a distinctly Western and modern form of "aversion" (*pratigha*).

The Two Dimensions of Basic Buddhism

Our first task, then, shall be to consider whether there is any aspect of traditional Buddhism that might warrant being called "hierarchical." While it is imperative that one remember the diversity within the different cultural expressions and traditions of Buddhism, it is nonetheless possible to identify a set of basic Buddhist teachings that remains at the core of the later variations. I am thinking of the basic doctrines of conditionality or dependent arising (*pratītya-samutpāda*), *karma*, the middle path, impermanence, and non-substantiality (*anātman*), among others. One quite useful approach I have found for gaining a more comprehensive understanding of "Basic Buddhism" in this sense is to recognize, running throughout Buddhist history, two fundamental aspects of the tradition: a developmental dimension and a relational dimension. While we shall see that each of these two dimensions is clearly distinct, we must also recognize that each complements the other in a way that is crucial to the integrity of the tradition.

Let us first consider these dimensions separately. When we speak of the developmental dimension or aspect of Buddhism, we are focusing on the transformational intent of the tradition, on the Buddhadharma as a practical means of spiritual growth and development. Buddhism, in all of its forms, sees the spiritual life as the transformation of delusion and suffering into enlightenment and liberation. Even the so-called nondual forms of Buddhism—Zen and Dzogchen, for example—acknowledge an experiential distinction between delusion and enlightenment, and certainly neither would trivialize the existential reality of suffering.[3] The second crucial aspect of basic Buddhism—what I have called the relational dimension of the tradition—comes to the fore, by contrast, whenever we note the distinctly Buddhist conception of the interrelatedness of all things. And "things" here may be taken to encompass not just all sentient beings but every aspect of the ecosystems in which they participate—ultimately, the ecosphere in its totality.[4]

Looking at Buddhism historically, we will quickly note that these two dimensions are rarely given equal stress in any given expression of the tradition. My argument here rests only on the assertion that both will always be present to some degree—that indeed there is a necessary complementarity between the two—even when one appears more prominent than the other. The fact that one dimension or the other will, within the context of a particular form of Buddhism, frequently receive relatively more or less emphasis thus raises no problem, since the basic complementarity is not thereby negated. Indeed, by noting in different schools of Buddhism the relative difference in emphasis given to the developmental or the relational dimension, we have one useful way of charting the complex and fascinating permutations that the basic *Dharma* manifested as the tradition made its way through the various cultural encounters of its twenty-five-hundred-year history.

To clarify the variable relationship between these two dimensions of basic Buddhism, we might think of the two axes of a graph, with the vertical axis indicating the developmental dimension of the tradition and the horizontal axis indicating the relational dimension (see figure 1). We have then a useful heuristic tool we can use to explore the rich elaboration of different Buddhist schools and teachings, plotting each in reference to the others by noting the relative degree of emphasis given to the developmental and relational dimensions respectively. While this approach is helpful in highlighting and understanding the diversity within Buddhism, the tool I am suggesting here will also help us recognize how the differences revealed indicate not so much a fundamental divergence among the forms of Buddhism as differences in approach and emphasis—expedient means (*upāya*) that reflect the ability of the tradition to adapt to the

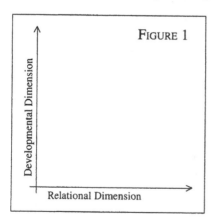

FIGURE 1

Developmental Dimension

Relational Dimension

needs and dispositions of different historical and cultural settings. One could, no doubt, even write a history of Buddhism by charting the various permutations of emphasis revealed by this simple x-y graph, but that would go well beyond the task at hand.

For our present purposes a few basic generalizations should suffice, both to illustrate the basic distinction between "vertical" and "horizontal" or "developmental" and "relational" within the tradition and to demonstrate the usefulness of this interpretative approach. Considering the two major divisions that arose within the history of Buddhism, Theravāda Buddhism (often called Hīnayāna), on the one hand, and Mahāyāna (including the later developments of Vajrayāna, Zen, etc.), on the other, we could, for example, note that the former places relatively more emphasis on the developmental dimension, while in the latter the relational aspect often comes more to the fore. Similarly, it would not be too rash to observe that, on the whole, the South Asian Indo-Tibetan forms of Buddhism tend to plot out higher on the developmental (the vertical axis), whereas East Asian forms on the whole tend to move further out on the horizontal or relational axis. As with all such generalizations, the exceptions are often all the more significant and more interesting than the instances that conform. And even more importantly, we must remember that what we are noting here is simply a matter of the *relative degree of emphasis* given each of these aspects, which does not assume any mutual exclusion between the two. Instances of a totally one-dimensional form of Buddhism would in fact be very difficult to find in the historical record, so much so that we would be justified in asking whether such a case was still legitimately Buddhism even if it referred to itself as such.

Working at this level of generalization and abstraction is unlikely to remain satisfying for very long, however. Now that we have the basic distinction between the two dimensions of Buddhism in mind, let us consider more specifically where we can locate these two general aspects within actual Buddhist teachings. This will help us to see just how deeply embedded in basic Buddhism these two dimensions are, and it will also reveal more clearly their mutual complementarity. The developmental dimension of Buddhism is perhaps most readily evident in the very conception of the *Dharma* as a path (*mārga*), whether presented in the elaborate sequence of steps the Buddha describes in the *Sāmaññaphala Sutta* of the *Dīgha*

Nikāya or in the perhaps more familiar early doctrines of the "threefold teaching" (morality-meditation-wisdom) and the "eightfold path." Here we can see the spiritual life advocated by the Buddha presented clearly in terms of a transformational soteriology, one that begins in a problematic state which is ultimately overcome, typically through the systematic cultivation of a variously detailed progression of positive mental and spiritual states or attainments. In this sense, Buddhism offers an interesting parallel to the "virtue tradition" of early and medieval Western thought.

We could explore many other expressions of this same vertical or developmental dimension of early Buddhism, looking for examples at the four levels of meditative absorption (*dhyāna*), the five spiritual faculties (*indriya*), the seven limbs of enlightenment (*bodhyanga*), the stages of *arhat*-hood, or the path of the twelve "positive" causes and conditions (*nidāna*) taught by the Buddha in the *Saṃyutta Nikāya*.[5] But all of these are examples of the developmental dimension seen in terms of different aspects of the development of the individual practitioner. We will understand better how deeply this vertical axis runs, however, if we recognize, in addition, a more systemic level at which this dimension is also evident. Basic Buddhist cosmology provides the best illustrations of this second form of the developmental dimension. Consider, for example, the vertical array of the "three world-levels" (*triloka*), which is further elaborated into a hierarchical taxonomy of six (or sometimes five) life-forms (*gati*): the gods, titans, humans, animals, *preta*s (hungry ghosts), and hell-beings. Not only does the spiritual life or path pursued by the individual have a crucial vertical dimension, but this verticality is also built into the very structure of the Buddhist conception of the cosmos itself.

Many of the instances of the developmental dimension of Buddhism that I have cited so far originated in and are often given more prominence in the early Buddhism of the Elders (*Theras*), which is consistent with the generalization I noted above regarding a relative difference of emphasis on the developmental and the relational between the two main divisions of Buddhism. I have also stressed, however, that these two dimensions are not mutually exclusive, and this will become more clear if we look also at instances of this verticality in the Mahāyāna tradition. First, we must remember that all of the doctrines discussed so far retain their place (if not necessarily the same degree of emphasis) within the Mahāyāna. The vertical dimension is never simply discarded: even when the Zen and Pure Land schools explore the dangers of taking "developmental" language in any overly literalistic way, they still maintain the crucial—and essentially vertical—distinction between the experience of enlightenment and the perpetuation of suffering. The Mahāyāna thus retains the verticality of the earlier tradition, but its recognition of this dimension is hardly limited to a residual carry-over of themes from the earlier tradition.

Many doctrines considered distinctly Mahayana reflect the same vertical perspective of a developmental path. One sees this in the *bodhisattva* ideal, which actually extends the older conception of the path in a spiritually significant way by stressing the importance of an altruistic motivation. The doctrines of the ten *bodhisattva* stages (*bhūmi*) and the six (or ten) *bodhisattva* virtues or perfections (*pāramitā*) are central Mahāyāna themes, both of which figure importantly in the Yogācāra elaboration of the spiritual map into a path of vision (*darśana-mārga*) followed by a path of cultivation or transformation (*bhāvana-mārga*). For all of its exploration of the relational axis, Mahāyāna thus remains just as fundamentally developmental, and this is true even of Zen where "sudden enlightenment" is expected to require a period—often quite a long period—of especially intensive practice.[6]

Turning next to the relational aspect, the horizontal axis of our grid, it will no doubt be teachings associated with the Mahāyāna that first come to mind. Ethically, this dimension is obvious in the transpersonal and altruistic focus of the *bodhisattva* ideal and, ontologically, in the notions of interrelatedness derived from the emptiness doctrine (*śūnyavāda*) richly elaborated in the Perfection of Wisdom literature, the *Avataṃsaka*, and other key Mahāyāna *sūtra*s. One key feature of the Mahāyāna was its insistence that the Buddha's enlightenment was not so much a combination of wisdom *and* compassion as the realization of a wisdom that must *be* compassion, by virtue of its insight into the fundamental interrelatedness of all existence. The very nature of the Buddha's enlightenment was thus seen to be interrelational, something that could only exist in the context of compassionate, altruistic activity.

But again, we must be careful not to assume that recognition of this relational dimension of the Buddha's enlightenment was a purely Mahāyāna innovation.

First of all, the roots of the *bodhisattva* ideal are well represented in the earlier tradition of the elders. And the early teachings on impermanence and *anātman* were already sufficient to establish a basic insight into the ultimate nonsubstantiality of any putative dichotomy of self-interest versus other-interest.[7] Even more revealing is the fact that the pre-Mahāyāna roots of the relational dimension are implicit in some of the very developmental teachings we have already considered above. An indispensable relational aspect is literally built right into even the most seemingly hierarchical doctrines of the early tradition. While the vertically arrayed taxonomy of life-forms recognized by all schools of Buddhism asserts an explicit hierarchy of levels of consciousness—adding still a higher level reached with the attainment of Buddhahood—the hierarchy here is nonetheless quite different from what we, as products of Western culture, might expect or fear. In Buddhism the point of these vertical distinctions is not to establish a hierarchy of privilege and subjugation. Quite the contrary. The hierarchy here is neither absolute nor does it justify the dominion or domination of one class of beings over another. In fact, as we shall see more clearly below, the vertical distinction here is a matter of compassion rather than of control.

In the religions of Abraham (Judaism, Christianity, and Islam), God is intrinsically superior to humankind, as is the creator to his creation. Similarly, humankind, which alone was created in God's image, is intrinsically and (unalterably) superior to the animals and all the rest of creation as well. The Buddhist taxonomy of life-forms (including Buddhahood) presents a crucial contrast. It too is thoroughly and incontrovertibly hierarchical in structure, yet in a fundamentally different way. All of the levels in the Buddhist "chain of being" are both dynamic and interpermeable. A given life-form moves up, and often down, in this deadly serious cosmic game of "chutes and ladders." The different levels in the Buddhist cosmology, while indicating spiritually significant differences in awareness and consciousness, do not entail the theocentric and anthropocentric perspective and privilege so familiar in our own cultural tradition. They represent, rather, the range of progressively greater degrees of awareness and ethical sensibility available to all life-forms. We might say that this is an ethically dynamic array of possibilities rather than an ontologically static hierarchy of privilege and status.

This is a crucial distinction, and one that is very easy for us to overlook, especially those of us who are the most disenchanted with and critical of the Western notions of ontological hierarchy. Indeed, there is an objection that invariably arises at this point in the minds of many contemporary Buddhists. How and why is the vertical, developmental dimension so complementary—and thus so necessary—if, as Buddhism asserts, all of existence is already by its very nature inherently interrelated? If everything is already the way it needs to be, what possible need is there for something to be done? If we have the relational dimension of the *Dharma*, what need is there for development, for doing?—especially since it is precisely "human doing" that has brought about the environmental crisis we now face. The anger and frustration that give rise to these questions, expressed often with a palpable tone of indignation, are feelings we have all no doubt shared at one time or another, and our tendency to feel this impatience is understandable. Yet these questions reflect a grave misunderstanding of the Buddhist teaching of interrelatedness and of enlightenment as a developmental process. We should note, especially, the tone of righteous indignation in which these questions are often expressed, moreover, for it betrays, I fear, the ultimate despair of an ethical scepticism, even cynicism, that is fundamentally at odds with the basically positive conception of human potential that characterizes the *Dharma*. In the West we have come to fear that the presence of any vertical, developmental perspective is antithetical to our newly gained recognition of horizontal relatedness. Thus we miss the point that for Buddhism neither is possible without the other. The developmental and the relational are not only complementary, they are inseparably interrelated. This last point is central to the concerns I expressed above that those of us most attracted to Green Buddhism may also be the most prone to seriously misunderstand Buddhism in our very effort to see it as part of the solution to the environmental question.

Green Buddhism and the Loss of the Vertical Dimension

I have argued that the developmental and the relational are inextricably linked in Buddhist ethics. Yet I have also suggested that contemporary Buddhists are strongly inclined to ignore or even deny that this could be true. We need to consider more closely how this peculiar circumstance has come about. What I wish to demonstrate is that, for all its laudable articulation of the environmental ethical themes within the Buddhist tradition, Green Buddhism at present also shows a subtle tendency that threatens to distort significantly the assimilation of the *Dharma* into the West, a tendency to reduce Buddhism to a one-dimensional teaching of simple interrelatedness. And the dangers of this tendency are all the more ironic and all the more insidious, I would further argue, because it is a tendency that arises out of our own cultural conditioning. It is a problem we are bringing to Buddhism rather than one inherent in the tradition. As such, it is a tendency that may well subvert the very potential Buddhism does have to contribute to the more environmentally ethical perspective we are currently struggling so hard to realize.

Hence my concern: we may, in our efforts to adopt Buddhism as an alternative to the worst in our own culture, end up divesting Buddhism of one of its most essential aspects. In doing so we may coincidentally and quite unwittingly denude Western Buddhism of the very aspect of Buddhism that we need to confront the magnitude of the present environmental crisis. But why, we may well ask, would contemporary Buddhism, especially Green Buddhism, develop this tendency to disavow or even deny a crucial element of traditional Buddhism? Part of the answer to this question lies, no doubt, in the historical fact that the forms of Buddhism that initially attracted the widest popularity in the West, and especially in North America, were forms in which we see a relatively greater emphasis on the horizontal, relational dimension of the tradition, forms in which one might initially overlook the importance of the developmental aspect. This is most obvious in the Western appropriation of Zen, for example, especially in its most popularized forms, those based on the writings of D. T. Suzuki and Alan Watts. It is, however, no historical accident that it was these particular forms of Buddhism that initially prevailed in much of the West; consequently, I see this as simply another symptom of a deeper circumstance, which has more to do with our own cultural history than with that of Asian Buddhism. What I am suggesting is that the Western cultural sensibility driving the critique of our own history of environmental practice is also significantly shaping how we see Buddhism, even influencing which forms of Buddhism strike us as the most attractive. This same Western sensibility, moreover, is also driving us toward a significantly distorted view of Buddhism, one which in its fear of hierarchy leads us to imagine the solution of our problems in a "Buddhism" free of any vertical or hierarchial structure.

The key to my argument lies in the degree to which many of us within the circle of Green Buddhism are extremely uncomfortable, even mortified, by any aspect of Buddhism that is in any sense hierarchical, so much so that some of us feel the need to redefine Buddhism, to purge it of anything that even vaguely resembles the Western forms of environmentally callous elitism and privilege we seek so desperately to flee. The motivation here is understandable and, in part, even commendable, yet its excesses are nonetheless deluded and the outcome may well be disastrous—for Western Buddhism, certainly, and perhaps even for Western environmental ethics more broadly. How has this come about? We have identified in our own cultural history an unquestionable tendency toward attitudes of exploitation and domination of nature, and we have rightly associated those attitudes with cultural institutions of hierarchy and privilege. The unwitting and often quite unconscious mistake we make, however, comes when we assume that all forms of hierarchy are the same. We assume that any and every manifestation of hierarchy leads inevitably to the dead end of domination and exploitation, and so we have even banished that now dreaded "h-word" from all forms of polite conversation. And, as Western Buddhists, we reassure ourselves that any apparently hierarchical element in our cherished Buddhism must be a mistake, perhaps the later corruption of some monastic elitists. Or perhaps we see it simply as a historical anomaly, one that can and indeed should be quickly swept under the carpet. But is this unconsidered assumption that all forms of hierarchy lead to attitudes of domination and exploitation actually true? And, even if it appears to be true within the (limited) context of our own cultural history, can we simply assume that it is true in other cultural traditions as well? Is this not

actually the height of cultural arrogance? And are we not overlooking the very difference between Western and Buddhist traditions
that I noted when discussing the fundamental "permeability"
Buddhist hierarchial thinking has in the context of the six saṃsāric
life-forms? I would answer affirmatively to all of the above, and I
would submit that our fear of any vertical dimension to the spiritual
life has become so strong that we are literally terrified of being
confronted by the fact that Buddhism is integrally hierarchical.

Consider the following passage written by Gary Snyder, one of
the most influential and respected Green Buddhists and someone
who has influenced much of my own appreciation for the "Green"
implications of Buddhism. Feeling the need to distinguish a
Buddhist sense of spiritual "training" from what he sees as a more
artificial notion of spiritual cultivation, Snyder observes that:

> The word *cultivation*, harking to etymologies of *till* and *wheel
> about*, generally implies a movement away from natural process.
> In agriculture it is a matter of "arresting succession, establishing
> monoculture." Applied on the spiritual plane this has meant
> austerities, obedience to religious authority, long bookish scholar-
> ship, or in some traditions a dualistic devotionalism (sharply
> distinguishing "creature" and "creator") and an overriding image
> of divinity being "centralized," a distant and singular point of
> perfection to aim at. The efforts entailed in such a spiritual practice
> are sometimes a sort of war against nature—placing the human over
> the animal and the spiritual over the human. The most sophisticated
> modern variety of hierarchical spirituality is the work of Father
> Teilhard de Chardin, who claims a special evolutionary spiritual
> destiny for humanity under the name of higher consciousness. Some
> of the most extreme of these Spiritual Darwinists would willingly
> leave the rest of earthbound animal and plant life behind to enter
> an off-the-planet realm transcending biology.[8]

While this may be an effective and appropriate critique of certain
Western religious attitudes, it is so heavy-handed in its blanket
condemnation of any notion of verticality, of any notion of the
development and evolution of consciousness, that it rejects, however
unintentionally, most of Buddhism as well. Snyder, in this passage
at least, implies that all notions of the evolution of consciousness

lead inevitably to the rejection of nature and the "natural" by an oppressive hierarchy of "Spiritual Darwinists." But what is the developmental dimension of Buddhism if not a teaching of the evolutionary transformation of consciousness? The very definition of Buddhahood asserts the developmental realization of a higher ethical sensibility expressed as compassion for all of existence.

I readily share Synder's concern to avoid any world-denying dualism that sets spirit off against nature. My concern is that his solution is too drastic. His cure may be as harmful as the disease, in that it compels the Western Buddhist to renounce not just the worst of Western religion but also the best of Buddhism, even as Snyder advocates the latter as one of the few established alternatives to the former available to us. What is it that is being overlooked here? I suggest that Western Buddhists can resolve this problem within our own cultural history only to the extent that we openly acknowledge and affirm the way in which the developmental aspect of Buddhism is hierarchical, while simultaneously continuing to criticize the specific hierarchical forms that have clearly misshaped Western attitudes toward nature and the environment.

It is thus central to my argument to establish that there is, in fact, a crucial difference that distinguishes the Buddhist conception of verticality or hierarchy from those forms of hierarchy that have dominated Western cultural history. Only once that difference is clear will I be able to argue my central thesis that we need actively to endorse this Buddhist notion of developmental verticality precisely for the sake of better environmental ethics, just as we strive to abandon the most familiar Western notions of hierarchy for the very same reason. The difference is not immediately obvious, however, and even the reader who is sufficiently sympathetic to consider that there might be a difference is no doubt wondering why I would choose, even insist, on contaminating whatever I have to say by using this dreaded "h-word" when I could just as easily have conformed to the prevailing cultural taboo and surreptitiously slipped in some more innocuous synonym for "hierarchy" when speaking of the vertical dimension of Buddhism. While it is true I could thereby avoid the risk of being dismissed as hopelessly atavistic even before I am able to make my case for the difference, there is a reason why I have chosen not to do this, one which I hope will soon become clear.

The first task, however, is to distinguish the two fundamentally different forms of hierarchy. Thinking, for the moment, not just historically but more theoretically in terms of a Weberian "ideal typology," I am suggesting that there are two forms of human practice that are sufficiently related one to the other to fall under the same general designation of "hierarchy," even though their respective outcomes are nonetheless diametrically opposite.

The Hierarchy of Oppression

To illustrate the two types of hierarchy we can imagine each form encompassing again both a developmental and a relational dimension of human experience, each of which we can plot on an x-y graph similar to the one we considered above. It is important to note the difference in what we are graphing now, however. Earlier, in figure 1, we were noting the relative emphasis given to the developmental versus the relational dimension of the *Dharma* in different forms of Buddhism, whereas now we shall be using the same axes to explore a rather different issue. In the next two figures we shall be plotting the relative balance between the developmental and relational dimensions of our existence in each of two different models of hierarchy. In each of these two figures, the further away from the center point we move horizontally (in either direction), the greater is the degree of interrelatedness. And the further we move up the vertical axis, the greater the degree of developmental progress. We shall see, however, that what constitutes vertical movement differs drastically in each of the two cases, and it is that difference that makes all the difference.

The first type of hierarchy or hierarchical structure we can designate a "hierarchy of oppression." We can understand its distinctive mechanisms by imagining superimposed on our x-y axes a triangle or a cone rising from a wide base to a single point at the apex (see figure 2). Imagine now that, as we move up the vertical axis, each horizontal section of the cone corresponding to the present vertical location represents a circle of interrelatedness. By "interrelatedness" here I mean not just any sense of relationship but, specifically, an understanding of the sense in which all beings share a communality of interests. The nature of a "hierarchy of oppression" is such that as one advances vertically, one's "circle of interrelatedness" becomes increasingly smaller. This is so because one advances in a hierarchy of oppression by exercising one's control over and domination of all those below. As a result of one's vertical progress, one necessarily becomes less and less aware of one's interrelatedness with them.

From the Buddhist perspective, of course, one's actual inter-relatedness remains constant and absolute. What in fact changes as one moves upward in figure 2 is not how interrelated one actually is but, rather, the extent to which one realizes and expresses that interrelatedness in one's actions. In other words, "progress" in a hierarchy of oppression requires that one actively deny and suppress any recognition of relatedness to those that one seeks to dominate. As one claws one's way to the top of the pyramid, submissively accepting subjugation from those above in return for the privilege and right to dominate those below, the extent of one's expressed interrelatedness, as plotted on the horizontal axis, becomes increasingly more narrow and circumscribed. For one cannot successfully dominate what is below except to the extent that one actively rejects any fundamental communality of interest and needs.

In the hierarchy of oppression, one moves upward only by gaining power over others, and to safeguard one's power and

FIGURE 2:
A Hierarchy of Oppression

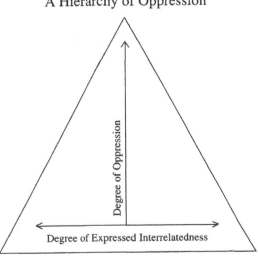

security one must seek ultimately to control all of existence, however unrealistic and deluded that aspiration inevitably turns out to be. One is able to sustain this aspiration, moreover, only to the extent that one actively suppresses and denies any sense of meaningful connection to all that is below. Reaching the apex of the cone in figure 2 would thus represent, in the terms of this model, the ultimate "success" to which one could aspire, but that ultimate "success" would, of course, be a state of total alienation—alienation not just from others but from oneself as well—because one can "succeed" only by rejecting one's actual nature of interrelatedness. If the folly of this approach to life is not schematically clear from the diagram, one need only reflect on the course of human history, especially (though not exclusively!) the history of the modern West.

The Hierarchy of Compassion

Imagine now the same image turned upside down, stood literally on its head as in figure 3. Here we find the apex point at the bottom, and we see that the cone broadens as it rises. This is a model of what I would call a "hierarchy of compassion." Note the fundamental difference. As one ascends the vertical, developmental axis in this case, something quite different happens, something that is precisely the inverse of the previous case. As one moves upwards, the circle of one's interrelatedness (or rather of one's expressed interrelatedness) increases. In fact, the only way one can move up is by actively realizing and acting on the fundamental interrelatedness of all existence. But the line of vertical ascent needs to be plotted somewhat differently in this case, because vertical movement now is not the simple, linear upward assertion of control over gradually more and more of the rest of existence. In the hierarchy of compassion, vertical progress is a matter of "reaching out," actively and consciously, to affirm an ever widening circle of expressed interrelatedness. Such an ever broadening circle plotted as a developmental line becomes the spiral path illustrated in figure 3.

Unlike the previous case, moreover, progress along this spiral path confers no increasing privilege over those who are below on the path. Quite the contrary, it entails an ever increasing sense of responsibility. This profoundly ethical sense of responsibility for an ever greater circle of realized relatedness is what is expressed by the Buddhist term *karunā*—compassion or "wisdom in action." Perhaps now it is beginning to become clear why I am so concerned about attempts to formulate Western Buddhism in any way that does not fully appreciate the vital complementarity of both the developmental and the relational dimensions of the tradition. Buddhism does offer an ethic

FIGURE 3:
A Hierarchy of Compassion

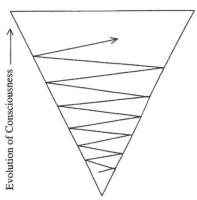

← Degree of Expressed Interrelatedness →

that might be capable of transforming our current deluded environmental practice, but the developmental dimension of the tradition is crucial to that ethic, because the Buddhist virtue of compassion is something one can cultivate only by progressing up the spiral path of the hierarchy of compassion. Before looking at this last assertion more closely, however, we must first consider a question I raised in the introduction to this article.

The two models I have just presented each have a vertical dimension, yet I have argued that there is a crucial difference. Why, if these two forms of "progress" or individual development are so different, do I feel so strongly that both models should be called "hierarchies," especially since that word sounds so objectionable to many modern ears? My point is to stress the close, yet decisively different, relation between the two, and that crucial point would be missed if we were to suggest that these two ways of living one's life are completely unrelated. Relating to others and to the environment as a whole in accord with the hierarchy of compassion is not just better than climbing the hierarchy of oppression: it is the very antithesis. To the extent that we do one, the other is literally impossible—and this is what is lost if we fail to stress the inherent relationship between the two. Hence the importance given in traditional Buddhism to the notion of "going forth." One can advance on the spiral path of compassion only to the extent that one has effectively gone forth away from pursuing the rewards of the hierarchy of oppression. Unlike some "new age" thinking, Buddhism does not suggest that we can have it all. On the contrary, it asserts that progress up the hierarchy of compassion becomes possible only to the extent that we "go forth" from the aspiration to have it all. For "having" in this sense is an expression of control and is possible only within the context of the hierarchy of oppression. Without seeing how the two hierarchies are related, one might still imagine that it might be possible to pursue simultaneously elements of both.

There is another reason to stress their relationship. Both the forms of hierarchy share a crucial feature in that both are about power. Or, perhaps we should say the one is about power and the other is about empowerment, the transformative power of compassion.[9] The first offers the power to control all, while the second cultivates the empowerment to transform oneself in order truly to benefit all life (including ourselves). It is this empowerment that we cannot afford to jettison in our desperate efforts to flee from the oppressive legacy of our past and present.

Reaffirming the Developmental Dimension of Traditional Buddhism

If the theory and the structure of the Buddhist hierarchy of compassion are now clear, one might well still wonder what this would look like in actual practice. This is the point at which the danger of overlooking the vertical, developmental aspect of Buddhism becomes most evident, for it is in the context of its developmental dimension that the tradition provides quite concrete suggestions as to how to put the insight of interrelatedness into actual practice. Without its developmental dimension, all that Buddhism has to offer contemporary environmental ethics is the metaphysical assertion that all things are interrelated. Lost is the fact that Buddhism offers also a systematic and comprehensive set of techniques by which one can actually realize that relatedness in practice.

I have already surveyed the doctrinal roots of the developmental aspect of the tradition, but the question we are currently addressing requires that we now focus on this aspect of the teaching as an actual path of practice. Consistently favoring pragmatism over metaphysical speculation, the Buddha would point out that the only way we can realize what a hierarchy of compassion would look like in practice is by actually doing the practice of *Dharma*, and this of course involves much more than just being more environmentally correct or sensitive, important as that may well be. Buddhism is saying, quite literally, that we cannot expect to act in an environmentally more ethical manner until we cultivate a much broader ability to act with compassion and wisdom. How we are to do that is the subject of a vast body of traditional teachings and techniques, but it is frequently summarized under the rubric of the "threefold learning" (*triśikṣā*): the systematic cultivation of morality, meditation, and insight into the actual nature of existence. Each of these three is widely explored by the various schools of Buddhism, and a full exposition of what is entailed goes well beyond the space available here. For our present purposes it will suffice to note simply how these three elements of Buddhist practice are related to one another and what implications this has for a contemporary environmental ethics based on Buddhist principles.

This threefold formulation of the Buddhist path is presented as clearly sequential, in that each step builds on the previous one. The three phases of the path do overlap, however, so the point is not that one cannot begin meditation before completing the practice of morality, for example. The point rather is that one cannot expect to make progress in one phase except on the basis of substantial progress in the previous phase. In other words, effective insight into the actual nature of existence requires real progress in the cultivation of higher states of awareness through meditative practice. And that, in turn, is possible only on the basis of a practice of the ethical precepts and a cultivation of the primary virtues. This may seem a simple point, but it has significant implications when we ask what a Buddhist environmental ethic would be like.

Buddhism says that we can expect to act in accord with the basic interrelatedness of all existence only once we have cultivated a significantly different state of awareness. Simply attempting to change specific environmentally detrimental behaviors will not work. Efforts to change our environmental behavior may well be part of the ethical practice that creates the necessary foundation for experiencing states of higher meditative awareness and ultimately for realizing transformative insight, but these efforts will be effective only to the extent that they are undertaken as part of the whole three-step program. The Buddhist solution to the environmental crisis is thus nothing short of the basic Buddhist goal of enlightenment. That may seem like an unimaginably distant and lofty goal, and indeed it does involve a fundamental and total transformation of what we are—nothing less. At the same time Buddhists need not feel overly daunted by the immensity of this undertaking, for enlightenment is, in one sense at least, simply (if not easily) a matter of becoming more fully human, in that this radical transformation is the potential of all humans, indeed of all beings. The solution to the problem is thus imminently possible, although that potential can only be actualized on the basis of both a clear vision of the goal and a well-defined path to reach it, coupled with a sustained effort to pursue that path to its completion.

A Buddhist environmental ethic is hence a "virtue ethic," one that asks not just which specific actions are necessary to preserve the environment but, more deeply, what are the virtues (that is, the precepts and perfections) we must cultivate in order to be able to act in such a way.[10] The relational dimension of Buddhism is necessary to secure an ecologically sound vision of the goal, but the developmental dimension of the tradition is every bit as necessary in that it provides the path that will enable us actually to reach that goal. Is there, then, truly a danger that Western Buddhists might overlook the central place of basic Buddhist ethics in formulating a new, "green" Buddhism? Not consciously, I suspect, but perhaps quite unintentionally as part of the effort to discard our own cultural legacy of hierarchies of oppression.

Consider the following comment made by yet another prominent and respected Green Buddhist. In "The Greening of the Self" Joanna Macy discusses the notion of "self-realization" that lies at the heart of Arne Naess's Buddhist-inspired sense of deep ecology, proclaiming it the foundation of what will become a new, environmentally benign conception of the self.[11] Citing his view that the process of self-realization, properly understood, involves leaving behind "notions of altruism and moral duty," Macy succumbs to a very dangerous, if seductive sentiment. Naess seeks to make a quite specific, if nonetheless ambiguous, point when he argues that the ethic of "self-realization" he envisions will not require that one act for the sake of others out of a sense of self-abnegating "duty." He takes "altruism" here very literally to mean something done "for others" *in contrast to* one's own self-interest. "Altruism" in this sense will become unnecessary, he asserts, when one reaches the point at which one's "self-interest" and the interests of others naturally converge. What he fails to clarify is that some form of ethical (and Buddhists would add meditative) practice is still necessary in order to reach that point, and the danger of this ambiguity is borne out by Macy's extension of his argument.

Naess's basic point may be sound enough, as far as it goes. We need an expanded sense of self, one in which acting on behalf of others and the ecosphere is ultimately acting in terms of "enlightened self-interest" and not out of some sense of moral obligation, or duty, or even the rights of others perceived as separate from our own interests.[12] Macy concurs but, falling prey to the implicit ambiguity, she is led seriously astray. She insists that "virtue is *not* required for the greening of the self or the emergence of the ecological self" (her italics).[13] In this formulation there is no ambiguity, and we are surely on ethical quicksand. She is clearly speaking not of the eventual goal but of the path itself, of the practice by which she feels the ecological self will "emerge." Apparently, thinking that the rejection of an ethic of duty entails rejecting all moral judgment and discernment—all effort to cultivate virtue—she arrives at the conclusion that ethical discipline and development have no place in the "new Buddhism" she envisions. If one simply has "self-realization" as one's goal, no further ethical effort is required. No practice is necessary, only an opening to what she concedes is something very close to the Christian concept of "grace." Let us hope that what she says, in this instance at least, is not actually what she intends, for this would surely be a case of throwing out one crucial aspect of Buddhism in the very act of professing another.

Conclusion

We have explored how some Green Buddhists, uncomfortable with any notion of hierarchy or developmental verticality, are moving, intentionally or not, toward a kind of unidimensional Buddhism, one in which the inverted cone of the hierarchy of compassion is simply collapsed into a single flat circle of relatedness. In doing this they very aptly stress the relevance of the horizontal, relational dimension of Buddhism to environmental ethics, but they overlook or even deny the equally vital vertical dimension, that aspect of the *Dharma* that sees enlightenment as a process involving the evolution of consciousness. This development of consciousness in Buddhism is expressed practically as an ever greater sense of responsibility to act compassionately for the benefit of all forms of life; hence its relevance to any discussion of Buddhist-inspired environmental ethics. Failing to distinguish between the two types of hierarchy outlined above, and obsessed with the need to dump out the dirty bath-water of Western hierarchies of oppression, some Green Buddhists fail to note that they are also discarding the "baby" of all potential for development—of the potential for meaningful growth toward a greater expressed sense of interrelatedness, toward a greater sense of environmental ethics in the most profound sense of the term.

There are thus two reasons why reaffirming the vertical dimension of Buddhism is so important: first, because it is central to the integrity of the tradition; and, second, because it is precisely that part of the tradition that has something useful to add to contemporary environmental ethics. This latter point may seem less than clear, even if one is prepared to concede the former. Could we not do as well or even better with just the circle of ultimate interrelatedness, even if it does seem a bit flat or one-dimensional? Is the loss of the vertical dimension not a relatively small price to pay at this particular moment in history, in order to secure thoroughly the long-neglected horizontal axis of relationship? Why, after all, should Buddhism need to assert, as it does, that we all too often perfidious human beings are somehow a "higher form of consciousness" than the loyal and faithful dog, for example, or even than a banana slug for that matter? The slug, at least, is content to mind his own business.

Given the dire situation of the environment, and given the human role in bringing about that crisis, the position suggested by these last few questions is indeed attractive, beguilingly so. Nonetheless I do see this newly emerging, unidimensionally horizontal form of Green Buddhism to be fundamentally flawed, flawed not just in that it misrepresents the actual nature of the Buddhist tradition, but even more seriously flawed in that it abdicates, however unwittingly and unintentionally, both the ethical responsibility and the ethical potential that might actually be just what we need to solve the predicament in which we find ourselves. If we deny the vertical dimension of the *Dharma*, we are denying the possibility of developing precisely the higher ethical sensibility that we are currently so manifestly lacking. And in denying that potential, we consign ourselves to wait helplessly, watching as the forces of human greed, hatred, and delusion proceed to destroy the ecosphere, watching either in disempowered rage and despair or perhaps in hope that some higher being will step in to save us from our sins.

Without an explicit recognition of the vertical challenge fundamental to Buddhist practice, the developmental quest for enlightenment with its concomitant increase in ethical sensibility is lost in favor of a view suggesting that there is really nothing we need do—indeed, nothing we can do beyond trusting in providence. This is not a Buddhist environmental ethic. What Buddhism offers is in fact quite a different message. And it is not just a message that the *Dharma* offers, it is a method. Herein lies the crucial difference. If we adopt only the relational teaching of the Buddha, then insight into the interrelatedness of all existence becomes simply an article of faith, something in which one is ardently to believe. The implicit message, one well embedded in our own cultural history, is that if one just believes in the right revelation faithfully enough, then all will turn out just fine—through the agency of some benign higher power. Stripped of the old theocentric "God-talk," this updated gospel of grace may seem both comfortable and familiar, but this must not obscure the fact that it is not the Buddhadharma. For Buddhism, the relational dimension of existence is not an article of faith; it is a reality to be experienced directly through the active cultivation of higher states of consciousness. Simply to affirm the interrelatedness of all things, whether as an article of faith or as an intellectual inference, has in the Buddhist perspective no transformative power. It is only through undertaking the ethical and meditative practice charted in the developmental dimension of the tradition that one's actual behavior begins to change to conform with the insight of interrelatedness.

Western ecology has given us an adequate model for understanding the ethical implications of how all things are interrelated.

It is nice that Buddhism confirms that insight, but we gain little from Buddhism if that is all we see in the tradition. And we gain even less if we feel that simply affirming this view of interrelatedness will, of itself, be sufficient to bring about the necessary changes in our ethical practice. Thus, the real value of Buddhism for us today lies not so much in its clear articulation of interrelatedness as in its other crucial dimension, in its conception of the ethical life as a path of practice coupled with its practical techniques for actually cultivating compassionate activity. The tendency in Green Buddhism to focus exclusively on the horizontal circle of interrelatedness thus endangers the very part of the tradition that we are most sorely lacking. What Green Buddhism needs to explore more thoroughly is the Buddhist principle that meaningful change in our environmental practice can come about only as part of a more comprehensive program of developing higher states of meditative awareness, along with the increased ethical sensibility which this evolution of consciousness entails. Otherwise, it seems, we are simply spinning our wheels.

Notes

1. This chapter was originally published in *Western Buddhist Review* 1 (December 1994):131–55. It was previously reprinted as a pamphlet in celebration of Earth Day, 20 April 1996, by the Friends of the Western Buddhist Order, Richland, Washington.

2. See, for example, *Buddhism and Ecology*, ed. Martine Batchelor and Kerry Brown (London: Cassell, 1992); *Dharma Gaia: A Harvest of Essays in Buddhism and Ecology*, ed. Allan Hunt Badiner (Berkeley: Parallax Press, 1990); Gary Snyder, *Practice of the Wild* (San Francisco: North Point Press, 1990); *Nature in Asian Traditions of Thought*, ed. J. Baird Callicott and Roger T. Ames (Albany: State University of New York Press, 1989); and Deane Curtin, "Dōgen, Deep Ecology, and the Ecological Self," *Environmental Ethics* 16, no. 2 (summer 1994):195–213.

3. Actually, to suggest that there are "nondual" forms of Buddhism in contrast to "dualistic" forms is a misnomer. All forms of Buddhism are nondualistic in that enlightenment is understood ultimately to transcend all ontological duality. Similarly, all Buddhist schools unavoidably adopt, in some form or another, an "operational dualism" reflected in the very distinction between delusion and enlightenment. There is a significant difference of emphasis in the way different schools speak of enlightenment and its relation to the state of suffering, but it is likely that this reflects more a difference of practical approach than of substantial ontological divergence. The difference between the gradualists and subitists within the tradition is thus best seen, in my view, as largely rhetorical, though part of the point, of course, is precisely that we often become trapped within the language we use.

4. The history of Buddhist views on whether plants and non-animate things have ethical standing is quite complex; see Lambert Schmithausen, *The Problem of Sentience of Plants in Earliest Buddhism* (Tokyo: International Institute of Buddhist Studies, 1991); and William R. LaFleur, "Saigyō and the Buddhist Value of Nature," parts 1 and 2, *History of Religions* 13, no. 2 (November 1973):93–128, and no. 3 (February 1974):227–48.

5. *Samyutta Nikāya*, 12:3, §23.

6. This is true at least of historical Zen, even if not of some of the modern-day versions of "Zen" promulgated in the West.

7. There is a logical and historical line linking the early doctrines of dependent co-arising (*pratītya-samutpāda*), impermanence (*anitya*), and the nonsubstantiality of the self (*anātman*) with the later Mahāyāna notions of emptiness and inter-relatedness, but tracing those links adequately would require more space than is available here.

8. Snyder, *Practice of the Wild*, 91.

9. My distinction between the hierarchy of oppression and the hierarchy of compassion is inspired in part by a similar distinction between the "power mode" and the "love mode" suggested by the Ven. Sangharakshita in "Mind—Reactive and Creative," *Middle Way*, August 1971. In Sangharakshita's distinction, however, the positive sense of empowerment (i.e., spiritual or ethical power) that I wish to stress here is not as evident.

10. Cf. Geoffrey B. Frasz's "Environmental Virtue Ethics: A New Direction for Environmental Ethics," *Environmental Ethics* 15, no. 3:259–74.

11. Joanna Macy, "The Greening of the Self," in *Dharma Gaia: A Harvest of Essays in Buddhism and Ecology*, ed. Allan Hunt Badiner (Berkeley: Parallax Press, 1990), 53–63.

12. Śāntideva provides a traditional Buddhist parallel to Naess's notion of "enlightened self-interest" (ibid.) when he points out that the hand helps the foot (by removing a thorn) even though the pain of the foot is not a pain of the hand; see the *Bodhicaryāvatāra*, 8:91–99.

13. Macy, "The Greening of the Self," 62.

Buddhism and the Discourse
of Environmental Concern:
Some Methodological Problems Considered

Ian Harris

Erosion of traditional cosmological thinking is a well-attested and significant strand in the recent history of religion in Europe and America. Undoubtedly, all of the major traditions have retained well-defined zones of resistance against the prevailing current of modernity, Christian creationism being a good example in this connection. However, as the current has grown in vigor, religious modernists have, at times reluctantly though often with enthusiasm, abandoned long-standing views on the place of the earth and the position of humanity within the created order—some of the most cherished beliefs of their tradition—and accepted, with few modifications, the modern scientific picture of the universe. Such capitulations are now, by and large, accepted and consigned to the historical past. However, the battle over humankind's position in the natural order, an order rendered incompatible with any conscious sense of meaning or responsible agency by the inexorable logic of the modern scientific method, has not yet been conceded by theologians. Under such circumstances it is perhaps unsurprising that a discourse of environmental concern, in part aimed at reintroducing meaning and purpose back into the bleak vastnesses of the modern cosmos, has taken such a prominent place in the pronouncements of leading theologians the world over.

Of course, Christianity is not the only religious tradition engaged in this rearguard action. Buddhism, too, has its eco-advocates. Indeed, Buddhism is often invoked as a far more environmentally beneficial set of beliefs and practices than Christianity could ever be, some writers going so far as to suggest that, of all the major religious traditions, Buddhism is the best equipped to form the heart of a new global environmentalist ethic. Now, positive environmentally oriented discourse does not have its origins in any specifically religious domain, although it is beyond the scope of this essay to discuss the romantic movement's repudiation of the scientific project that so clearly contributed to its emergence.[1] Nevertheless, the politization of this discourse has become a significant theme, particularly in the latter part of the twentieth century, and no world-historical religious movement would wish to jeopardize its standing by failing to endorse such a "self-evident" collection of truths about the world and our place within it. It is clear that the benefits of taking such a stance will be considerable.

There is now much good evidence that a significant number within Buddhism[2] itself, plus those who give intellectual assent to selected elements of the Buddhist tradition as part of their armory in the fight against the worst excesses of "technological society," have declared themselves favorably disposed to ecologically motivated activity, whether it be of the shallow or deep variety. Organized Buddhism undoubtedly embodies virtues that appear, at least from the superficial perspective, in tune with the discourse of environmental concern.[3] The task of this essay will be to assess the tradition as a whole, and the methodological presuppositions underlying ecoBuddhism, and to confirm or deny the truth of these impressions. My central contention will be that, with one or two notable exceptions (Schmithausen[4] springs to mind here), supporters of an authentic Buddhist environmental ethic have tended toward a positive indifference to the history and complexity of the Buddhist tradition. In their praiseworthy desire to embrace such a "high profile" cause, or, to put it more negatively, in their inability to check the influence of a significant element of modern globalized discourse, Buddhist environmentalists may be guilty of a *sacrificium intellectus* very much out of line with the critical spirit that has played such a major role in Buddhism from the time of the Buddha himself down to the modern period.

A fundamental problem confronting any serious examination of the Buddhist tradition's "attitude to nature" is philological. The most obvious starting point ought to be the identification of a Buddhist term or terms equivalent in range of meaning to our word "nature."

However, this is more complex than it seems on the surface. In the first place, there are many canonical languages to choose from. We could simply choose to differentiate between Indic terms, on the one hand, and those originating in the East Asian area, on the other, but even if this was deemed a suitably sophisticated methodology, and I am not sure myself that it would be, a further difficulty presents itself. Each of these languages is bound to cultures that possess their own specific modes of development. Indeed, the original attempts to translate Sanskrit technical jargon into Chinese are known to have encountered many intractable difficulties, not least because of the existence of a sophisticated philosophical vocabulary in China prior to the arrival of Buddhism. Moving to the contemporary setting, we must not forget that the interpretation of textual material can never be a culture-free exercise, whether it be done by contemporary Buddhist themselves or by those who seek corroboration of their own ideas from the Buddhist tradition. As Hans Georg Gadamer has pointed out, we must be aware of the prejudgments we bring to the understanding of a text and must acknowledge the distance in historical terms between us and the text's author. Without this we are likely to deceive ourselves into thinking that we can uncritically "stand in immediate relation with the past."[5] Also, let us not ignore the fact that the languages of canonical Buddhism reflected the concerns of a segment within the wider culture and, by and large, are to be identified with the worldview of small but influential elites. The question must arise as to how far the sacred writings and their commentaries represent the understandings and practices of ordinary people who, after all, will be the prime agents in the interaction of Buddhism and the natural world, for monks, by virtue of their disciplined existences, are practically restrained from most potentially damaging activities, such as agriculture and the like. It is clear, then, that all of these matters must be examined more rigorously than has been done to date before we can confidently assert that Buddhism, of whatever form, possesses the necessary philological, cultural, and philosophical structures to accept the imposition of a discourse of environmental concern without undue distortion.

Another element, this time relating to the range of meanings the term "nature" has come to represent in the West, must also be considered. Kate Soper[6] identifies three ways in which nature has been conceptualized in modern environmentalist discussions, of which the first, or metaphysical, relates to that part of the world which lies beyond the human or merely artificial. The nature/culture dichotomy is clearly at the heart of this definition. The second meaning is associated with "the structures, processes and causal powers. . .operative within the physical world" and therefore represents that sector of existence understood as the proper object of study in the natural sciences. The final "lay" or "surface" concept is concerned with the distinction between the "natural" as opposed to urban or industrial landscapes and is intimately bound up with aesthetic judgment. Soper accepts that the third meaning dominates the discourse of the green movement, although it is clearly dependent on and interrelated with the others.

The evolution of the modern ecological definition of "nature" and "the natural" can only be fully understood against the background of the history of Western thought itself. With this in mind, it would be unwise to neglect two other crucial distinctions: the Aristotelian tension between "nature" understood as the totality of all that exists and "nature" as the essence or active principle of things; and the medieval nature/supernature dichotomy. Although the term *supernaturalis* only seems to have emerged fairly late in the history of Christian thought, most notably in the work of Thomas Aquinas, the modern manner of construing reality entails assent to, or at the least criticism of, the notion that nature lacks many of the clues necessary for a full understanding of things. The scientific worldview, then, is clearly a rejection of the supernaturalist claims of theism, but, intriguingly, environmentalism—particularly of the ecospiritual type,[7] a form that has had a sizable impact on contemporary ecoBuddhism—represents a reappropriation of prescientific modes of thinking with its Spinozist insistence on *natura naturans* as an almost pantheist power of nature.

Buddhist scholars and activists have, in recent times, offered a range of Buddhist technical terms that they deem to correspond with the English term "nature." An obvious question in this context is, what sense of this richly nuanced term are they thinking of and are they all in agreement on the matter? I do not believe that this question has even begun to be answered, and this essay may be seen as a humble and highly provisional attempt to get such a debate off the ground. A list of the most commonly mentioned Indic equivalents of the term "nature" includes *saṃsāra, prakṛti, svabhāva, pratītya-samutpāda, dharmadhātu, dharmatā,*[8] and *dhammajāti.*[9] The range of significances covered by such terms is vast and detailed analysis is beyond the scope of our present discussions, although sustained work on the topic would undoubtedly do much to advance our present understanding. One example will have to suffice. *Saṃsāra* in its usual sense denotes the totality of sentient beings (*sattvaloka*) caught in the round of life after life, although it may also encompass those parts of the cosmos that fall below the level of sentience and, as such, act as the stage or receptacle (*bhajanaloka*) on which the beginningless cycle of life on life unfolds. However, even in this extended manner, *saṃsāra* can hardly be regarded as *natura naturata* in any obviously Western sense for it contains hell-beings, gods, and ghosts quite apart from its human and animal residents. Indeed, above this region of physicality and gross desire lie two other more subtle regions of reality, the whole comprising the traditional Buddhist triple-decker universe. Built into this model is the possibility of movement from one level to the other through the activation of mental powers gained in meditation. *Saṃsāra*, then, incorporates elements which, from a Western perspective, encompass both the natural and the supernatural. Consideration of other terms offered by scholars as Buddhist equivalents of "nature" tend to reveal similar mismatches.

Statements of the kind "Buddhism is. . ." are problematic in that they very often fail to take account of the historical, doctrinal, and cultural diversity of the tradition. For instance, a fundamental distinction needs to be maintained between Buddhism in its Indic forms (in this category I include the Theravāda traditions of South and Southeast Asian as well as the Mahāyānist Tibetan forms of Buddhism) and the Chinese and East Asian transformations of the Indic tradition. It also makes good sense to distinguish between the historical phases in the development of Buddhist thought and practice. Heinz Bechert, for instance, chooses to divide Buddhist history into canonical, classical, and modern phases,[10] while Charles F. Keyes, in a manner possibly more conducive to our investigation of Buddhism's understanding of the "environment," distinguishes between a premodern cosmological Buddhism, on the one hand, and modernist forms, influenced by aspects of Western thought and social organization, on the other.[11] Whatever classificatory scheme we choose to use, the generalization of ideas or practices from one historical, geographical, or cultural phase of the tradition, in an attempt to justify some monolithic Buddhist position, will be largely illegitimate.

An example should give a good illustration of this point. Frank E. Reynolds, in an important discussion of the three overlapping types of cosmological thinking present in the traditional Buddhist countries of Southeast Asia, points to the *karma/saṃsāra* complex of doctrines—his "saṃsāric cosmogony"[12]—as the point from which laypeople and monks orient themselves ethically one to another. Such interactions generate a "total field"[13] system in which one's present existence is ethically enmeshed in a vast, causally connected, and highly stratified cosmic order encompassing humans, animals, gods, and so forth, arranged hierarchically from the realms of the gods all the way down to the infernal regions. In the *Saddharmasmṛtyupasthāna Sūtra* (Sūtra of the remembrance of the good law),[14] classified by Chinese tradition as a work of the Hīnayānist Abhidharma and mainly important because it provided the basis for Genshin's (942–1017) famous description of hell, the Ōjōyōshū,[15] the eight levels of hell are further subdivided. Thus, a subregion of the hell of repetition (*saṃjiva*) is called the "place of excrement" because this is the place in which sinners who have killed birds and deer without regret are punished by being forced to eat dung that is crawling with flesh-eating worms. The "hell where everything is cooked," a sublevel of the burning hell (*tapana*), is reserved for those who have deliberately destroyed forests by fire, while the "bird hell" in the hell of no interval (*avīci*) contains malefactors who deliberately caused famines through the disruption of water supplies.[16] It may well be that the moral implications of these doctrines did serve to inhibit environmentally destructive behavior in the premodern period, but we should be aware of two issues before we try to import them into a modern context. First, one of the cardinal features of modernist Buddhism is precisely its embarrassment about traditional (mythological or prescientific) cosmologies. As such, it represents an erosion of tradition and an accommodation to the prevailing current of scientific thinking. Indeed, the majority of social activist, including environmentalist,[17] forms of Buddhism today can be seen to have arisen as a result of these changes in emphasis. How paradoxical, then, that the claims of modernist Buddhists to stand in good harmony with nature seem to be premised on the scientism of the Enlightenment, a movement in European history that did so much to liberate the individual from the "thrall of nature"[18] and opened up the forces that have now led to its potential destruction. Second, until evidence is offered to the contrary, we shall have to remain skeptical of the inhibitory power of the Buddhist conception of hell, at least from the environmentalist perspective, in a premodern Asian world that was fundamentally unaffected by the factors that may have rendered large-scale ecological degradation a realistic possibility.

Reynolds terms the traditional Buddhist world system the "rūpic,"[19] or devolutionary, cosmogony. However, any positive interpretation of this hierarchically organized and interrelated vision of the universe—one is tempted to employ the term "nature" in this context—is rather undermined by the tradition's own assessment of the radically unstable nature of all conditioned things. The Indic, and specifically early Upaniṣadic and hence pre-Buddhist, roots of this way of thinking now become plain. For traditional Theravāda Buddhism, the universe is a vast unsupervised recycling plant in which unstable entities circulate from one form of existence to the next—a Joycean *"commodius vicus* of recirculation." This seems an ideal metaphor from the environmentalist perspective, for, if Buddhists envisage the world process in this manner, there is some justification in the conclusion that we should seek to replicate the processes of which we are such an intrinsic part. Two objections immediately arise, however. In the first place, environmentalists are certainly committed to the principle of the recirculation of inanimate materials, such as wood products and the like, but how far are they prepared to go in the direction of the recycling of sentiency itself? It seems to me that there are few intellectual resources in the Western thought universe to support such a move! In the second place, and from the perspective of the "ultimate evaluation of existence,"[20] the Buddhist universe lacks any genuine *telos*. It is dysteleological.[21] As we have already noted, Reynolds employs the term "devolutionary" in his discussion of the rūpic cosmogony, a term that implies a regular, though lengthy, degeneration of the physical world, a process mirrored in the inevitable moral decline of humans. The outworldly character of Theravāda cosmology is now apparent, although, to give a full account of this particular interpretation of existence, we must introduce a final element into the equation, the *mokṣa/nirvāṇa* complex. If we now return to the environmentalist perspective, it becomes clear that recycling is connected with *saṃsāra*. This is the positive part of the message. However, it is somewhat compromised by the fact that, ultimately, the Buddha's teachings point to a goal that represents the overcoming of the restrictions entailed by *saṃsāra*.

Ecology, even in its so-called deep form, must be premised on some distinction between nature and humanity, for without it our activities become, by definition, "natural" and, under such circumstances we can be held no more responsible for the adverse effects of our activities than can any other species. However, Martin Heidegger, among others, has pointed to the difficulties inherent in this fundamental distinction. For him, the problem of "construing the humanity-nature relationship as a Subject-Object antithesis is that it already presupposes a division between 'subjects' and 'objects' that is, strictly speaking, illegitimate."[22] Heidegger's point is that scientific modes of thinking, while "deeply counter-intuitive"[23] have accustomed us to regard the things of the world as "objects," with the result that we, as heirs to the Western intellectual tradition, have become alienated from an earlier, premodern "pre-understanding of the world." This is interesting because it seems to tie in with the ·Buddhist Yogācāra/Vijñānavāda view that the imagination of the subject/object dichotomy (*grāhya-grāhakakalpanā*) is a function of mental processes contaminated by ignorance (*avidyā*). The attainment of *nirvāṇa* as a return to this primitive mental purity, then, represents the uprooting of saṃsāric addiction. In Vasubandhu's words:

> From the non-perception of the duality [of subject/object] there arises the perception of the *dharmadhātu*. From the perception of the *dharmadhātu* there arises the perception of splendour.[24]

The term *dharmadhātu* represents the "realm of *dharmas*," those elements of existence that are held to comprise the totality of things, including human knowledge, culture, artifice, and so on, that make up the Buddhist universe, and we might, therefore, be tempted (as indeed some contemporary Buddhists are) to translate *dharmadhātu* as the "natural realm." The Japanese philosopher Nishida Kitarō (1870–1945) seems to adopt a Yogācārin line in his distinctive development of a doctrine of pure nondual experience. He is careful to note, however, that this experience will be "incompatible with Western naturalism."[25] I take this to mean that Nishida understands Buddhism's ultimate goal as a pure, nature-transcending subjectivity. This certainly meets the criteria of Heidegger's antitechnological vision of reality, but it hardly qualifies as the kind of concept to act as the basis for an authentically environmentalist ethic. Indeed, the splendid perceptions of the enlightened saint are discussed at some length in Yogācārin sources and they are not of the kind that offer much comfort for the environmentalist. The Yogācāra scholar Sthiramati (ca. 510–570), for instance, tells us that for a Buddha whose vision is purified in this way "the external world is perceived as consisting not of clay, pebbles, thorny plants, abysses, etc. but of gold, jewels, etc."[26] Of course, we may choose to interpret claims like this in an entirely metaphoric light, but it is surprising how well the purified vision of the Mahāyānist saint does correspond with Reynolds's third and final Theravādin "dhammic" cosmological type.[27] There is undoubtedly some overlap here with the later Tantric notion that, while the things of the world may appear to be conventionally "natural," from the ultimate perspective, they are merely parts of the body of the cosmic Buddha (*dharmakāya*) in one of its many forms, for example, as Vairocana.[28] Indeed, the Tantric view of the world, with its origins deep within the Indic tradition, contains much that appears to be rather inimical to the environmentalist project, not least its emphasis on the subjugation of—or, at any rate, the gaining of power over—nature.[29] In this way Tantricism, and perhaps the whole of the Buddhist dhammic cosmology, focusing as it does on the otherworldly vision of the completed saint, has something in common with the dominion ideal[30] that has been seen from the ecological perspective as such an unhelpful strand within the Judeo-Christian tradition.

Just to add one further complication, let us now turn to Buddhism in its East Asian forms. It is clear that the outworldly character of the Indic *karma/saṃsāra* complex of doctrines had some difficulty in being accepted in China during the period of the initial diffusion of Buddhism, not least because of its apparent conflict with established Confucian social ethics. The "morbid nihilism" associated with the new ideas in the minds of the Chinese intellectual elite has led to a tendency within East Asian Buddhism to characterize the "natural world" in a manner distinct from that found, for instance, in the Hinduized states of Theravādin Southeast Asia. Of course, concern for the welfare of animals, for example, is attested in the earliest Indic canonical sources, as it is in the edicts of Aśoka, and this attitude transplanted itself easily in the Chinese context, no doubt because it harmonized with indigenous traditions. It also seems to have counteracted the negativity of Indic otherworldliness. Thus, the Liang emperor, Wu Ti (502–550), is said to have fed fish held in a monastery pond as part of his Buddhist devotions, while, in 759, the T'ang emperor is reported to have donated a substantial sum toward the construction of eighty-one such ponds (*fang sheng ch'ih*) for the preservation of animal life. Johannes Prip-Møller,[31] in his classic account of Chinese Buddhist monasteries, reports that, as late as the mid-1930s, the National Buddhist Association broadcast radio lectures on the need for animal protection, particularly around the period of "animal day," a date that traditionally coincided with the Buddha's birthday festivities. Even today, after the traumas of Buddhism's recent past in China, ethno-botanical evidence[32] exists to support the notion of monastery as nature reserve. However, not all of the evidence points in the same direction. We know, for instance, that during the high-water mark of Chinese Buddhism in the T'ang period, monasteries "engaged in multifarious commercial and financial activities"[33] that may very well have had an adverse influence on the natural environment. So, a monastery near Ningpo, having fallen on hard times around 836, was able to recoup its losses by large-scale deforestation of surrounding hillsides, while a few years later, in 841, another monastery connived with commercial fuel-gatherers to exploit timber and other forest resources for financial advantage.[34] It seems that at least some of these environmentally damaging commercial enterprises may have been associated with entrepreneurs already engaged in environmentally dubious undertakings—we could call them "monks of convenience," who seem to have opted for the monastic life as a kind of tax-avoidance strategy. Still, it would be unwise to jump to general conclusions about the activities of the monastic order on the evidence of a few bad apples.

There can be little doubt that the environmentalist discourse of Westernized cultures forms part of a broad critique of negative aspects of the capitalist/technological nexus and, in particular, of the twin system of mass-production and consumption wholly oriented toward the satisfaction of material desires that has emerged most fully in recent times.[35] It is not unreasonable to suppose that the genealogy of this critique will be located within the broad pastures of European intellectual history. To illustrate this point we need only look to a figure like Arne Naess,[36] who, while nodding sympathetically but rather uncritically in the Buddhist[37] direction, has successfully erected his system of "deep ecology" on almost purely Spinozan foundations. This is not surprising, for the classical forms of Buddhism emerged as the result of social and economic factors that were uniquely Asiatic. Of course, we shall have to admit that Asia has lacked any overarching homogeneity in terms of its means of production, and this should make us suspicious of terms, such as "the Asiatic mode of production," "semi-feudalism,"[38] or, indeed, "oriental despotism,"[39] employed to describe the premodern economies of India and China. Nevertheless, there is little hard evidence to suggest the presence of indigenous economic systems that depended on high levels of industrial production in premodern Buddhist cultures, although the situation has been drastically different since the advent of the modern period.

In this light, it would be unwise to claim, as do many exponents of an environmentally engaged Buddhism, that Buddhism contains the intellectual and practical resources necessary to counteract the adverse effects of modernity. My response to such high levels of confidence is to raise two further questions: Can the supporters of Buddhism's claim to represent an authentic environmental ethic be certain that they have not fallen prey to "the myth of primitive ecological wisdom"[40] that seems a common ingredient of some recent critiques of industrialism? And, have they given sufficient thought to the genealogy of modernist Buddhism, of which they are generally a part? For, when this is done, it becomes clear that a range of features alien to the abiding character of classical Buddhism—features that tend to be connected with the arrival of Westernized forms of religion and socioeconomic organization—is deeply embedded in the contemporary Asian Buddhist heartlands. Thus, if we turn to recent Thai Buddhist critiques[41] of the negative environmental consequences of multinational logging activities and the like, we can observe that the arguments have no discernibly Buddhist character. The rhetoric employed is actually a blend of the sort of globalized environmental discourse we might meet with in any part of today's world—in effect a romantic "summons to. . . discover in 'nature' both inner and outer, the source of redemption from the alienation and depredations of industrialism and the 'cash nexus' deformation of human relations,"[42] leavened with a good dose of nineteenth-century nationalism.

Japan provides a particularly apt illustration of the ways in which Buddhism, nationalism, and environmental discourse can mesh together. In a revealing passage, D. T. Suzuki, probably the greatest of all modern Buddhist propagandists, contrasts the occidental and oriental attitudes to mountains, concluding that Europeans have characteristically sought to "conquer" them on climbing expeditions and the like, while the Japanese treat mountains, indeed the whole of the natural realm, in a far more respectful manner. He writes:

> The idea of the so-called "conquest of nature" comes from Hellenism. . .in which the earth is made to be man's servant, and the winds and the sea are to obey him. Hebraism concurs with this view, too. In the East, however, this idea of subjecting Nature to the commands or service of man according to his selfish desires has never been cherished. For Nature to us has never been uncharitable, it is not a kind of enemy to be brought under man's power. We of the Orient have never conceived Nature in the form of an opposing power. On the contrary, Nature has been our constant friend and companion, who is to be absolutely trusted in spite of the frequent earthquakes assailing this land of ours. The idea of conquest is abhorrent.[43]

Let us note that Suzuki uncritically conflates a heterogeneous collection of cultures, both Buddhist and non-Buddhist, under the heading of the "Orient," a sort of reverse orientalism. However, we should not judge him too harshly, for such lack of precision is a common foible and, in fact, Suzuki means something far more specific by the term "Orient" than appears on the surface. For him, the essence of the Orient is nothing other than the spirit of Zen. Perhaps Zen, then, with its insistence on "naturalism," particularly in the arts, may hold the key to the development of an authentically Buddhist ecological ethic.

In order to pursue this question in a more informed manner, it is necessary to place Suzuki's literary career as a Zen propagandist in its sociohistorical context. In the early part of Suzuki's life Japanese Buddhists were still coming to terms with the trauma induced by the Meiji (1868–1912) persecution of Buddhism. In order to reassert itself in the face of official hostility, a modernist and nationalistic New Buddhism (*shin bukkyō*) emerged that placed great emphasis on the essential dissimilarities between "oriental" and "occidental" ways of thinking. The fundamental uniqueness of the Japanese character (*nihonjinron*) came to be stressed, particularly by members of the influential Kyoto school of thought, such as Nishida. In a recent discussion of these *nihonjinron* thinkers, Robert Scharf observes that they:

> would assert that the Japanese are racially and/or culturally inclined to experience the world more directly than are the peoples of other nations.[44]

It is clear from our earlier quotation that Suzuki eagerly embraced this style of thinking, and his significance, particularly for the reception of Buddhist ideas in the West, is twofold. In the first place, he was an active promoter of the notion that the Japanese uniquely respond to nature along lines that now seem entirely compatible with the aims and ideals of modern ecology. In the second, he identified Zen as the prime factor in this attitude. Echoes of these ideas are still found in the scholarly literature with social scientists and art historians, for instance, regularly claiming that Japanese culture promotes a "relative minimization of the importance of the subject as against the environment. . . ."[45] This is said to result in a valorization of nature, or, as Augustin Berque observes:

> Japanese culture. . .persistently placed nature and the natural at the acme of culturalness. . .a sense of place (*bashosei*) is particularly pronounced in cultures which, as in the Japanese case, do not enhance the subject's pre-eminence to the degree that European culture has done.[46]

This is an interesting corruption—"orientalization" is perhaps a better term—of Nishida's position as discussed above.[47] Nevertheless, it is generally agreed that the belief that all things, including those associated with the "realm of nature," possess the capacity to gain *nirvāṇa* is a distinctive feature of East Asian

Buddhism. The idea that trees and grasses, indeed the land itself, are destined for enlightenment is probably not found in Indic sources, although a belief in the partial sentience of plants may have been a feature of popular Buddhism from the earliest times.[48] The doctrine is variously claimed to have its source either in the Mahāyānist *Mahāparinirvāṇa Sūtra* or in the chapter entitled "Medicinal Herbs" of the *Lotus Sūtra*.[49] The former text, concerned primarily with the teaching that all beings are possessed of an embryo of the *Tathāgata* (*tathāgatagarbha*), is claimed to have been translated into Chinese in about 417 C.E. by Fa-hsien and Buddha-bhadra. However, since no Sanskrit version is known, some scholars believe that it may be a uniquely Chinese work without an Indian counterpart. Now, while the idea of the "attainment of Buddhahood by nonsentient beings" (Japanese, *hijō jōbutsu*) may plausibly be traced to the previously mentioned Mahāyāna Sūtras, the first explicit reference to the doctrine is found in disputations between masters of the Sui period (581–617 C.E.), such as Hui-yuan and Chih-i. These debates were further developed by Chan-jan, a T'ien-t'ai writer of the T'ang (624–907 C.E.). Saichō (767–822) and Kūkai (774–835) seem to have been the first to have imported the doctrine into Japan, although it is to Annen (841–915), a prominent Tendai Esotericist, that we should look in order to find full systematization and defense of the doctrine of the innate enlightenment (*hongaku shiso*) of all things. His *Private Notes on Discussions of Theories on the Realization of Buddhahood by Grasses and Trees (Shinjo sōmoku jōbutsu shiki)*[50] provides the most detailed presentation of the notion, with a defense undergirded by appeal to the esoteric teaching that "this phenomenal world is nothing but the world of Buddhas."

In this connection, consideration of a painting entitled *Yasai Nehan* (Vegetable Nirvāṇa) by the Japanese artist Itō Jakuchū (1716–1800) may be instructive (see figure 1). At present housed in the collection of the Kyoto National Museum, this scroll once belonged to the Seiganji, a Kyoto temple of the Nishi Honganji form of the Pure Land or Jōdo Shin sect. Clearly Buddhist in one obvious sense, then, the painting shows a variety of vegetables arranged around a central image which happens to be a large radish (*daikon*) laying on a mat or bed of some sort. A partial clarification of the meaning of the piece becomes apparent when we realize that the composition is a coded reference to the Buddha's death (*parinirvāṇa*) scene, which has customarily centered on a reclining Śākyamuni surrounded by mourners, all within a vaguely sylvan setting. A proper interpretation of the work is only possible once we have factored in the previously mentioned doctrine of the Buddhahood of plants (*sōmoku jōbutsu*).[51] We may also wish to know why it is that the artist has chosen to represent the Buddha by the humble—at least from the occidental perspective—radish. This makes sense when we understand more about the rise and subsequent ubiquity of the radish motif in Japanese painting from the early thirteenth century, a subject exhaustively discussed by Yoshiaki Shimizu.[52] The obvious conclusion is that the painting is a visual exposition of East Asian belief in the essential capacity of all things, including those within the vegetable realm, to reach the enlightened state. However, there is more to the painting than meets the eye. It is likely that the painting was donated to the Jōdo Shin temple in 1792 in commemoration of the death of the painter's eldest brother. The painting thus serves as a twin memorial to the Buddha and to Jakuchū's brother. The painter also happens to have been a fourth-generation member of a family of greengrocers.[53] The work can also be read, then, as a celebration of the hereditary occupation, an occupation with which Jakuchū, as the new head of the family, will have to become more fully involved.

Yoshiaki Shimizu concludes his memorable study of Jakuchū's work by noting that the complex metaphoric commemoration alluded to above tends to be absent in other cultures and must be regarded as "indigenously Japanese."[54] If this is so, the question arises for us as to how such works may best be categorized. Should they be considered mainly under the heading of "Buddhism" or are they primarily manifestations of Japanese culture? The answer to such a question has a bearing on how evidence from the East Asian cultural domain may be legitimately employed to advance the cause of an authentic Buddhist environmentalism. Indeed, this is precisely the point made by Ienaga Saburō in his consideration of the general question of the salvific role of nature in Japanese religious thought. In a discussion of such motifs in the work of Saigyō, the twelfth-century Shingon-oriented poet, Ienaga notes that the absolutization of nature as a religious category among some Buddhists of the time created a contradiction between the desire for union with a divinized

FIGURE 1: Itō Jakuchū (1716–1800),
Yasai Nehan (Vegetable Nirvāṇa), ca. 1792
(courtesy of Kyoto National Museum)

nature, on the one hand, and a suspicion of "nature's captivating beauty,"[55] on the other. Ienaga links the former desire very firmly with indigenous factors within Japanese culture, while the latter is the Buddhist ingredient in the mixture.

At this point it might be worth adducing a further piece of evidence that, to some extent, compromises the superficial interpretation of the *sōmoku jōbutsu* doctrine. Dōgen, the Sōtō Zen author of the *Shōbōgenzō*, though admittedly not an adherent of Tendai (although he initially trained in the school), seems to allow the doctrine only in a highly restricted sense. He argues that:

> Since the plants and trees exist in [our] consciousness as reality, they are part of the universal Buddha-nature.[56]

The idealism inherent in this pronouncement is hardly of much use in supporting any conventional environment ethic. Indeed, the ubiquity of statements like this in the East Asian Buddhist context seems to reinforce the antirealist Indic and Yogācāra-derived picture of a world radically transformed in the understanding of the purified saint.[57]

What is apparent from the discussion so far is that the vegetable world, as it appears in Japanese literary sources, may be read as the locus of shifting significances. Another example of this is the banana plant (*bashō*) motif. Matsuo Bashō (1644–1694) is Japan's most celebrated poet. His name, which may be literally rendered as "Master Banana Plant,"[58] derives from the fact that he lovingly tended such a plant, a gift from a disciple, in the garden outside his hut. For Bashō the banana plant is tender, exotic, and rare. Not native to Japan, it is easily damaged by autumn winds and rains:

> The banana in the autumn blast—
> the night I hear
> rain [dripping] in a tub.[59]

In a sense, then, the plant has been torn from its natural home in warmer climes and must stand alone and defenseless in an environment that renders it stunted and unable to set fruit. Tradition informs us that the poet himself was constitutionally weak and prone to various illnesses even though he conducted a life of rigorous asceticism. In this way the banana plant speaks to Bashō's condition and underlines the universal frailty of human existence. More

generally, in Japanese literature *bashō* is both a realistic manifestation of vegetable existence and the metaphorical symbol of
insubstantiality. Thus, the Nō text *Yōkyoku* talks of "the uncertainty
of human life, the way of this world of banana plants and foam,
yesterday's flowers are today's dream. . . ."[60] The connection
between the plant and evanescence derives from the fact that the
plant has a hollow core. On stripping away the outer leaves, the
center is revealed as devoid of solidity, a literary allusion that seems
to have its origin in the *Vimalakīrtinirdeśa Sūtra*[61] and, hence, in
the Indic tradition.[62] Bashō's composition—

> The garden
> Of this temple is full
> Of *bashō*.[63]

—rather nicely illustrates the two primary meanings of this term.

One of the most striking differences between Indic and East
Asian forms of Buddhism involves their attitudes to the fine arts.
Both have customarily employed art for didactic purposes, and most
of us are familiar with scenes of the Buddha's enlightenment and
death, celestial *bodhisattva*s, the realms of gods, *yakṣa*s, hell-beings,
and the like. However, it is significant that art depicting actual as
opposed to religious or imaginary subjects—that is, naturalist art—
is almost absent from Indian Buddhist sources, although one must
concede that naturalistic elements are sometimes employed to fill
in gaps between the main mythological elements of the work. On
the other hand, landscapes, perhaps the most celebrated of which
are associated with the Zen monk Sesshū (1421–1506), and related
forms of naturalistic art, like gardening, are almost a defining feature
of East Asian, and particularly Japanese, Buddhism.[64] We should
not neglect the fact that elements beyond the strictly Buddhist,
notably Taoism, may be an additional factor here. Nevertheless,
there is little doubt that Indian Buddhist artists were largely immune
to the beauty of the natural world. Ananda K. Coomaraswamy's
insistence on the primacy of iconography in Indian religious art
confirms this point. For him, the "Indian icon fills the whole field
of vision at once. . .the eye is not led to range from one point to
another"[65] in the manner demanded by the naturalistic artist. Instead
the work acts as a geometrical representation of a transfigured,
divine, and ultimately antinaturalistic realm, good examples here

being depictions of ideal worlds, such as Sukhāvati with its jewel trees, artificial birds, and absence of women, or Shambhala, whose landscape, at least in the Tibetan tradition, is subsumed into the highly geometric *maṇḍala* of Kālacakra.[66]

It is interesting that, while a considerable body of material on aesthetics is preserved in the East Asian Buddhist tradition, nothing of the kind seems to have been produced by Indian Buddhists, although Indic, and specifically Hindu, works focusing on technical as opposed to aesthetic matters are common.[67] Of course, this must be in part because of the early Buddhist teachings on the dangers associated with sense desires. Consideration of the beautiful was probably regarded as deeply suspect within a monastic tradition that inclined toward moderate displays of asceticism and, in any case, the world was seen as something to be abandoned rather than aesthetically contemplated.[68] If we turn to the forms of aesthetics that flourished in Hindu contexts during the Buddhist period, the same general conclusions can be drawn. Thus, the author of the fourth- to fifth-century *Naṭyasāstra*, the earliest work extant on the topic, and Abhinavagupta (late tenth century), the figure who did most to bring the discipline of Indian aesthetics to its zenith, agree that the perception of beauty is a function of the emotions (*rasa*). Of the eight or nine *rasa*s mentioned in the literature, none appear to be induced by contemplation of the natural world.[69]

In conclusion, we have seen how influential segments of the Buddhist world have responded to the challenge of modernity—in particular the erosion of traditional cosmologies—by presenting a positive ecological message for consumption both within and without the tradition. This puts Buddhism in line with most other major religions. While this is to be applauded in various ways, I have sought here to suggest that uncritical endorsement of aspects of a global environmentalist discourse rooted in the economic and intellectual thought of European and American culture raises a number of intriguing and difficult questions. The most important of these is connected with the indifference, probably unconscious, of ecoBuddhism to the historical, philosophical, and cultural diversity of the Buddhist tradition itself. I have attempted to show in this essay that a range of philosophical and philological issues relating to the richness of meanings attributed to the term "nature" inevitably emerge when the concept is translated into a Buddhist context. I have also pointed to the ambiguity of certain fundamental Indic concepts, such as *saṃsāra* or *nirvāṇa*—not least the anti-naturalistic flavor of the latter—when drawn into an environmentalist context. Aesthetically, and in a number of ways related to its history of doctrine formation, East Asian Buddhisms seem to offer more promise in this regard. However, this should not blind us to the equivocal nature of the East Asian historical record nor to the ways in which a sort of "proto-environmentalist" Buddhism has been employed in the service of Japanese and other Asian manifestations of nationalism.

Clearly there are difficulties involved in translating Western environmentalist discourse into an authentically Buddhist setting or, indeed, in calling on Buddhism to provide a rationale for ecological activity. This does not mean that the task is hopeless. I, for one, remain optimistic about the outcome. Nevertheless, it must be admitted that the work, for scholars and scholarship, is only just beginning.

Notes

1. On the way in which romanticism fed into New England transcendentalism and subsequently on to American ecoBuddhism, see my "Buddhist Environmental Ethics and Detraditionalization: The Case of EcoBuddhism," *Religion* 25, no. 3 (July 1995):199–211.

2. Manifestations of this are many and varied. Indeed, the literature on the topic is growing at a fairly rapid rate. Examples include the "tree ordination movement" in Thailand, environmental awareness programs among Tibetan refugee communities in India, and the work of socially engaged Western Buddhists.

On Thailand, see Buddhadāsa Bhikkhu, *Buddhasāsanik Kap Kān Anurak Thamachāt* (Buddhists and the conservation of nature) (Bangkok: Kōmol Khīmthong Foundation, 1990). See also J. L. Taylor, *Forest Monks and the Nation-State: An Anthropological and Historical Study in Northeastern Thailand* (Singapore: Institute of Southeast Asian Studies, 1993); Leslie E. Sponsel and Poranee Natadecha-Sponsel, "Buddhism, Ecology, and Forests in Thailand: Past, Present, and Future," in *Changing Tropical Forests: Historical Perspectives on Today's Challenges in Asia, Australasia, and Oceania,* ed. John Dargavel, Kay Dixon, and Noel Semple (Canberra: Centre for Resource and Environmental Studies, 1988), 305–25; Leslie E. Sponsel and Poranee Natadecha-Sponsel, "The Role of Buddhism in Creating a More Sustainable Society in Thailand," in *Counting the Costs: Economic Growth and Environmental Change in Thailand,* ed. Jonathan Rigg (Singapore: Institute for Southeast Asian Studies, 1995); Phra Depvedī, *Phra Kap Pā* (Monks and the forest) (Bangkok: Vanāphidak Project, 1992); and Kasetsart University, *Invitation to Tree Planting at Buddhamonton* (Bangkok: Public Relations Office, 1987).

Tibetan sources include: Tenzin Gyatso, His Holiness the Fourteenth Dalai Lama, "A Tibetan Buddhist Perspective on Spirit in Nature," in *Spirit and Nature: Why the Environment Is a Religious Issue,* ed. Steven C. Rockefeller and John C. Elder (Boston: Beacon Press, 1992), 109–23; Bstan-dzin rgya-mtsho, Dalai Lama XIV, *On the Environment* (Dharamsala: Department of Information and International Relations, Central Tibetan Administration of His Holiness the Fourteenth Dalai Lama, 1994).

The Buddhist Perception of Nature Project was initiated by its international coordinator, Nancy Nash, in 1985 and is influential in both Tibetan and Thai circles; see *Tree of Life: Buddhism and the Protection of Nature,* ed. Shann Davies (Hong Kong: Buddhist Perception of Nature Project, 1987).

For essays representing ecoBuddhist and related matters, see Allan Hunt Badiner, ed., *Dharma Gaia: A Harvest of Essays in Buddhism and Ecology* (Berkeley: Parallax Press, 1990); and *Inner Peace, World Peace: Essays on Buddhism and Nonviolence,* ed. Kenneth Kraft (Albany: State University of New York Press, 1992).

3. The most detailed examination to date of the evidence for and against may be found in Lambert Schmithausen, "The Early Buddhist Tradition and Ecological Ethics," *Journal of Buddhist Ethics* 4 (1997):1–42.

4. Ibid.

5. John C. Maraldo, "Hermeneutics and Historicity in the Study of Buddhism," *Eastern Buddhist* 19 (1986):23.

6. Kate Soper, *What Is Nature? Culture, Politics, and the Non-Human* (Oxford and Cambridge, Mass.: Blackwell, 1995), 155f.

7. See notes 17 and 36 below.

8. David J. Kalupahana claims that "Dependent arising [*pratītyasamutpāda*] is often referred to as dharmatā which is the Buddhist term for nature"; David J. Kalupahana, "Toward a Middle Path of Survival," in *Nature in Asian Traditions of Thought: Essays in Environmental Philosophy*, ed. J. Baird Callicott and Roger T. Ames (Albany: State University of New York Press, 1989), 252.

9. Donald K. Swearer, "The Hermeneutics of Buddhist Ecology in Contemporary Thailand: Buddhadāsa and Dhammapiṭaka," included in this volume, 24.

10. Heinz Bechert, "Sangha, State, Society, and 'Nation': Persistence of Traditions in 'Post-Traditional' Societies," *Daedalus* 102, no. 1 (1973):85–95 (reprinted in *Post-Traditional Societies*, ed. S. N. Eisenstadt [New York: Norton, 1972]).

11. Charles F. Keyes, "Communist Revolution and the Buddhist Past in Cambodia," in *Asian Visions of Authority: Religion and the Modern States of East and Southeast Asia*, ed. Charles F. Keyes, Laurel Kendall, and Helen Hardacre (Honolulu: University of Hawaii Press, 1994), 43f.

12. Frank E. Reynolds, "Multiple Cosmogonies and Ethics: The Case of Theravada Buddhism," in *Cosmogony and Ethical Order: New Studies in Comparative Ethics*, ed. Robin W. Lovin and Frank E. Reynolds (Chicago and London: University of Chicago Press, 1985), 203–24. The three types mentioned here are the saṃsāric, the rūpic, or devolutionary, and the dhammic. There may be some justification in regarding these as, respectively, psychological, mythological, and supramundane or purified visions of existence.

13. For a discussion of this phrase, see Charles F. Keyes, *The Golden Peninsula: Culture and Adaptation in Mainland Southeast Asia*, SHAPS Library of Asian Studies (Honolulu: University of Hawaii Press, 1995), 88f.

14. *Shōbōnenjokyō* in *Taishō shinshu Daizōkyō*, ed. Daizōkyō Kankōkai, 85 vols. (Tokyo: Taishō Issaikyō Kankōkai, 1924–1932), 17, text 721 (hereafter cited as T.).

15. T. 84, text 2682.

16. For a full discussion of the Buddhist hells, see Daigan Matsunaga and Alicia Matsunaga, *The Buddhist Concept of Hell* (New York: Philosophical Library, 1972), particularly 107–36.

17. I accept that not all Buddhist environmentalists are going to be modernist in their approach. Elsewhere, I offer four types of contemporary Buddhist environmentalism—ecospiritual, ecoconservative, eco-apologetic and ecojust—in which only the latter is strictly modernist. See my "Getting to Grips with Buddhist Environmentalism: A Provisional Typology," *Journal of Buddhist Ethics* 2 (1995):173–90.

18. Soper, *What Is Nature?* 29

19. Reynolds, "Multiple Cosmogonies and Ethics," 209f.

20. Schmithausen's term, in Schmithausen, "The Early Buddhist Tradition and Ecological Ethics," 4.

21. On this term in the context of the Buddhist understanding of the world, see Ian Harris, "Causation and 'Telos': The Problem of Buddhist Environmental Ethics," *Journal of Buddhist Ethics* 1 (1994):45–56.

22. Soper, *What Is Nature?* 47

23. Erazim Kohák, *The Embers and the Stars: A Philosophical Inquiry into the Moral Sense of Nature* (Chicago and London: University of Chicago Press, 1984), 11.

24. *Trisvabhāvanirdeśa* 37, discussed in Ian Harris, *The Continuity of Madhyamaka and Yogācāra in Indian Buddhism* (Leiden: E. J. Brill, 1991), 149.

25. Andrew Feenberg, "The Problem of Modernity in the Philosophy of Nishida," in *Rude Awakenings: Zen, the Kyoto School, and the Question of Nationalism*, ed. James W. Heisig and John C. Maraldo (Honolulu: University of Hawaii Press, 1995), 156. Also Kitarō Nishida, *An Inquiry into the Good* (New Haven and London: Yale University Press, 1990), 72. Intriguingly, Heidegger may have borrowed some elements in his later thought from Nishida and other Kyoto philosophers. On this important topic, see Graham Parkes, "Heidegger and Japanese Thought: How Much Did He Know, and When Did He Know It?" in *Martin Heidegger: Critical Assessments*, ed. Christopher E. Macann, (London and New York: Routledge, 1992).

26. *Sūtrālamkāravṛttibhāṣya* (Peking Tanjur, *Sems-tsam*, vol. Mi), 210 b8f; quoted in Lambert Schmithausen, "Buddhism and Ecological Responsibility," in *The Stories They Tell: A Dialogue among Philosophers, Scientists, and Environmentalists*, ed. Lawrence Surendra, Klaus Schindler, and Prasanna Ramaswamy (Madras: Earthworm Books, 1997), 71, n. 73.

This quotation seems to coincide with the deeply un-naturalistic decriptions of Buddhist Pure Lands, such as Sukhāvati, found in the early Mahāyāna Sūtras.

27. Reynolds, "Multiple Cosmogonies and Ethics," 213f.

28. Toni Huber, for instance, discusses the connection between the landscape of a region of southern Tibet and the *yidam* Cakrasaṃvara in Toni Huber, "Traditional Environmental Protectionism in Tibet Reconsidered," *Tibet Journal* 16, no. 3 (1991):70f.

29. Cf. David L. Snellgrove, *Indo-Tibetan Buddhism: Indian Buddhists and Their Tibetan Successors* (London: Serindia, 1987), 235f; and Schmithausen, "Buddhism and Ecological Responsibility," 68.

30. Gen. 9.2, for example.

31. Johannes Prip-Møller, *Chinese Buddhist Monasteries: Their Plan and Its Function as a Setting for Buddhist Monastic Life* (Copenhagen: G. E. C. Gads Forlag; and London: Oxford University Press, 1937), 161–63.

32. On the influence of Buddhist temples on the dispersal of certain plant species, see Sheng-ji Pei, "Some Effects of the Dai People's Cultural Beliefs and Practices on the Plant Environment of Xishuangbanna, Yunnan Province, Southwest China," in *Cultural Values and Human Ecology in Southeast Asia*, ed. Karl L. Hutterer, A. T. Rambo, and G. Lovelace, Michigan Papers on Southeast Asia, 24 (Ann Arbor: Center for South and Southeast Asian Studies, University of Michigan, 1985), 321–39.

33. D. C. Twitchett, "Monastic Estates in T'ang China," *Asia Major*, n.s., 5 (1956):123.

34. Ibid., 138; also D. C. Twitchett, "The Monasteries and China's Economy in Medieval Times" (a review of Jacques Gernet's *Les aspects économiques du bouddhisme dans la société chinoise du Vᵉ au Xᵉ siècle* [Saigon: École Française d'Extrême Orient, 1956]), *Bulletin of the School of Oriental and African Studies* 19, no. 3 (1957):536–37, 541.

35. Jan Patocka uses the term "prehistoric" to characterize such a culture: Jan Patocka, *Essais hérétiques sur la philosophie de l'histoire*, trans. Erika Adams (La Grasse: Éditions Verdier, 1981); quoted in Kohák, *The Embers and the Stars*, 21.

36. See Arne Naess, "Through Spinoza to Mahayana Buddhism or Through Mahayana Buddhism to Spinoza?" in *Spinoza's Philosophy of Man: Proceedings of the Scandinavian Spinoza Symposium 1977*, ed. J. Wetlesen (Oslo: University of Oslo Press, 1978), 136–58; Naess, *Spinoza and the Deep Ecology Movement* (Delft: Eburon, 1992); and Naess, "Spinoza and Ecology," in *Speculum Spinozanum, 1677–1977*, ed. Siegfried Hessig (London: Routledge and Kegan Paul, 1977), 418–25.

37. This is discussed in my "The American Appropriation of Buddhism," in *The Buddhist Forum Volume IV: Seminar Papers 1994–1996* (London: School of Oriental and African Studies, 1996), 125–39, particularly 133–34.

38. Barry Hindess and Paul Q. Hirst, *Pre-Capitalist Modes of Production* (London: Routledge and Kegan Paul, 1975), 182.

39. Cf. Karl August Wittfogel, *Oriental Despotism: A Comparative Study of Total Power* (New Haven: Yale University Press, 1963).

40. Kay Milton, *Environmentalism and Cultural Theory: Exploring the Role of Anthropology in Environmental Discourse* (New York: Routledge, 1996), 109f.

41. See, for example, Chaiwat Satha-Anand and Suwanna Wongwaisayawan, "Buddhist Economics Revisited," *Asian Culture Quarterly* 7, no. 4 (1979):37–45. See also Sulak Sivaraksa, "Buddhism and Contemporary International Trends," in *Inner Peace, World Peace: Essays on Buddhism and Nonviolence*, ed. Kenneth Kraft (Albany: State University of New York Press, 1992).

42. Soper, *What Is Nature?* 27.

43. D. T. Suzuki, *Zen and Japanese Culture*, 2d ed. (London: Routledge and Kegan Paul, 1959), 334.

44. Robert H. Scharf, "The Zen of Japanese Nationalism," in *Curators of the Buddha: The Study of Buddhism under Colonialism*, ed. Donald S. Lopez, Jr. (Chicago and London: University of Chicago Press, 1995), 124. See also his "Whose Zen? Zen Nationalism Revisited," in *Rude Awakenings: Zen, the Kyoto School, and the Question of Nationalism*, ed. James W. Heisig and John C. Maraldo (Honolulu: University of Hawaii Press, 1995), 48. For a detailed analysis of *nihonjinron* thought, cf. Peter N. Dale, *The Myth of Japanese Uniqueness* (London: Routledge, 1986).

45. S. N. Eisenstadt, "The Japanese Attitude to Nature: A Framework of Basic Ontological Conceptions," in *Asian Perceptions of Nature: A Critical Approach*, ed. Ole Bruun and Arne Kalland (London: Curzon Press, 1995), 190.

46. Augustin Berque, "The Sense of Nature and Its Relation to Space in Japan," in *Interpreting Japanese Society: Anthropological Approaches*, ed. Joy Hendry and Jonathan Weber, JASO Occasional Papers, 5 (Oxford: JASO, 1986), 103.

47. See note 25 above.

48. On this important topic, see Lambert Schmithausen, *The Problem of the Sentience of Plants in Earliest Buddhism*, Studia Philologica Buddhica, Occasional Paper Series, 6 (Tokyo: International Institute for Buddhist Studies, 1991).

49. See *Scripture of the Lotus Blossom of the Fine Dharma*, translated from the Chinese of Kumārājīva by Leon Hurvitz, Records of Civilization: Sources and Studies, 94 (New York: Columbia University Press, 1976), 101f.

50. Recently published for the first time in moveable type, together with modern Japanese translation and notes, by Fumihiko Sueki, *Heian shoki Bukkyo shiso-shi no kenkyu* (Shunjusha, 1995). Also see Fumihiko Sueki, "Annen: The Philosopher Who Japanized Buddhism," *Acta Asiatica* 66 (1994).

51. For a detailed discussion of *sōmoku jōbutsu*, see part 1 of William LaFleur, "Saigyō and the Buddhist Value of Nature," parts 1 and 2, *History of Religions* 13, no. 2 (November 1973):93–128; no. 3 (February 1974): 227–48. A condensed and revised version of this article may be found in *Nature in Asian Traditions of Thought: Essays in Environmental Philosophy*, ed. J. Baird Callicott and Roger T. Ames (Albany: State University of New York Press, 1989), 183–209.

52. Yoshiaki Shimizu, "Multiple Commemorations: *The Vegetable Nehan* of Itō Jakuchū," in *Flowing Traces: Buddhism in the Literary and Visual Arts of*

Japan, ed. James Sanford, William LaFleur, and Masatoshi Nagatomi (Princeton: Princeton University Press, 1992), 201–33, particularly 217f.

53. Ibid., 229.

54. Ibid., 233.

55. Quoted by LaFleur, "Saigyō and the Buddhist Value of Nature," in Callicott and Ames, *Nature in Asian Traditions of Thought*, 204.

56. T. 82:97c–98a, text 2582; quoted in Sanford, LaFleur, and Nagatomi, *Flowing Traces*, 214.

57. See note 24 above.

58. For consideration of the link between the poet and the plant, see Donald H. Shively, "Bashō—the Man and the Plant," *Harvard Journal of Asiatic Studies* 16, no. 1–2 (1953):146–61.

59. *Bashō nowaki shite / tarai ni ame o / kiku yo kana*; quoted in ibid., 152.

60. Quoted in ibid., 148.

61. William LaFleur, *The Karma of Words: Buddhism and the Literary Arts in Medieval Japan* (Berkeley: University of California Press, 1983), 68–69; and ibid.

62. There is no shortage of Indic references to the insubstantiality of the plant kingdom, for example, Candrakirti's frequent depiction of *saṃsāra* as a forest. See my "How Environmentalist Is Buddhism?" *Religion* 21 (April 1991):101–14, particularly 108f.

63. *Kono tera wa / niwa ippai no / Bashō kana*; quoted in R. H. Blyth, *Haiku* (Tokyo) 4 (autumn-winter 1952):127.

64. Cf. note 43 and ensuing discussion above.

65. Ananda K. Coomaraswamy, *The Transformation of Nature in Art* (New York: Dover, 1956), 29.

66. For more discussion on the depiction of the mythical kingdom of Shambhala, see Marylin M. Rhie and Robert A. F. Thurman, *Wisdom and Compassion: The Sacred Art of Tibet*, rev. ed. (London: Thames and Hudson, 1996), 378–79, 482.

67. For information on works concerned with the technicalities of Indian art, see P. Hardie, "Concept of Art," in *The Dictionary of Art*, ed. Jane Turner, vol. 6 (London and New York: Grove's Dictionaries, 1996), 633–35.

68. Having said this, one or two examples of "nature mysticism" may be detected, for instance, in early Pali texts. On this, see my "How Environmentalist Is Buddhism?" 107.

69. On the theory of *rasas*, see Raniero Gnoli, *The Aesthetic Experience according to Abhinavagupta*, 2d ed. (Varanasi: Chowkhamba Sanskrit Series Office, 1968), xvff.

Bibliography on Buddhism and Ecology

Duncan Ryūken Williams

Abe, Masao. "Man and Nature in Christianity and Buddhism." *Japanese Religions* 7, no. 1 (July 1971):1–10.

Abraham, Ralph. "Orphism: The Ancient Roots of Green Buddhism." In *Dharma Gaia: A Harvest of Essays in Buddhism and Ecology*, ed. Allan Hunt Badiner, 39–49. Berkeley: Parallax Press, 1990.

Aitken, Robert. *The Mind of Clover: Essays in Zen Buddhist Ethics*. San Francisco: North Point Press, 1984.

———. "Gandhi, Dogen, and Deep Ecology." In *Deep Ecology: Living As If Nature Mattered*, ed. Bill Devall and George Sessions, 232–35. Salt Lake City: Peregrine Smith Books, 1985. Reprinted in *The Path of Compassion:Writings on Socially Engaged Buddhism*, ed. Fred Eppsteiner, 86–92 (Berkeley: Parallax Press, 1988).

———. "Right Livelihood for the Western Buddhist" In *Dharma Gaia: A Harvest of Essays in Buddhism and Ecology*, ed. Allan Hunt Badiner, 227–32. Berkeley: Parallax Press, 1990. Reprinted in *Primary Point* 7, no. 2 (summer 1990):19–22.

———. *The Practice of Perfection: The Pāramitās from a Zen Buddhist Perspective*. New York: Pantheon Books, 1994.

Almon, Bert. "Buddhism and Energy in the Recent Poetry of Gary Snyder." *Mosaic* 11 (1977):117–25.

Anderson, Bill. "The Use of Animals in Science: A Buddhist Perspective." *Zen Bow Newsletter* 6, no. 2-3 (summer-fall 1984):8–9.

Ariyaratne, A. T., and Joanna Macy. "The Island of Temple and Tank. Sarvodaya: Self-help in Sri Lanka." In *Buddhism and Ecology*, ed.

Martine Batchelor and Kerry Brown, 78–86. London and New York: Cassell, 1992.

Badiner, Allan Hunt. "Dharma Gaia: The Green Roots of American Buddhism." *Vajradhatu Sun*, April-May 1988, 7.

———. "Is the Buddha Winking at Extinction?" *Tricycle* 3, no. 2 (winter 1993):52–54.

———, ed. *Dharma Gaia: A Harvest of Essays in Buddhism and Ecology.* Berkeley: Parallax Press, 1990.

Barash, D. P. "The Ecologist as Zen Master." *American Midland Naturalist* 89 (1973):214–17.

Bari, Judi. "We All Live Here: An Interview with Judi Bari." By Susan Moon. *Turning Wheel*, spring 1994, 16–19.

Barnhill, David L. "Indra's Net as Food Chain: Gary Snyder's Ecological Vision." *Ten Directions*, spring-summer 1990, 20–28.

———. "A Giant Act of Love: Reflections on the First Precept." *Tricycle* 2, no. 3 (spring 1993):29–33.

Batchelor, Martine, ed. "Even the Stones Smile: Selections from the Scriptures." In *Buddhism and Ecology*, ed. Martine Batchelor and Kerry Brown, 2–17. London and New York: Cassell, 1992.

Batchelor, Martine, and Kerry Brown, eds. *Buddhism and Ecology.* London and New York: Cassell, 1992.

Batchelor, Stephen. "Buddhist Economics Reconsidered." In *Dharma Gaia: A Harvest of Essays in Buddhism and Ecology*, ed. Allan Hunt Badiner, 178–82. Berkeley: Parallax Press, 1990.

———. "Images of Ecology." *Primary Point* 7, no. 2 (summer 1990):9–11.

———. "The Sands of the Ganges: Notes towards a Buddhist Ecological Philosophy." In *Buddhism and Ecology*, ed. Martine Batchelor and Kerry Brown, 31–39. London and New York: Cassell, 1992.

Birch, Pru. "Individual Responsibility and the Greenhouse Effect." *Golden Drum: A Magazine for Western Buddhists*, February-April 1990, 10–11.

Bloom, Alfred. "Buddhism, Nature, and the Environment." *Eastern Buddhist*, n.s., 5, no. 1 (May 1972):115–29.

————. "Buddhism and Ecological Perspective." *Ecology Center Newsletter*, December 1989, 1–2.

Brown, Brian Edward. "Buddhism in Ecological Perspective." *Pacific World*, n.s., 6 (fall 1990):65–73.

Buddhadāsa Bhikkhu. "A Notion of Buddhist Ecology." *Seeds of Peace* 2 (1987):22–27.

Burkill, I. H. "On the Dispersal of the Plants Most Intimate to Buddhism." *Journal of the Arnold Arboretum* 27, no. 4 (1946):327–39.

Byers, Bruce A. "Toward an Ecocentric Community: From Ego-self to Eco-self." *Turning Wheel*, spring 1992, 39–40.

Calderazzo, John. "Meditation in a Thai Forest." *Audubon*, January–February 1991, 84–91.

Chapple, Christopher Key. "Nonviolence to Animals in Buddhism and Jainism." In *Animal Sacrifices: Religious Perspectives on the Use of Animals in Science*, ed. Tom Regan, 213–35. Philadelphia: Temple University Press, 1986. Reprinted In *Inner Peace, World Peace: Essays on Buddhism and Nonviolence*, ed. Kenneth Kraft, 49–62 (Albany: State University of New York Press, 1992).

————. *Nonviolence to Animals, Earth, and Self in Asian Traditions*. Albany: State University of New York Press, 1993.

Codiga, Doug. "Zen Practice and a Sense of Place." In *Dharma Gaia: A Harvest of Essays in Buddhism and Ecology*, ed. Allan Hunt Badiner, 106–11. Berkeley: Parallax Press, 1990.

Colt, Ames B. "Perceiving the World as Self: The Emergence of an Environmental Ethic." *Primary Point* 7, no. 2 (summer 1990):12–14.

Cook, Francis. *Hua-yen Buddhism: The Jewel Net of Indra*. University Park: Pennsylvania State University Press, 1977.

————. "Dogen's View of Authentic Selfhood and Its Socio-ethical Implications." In *Dogen Studies*, ed. William R. LaFleur, 131–49. Honolulu: University of Hawaii Press, 1985.

————. "The Jewel Net of Indra." In *Nature in Asian Traditions of Thought: Essays in Environmental Philosophy*, ed. J. Baird Callicott and Roger T. Ames, 213–29. Albany: State University of New York Press, 1989.

Crawford, Cromwell. "The Buddhist Response to Health and Disease in Environmental Perspective." In *Radical Conservatism: Buddhism in the Contemporary World: Articles in Honour of Bhikkhu Buddhadasa's 84th Birthday Anniversary*, 162–71. Bangkok: Thai Inter-Religious Commission for Development/ International Network of Engaged Buddhists, 1990. Reprinted in *Buddhist Ethics and Modern Society,* ed. Charles Wei-hsun Fu and Sandra A. Wawrytko, 185–93 (New York: Greenwood Press, 1991).

Currier, Lavinia. "Report from Rio: The Earth Summit." *Tricycle* 2, no. 1 (fall 1992):24–26.

Curtin, Deane. "Dōgen, Deep Ecology, and the Ecological Self." *Environmental Ethics* 16, no. 2 (summer 1994):195–213.

Dalai Lama. "Buddhism and the Protection of Nature: An Ethical Approach to Environmental Protection." *Buddhist Peace Fellowship Newsletter*, spring 1988.

———. Foreword to *Dharma Gaia: A Harvest of Essays in Buddhism and Ecology*, ed. Allan Hunt Badiner. Berkeley: Parallax Press, 1990.

Darlington, Susan Marie. "Buddhism, Morality, and Change: The Local Response to Development in Northern Thailand." Ph.D. diss., University of Michigan, 1990.

———. "Monks and Environmental Conservation: A Case Study in Nan Province." *Seeds of Peace* 9, no. 1 (January-April 1993):7–10.

———. "Monks and Environmental Action in Thailand." *Buddhist Forum*, 1994.

Davies, Shann, ed. *Tree of Life: Buddhism and the Protection of Nature.* Hong Kong: Buddhist Perception of Nature Project, 1987.

De Silva, Lily. "The Buddhist Attitude towards Nature." In *Buddhist Perspectives on the Ecocrisis*, ed. Klas Sandell, 9–29. Sri Lanka: Buddhist Publication Society, 1987.

———. "The Hills Wherein My Soul Delights: Exploring the Stories and Teachings." In *Buddhism and Ecology*, ed. Martine Batchelor and Kerry Brown, 18–30. London and New York: Cassell, 1992.

De Silva, Padmasiri. "Buddhist Environmental Ethics," In *Dharma Gaia: A Harvest of Essays in Buddhism and Ecology*, ed. Allan Hunt Badiner, 14–19. Berkeley: Parallax Press, 1990.

————. "Environmental Ethics: A Buddhist Perspective." In *Buddhist Ethics and Modern Society*, ed. Charles Wei-hsun Fu and Sandra A. Wawrytko, 173–84. New York: Greenwood Press, 1991.

Devall, Bill. *Simple in Means, Rich in Ends: Practicing Deep Ecology.* Salt Lake City: Peregrine Smith Books, 1988.

————. "Ecocentric Sangha." In *Dharma Gaia: A Harvest of Essays in Buddhism and Ecology*, ed. Allan Hunt Badiner, 155–64. Berkeley: Parallax Press, 1990.

Devall, Bill, and George Sessions. *Deep Ecology: Living As If Nature Mattered.* Salt Lake City: Peregrine Smith Books, 1985.

Dhamma Bhikkhu Rewata. "Buddhism and the Environment." In *Radical Conservatism: Buddhism in the Contemporary World: Articles in Honour of Bhikkhu Buddhadasa's 84th Birthday Anniversary*, 156–61. Bangkok: Thai Inter-Religious Commission for Development/ International Network of Engaged Buddhists, 1990.

Donegan, Patricia. "Haiku and the Ecotastrophe." In *Dharma Gaia. A Harvest of Essays in Buddhism and Ecology*, ed. Allan Hunt Badiner, 197–207. Berkeley: Parallax Press, 1990.

Dutt, Denise Manci. "An Integration of Zen Buddhism and the Study of Person and Environment." Ph.D. diss., California Institute of Integral Studies, 1983.

Duval, R. Shannon, and David Shaner. "Conservation Ethics and the Japanese Intellectual Tradition." *Conservation Ethics* 11 (fall 1989):197–214.

Earhart, H. Byron. "The Ideal of Nature in Japanese Religion and Its Possible Significance for Environmental Concerns." *Contemporary Religions in Japan* 11, no. 1-2 (March-June 1970):1–25.

Ehrlich, Gretel. "Pico Iyer Talks With Gretel Ehrlich: Buddhist at the Edge of the Earth." *Tricycle* 5, no. 3 (spring 1996):77–82.

Einarsen, John, ed. *The Sacred Mountains of Asia.* Boston: Shambhala Press, 1995.

Eppsteiner, Fred, ed. *The Path of Compassion: Writings on Socially Engaged Buddhism.* Berkeley: Parallax Press, 1988.

Fields, Rick. "A Council of All Beings." *Yoga Journal*, November-December 1989, 52, 108.

————. "The Very Short Sutra on the Meeting of the Buddha and the Goddess," In *Dharma Gaia: A Harvest of Essays in Buddhism and Ecology*, ed. Allan Hunt Badiner, 3–7. Berkeley: Parallax Press, 1990.

Fitzsymonds, Sue. "Treading Softly on This Earth." *Golden Drum: A Magazine for Western Buddhists*, February-April 1990, 12.

Franke, Joe. "The Tiger in the Forest: A Walk with the Monk Who Ordained Trees." *Shambhala Sun* 4, no. 2 (November 1995):48–53.

Gates, Barbara. "Reflections of an Aspiring Earth-Steward." *Inquiring Mind* 7, no. 2 (spring 1991):18–19.

Getz, Andrew. "A Natural Being: A Monk's Reforestation Project in Thailand." *Buddhist Peace Fellowship Newsletter*, winter 1991, 24–25.

Giryo, Yanase. *O Buddha! A Desperate Cry from a Dying World*. Nagoya, Japan: KWIX, 1986.

————. *An Appeal for Your Help in Halting World Environmental Destruction Now for Future Generations*. (Obtainable from Jiko-bukkyo-kai, Okaguchi 2 chome 3-47, Gojo, Nara Prefecture, Japan 637.)

Grady, Carla Deicke. "Women and Ecocentric Conscience." *Newsletter on International Buddhist Women's Activities* 21 (October 1989). Reprinted as "Women and Ecocentricity," in *Dharma Gaia: A Harvest of Essays in Buddhism and Ecology*, ed. Allan Hunt Badiner, 165–68 (Berkeley: Parallax Press, 1990).

————. "A Buddhist Response to Modernization in Thailand: With Particular Reference to Conservation Forest Monks." Ph.D. diss., University of Hawaii, 1995.

Gray, Dennis D. "Buddhism Being Used to Help Save Asia's Environment." *Seeds of Peace* 2 (1987):24–26.

Grosnick, William Henry. "The Buddhahood of the Grasses and the Trees: Ecological Sensitivity or Scriptural Misunderstanding." In *An Ecology of the Spirit: Religious Reflection and Environmental Consciousness*, ed. Michael Barnes, 197–208. Lanham, Md.: University Press of America, 1994.

Gross, Rita. "Toward a Buddhist Environmental Ethic." *Journal of the American Academy of Religion* 65, no. 2 (summer 1997):333–53.

Halifax, Joan. "The Third Body: Buddhism, Shamanism, and Deep Ecology." In *Dharma Gaia: A Harvest of Essays in Buddhism and Ecology*, ed. Allan Hunt Badiner, 20–38. Berkeley: Parallax Press, 1990.

———. *The Fruitful Darkness: Reconnecting with the Body of the Earth*. San Francisco: Harper San Francisco, 1993.

Hannan, Pete. "Images and Animals." *Golden Drum: A Magazine for Western Buddhists*, August-October 1989, 8–9.

Harris, Ian."How Environmentalist Is Buddhism?" *Religion* 21 (April 1991):101–14.

———. "Causation and 'Telos': The Problem of Buddhist Environmental Ethics." *Journal of Buddhist Ethics* 1 (1994):46–59.

———. "Buddhist Environmental Ethics and Detraditionalization: The Case of EcoBuddhism." *Religion* 25, no. 3 (July 1995):199–211.

———. "Getting to Grips with Buddhist Environmentalism: A Provisional Typology." *Journal of Buddhist Ethics* 2 (1995):173–90.

Hayward, Jeremy. "Ecology and the Experience of Sacredness." In *Dharma Gaia: A Harvest of Essays in Buddhism and Ecology*, ed. Allan Hunt Badiner, 64–74. Berkeley: Parallax Press, 1990.

Head, Suzanne. "Buddhism and Deep Ecology." *Vajradhatu Sun*, April-May 1988, 7–8, 12.

———. "Creating Space for Nature." In *Dharma Gaia: A Harvest of Essays in Buddhism and Ecology*, ed. Allan Hunt Badiner, 112–27. Berkeley: Parallax Press, 1990.

Ho, Mobi. "Animal Dharma." In *Dharma Gaia: A Harvest of Essays in Buddhism and Ecology*, ed. Allan Hunt Badiner, 129–35. Berkeley: Parallax Press, 1990

Htun, Nay. "The State of the Environment Today: The Needs for Tomorrow." In *Tree of Life: Buddhism and the Protection of Nature*, ed. Shann Davies, 19–29. Hong Kong: Buddhist Perception of Nature Project, 1987.

Hughes, James, ed. *Green Buddhist Declaration*. Moratuwa: Sarvodaya Press, 1984. (Obtainable from 98 Rawatawatte Rd., Moratuwa, Sri Lanka.)

Ikeda, Daisaku. "Man in Nature." In *Dialogue on Life,* vol. 1, 26–56. Tokyo: Nichiren Shoshu International Center, 1976. Reprinted in *Life: An Enigma, A Precious Jewel,* trans. Charles S. Terry, 28–46 (Tokyo and New York: Kodansha International, 1982).

————. "Life and the Environment." In *Dialogue on Life,* vol. 2, 78–90. Tokyo: Nichiren Shoshu International Center, 1977.

Inada, Kenneth K. "Environmental Problematics." In *Nature in Asian Traditions of Thought: Essays in Environmental Philosophy,* ed. J. Baird Callicott and Roger T. Ames, 231–45. Albany: State University of New York Press, 1989.

Ingram, Catherine. *In the Footsteps of Gandhi: Conversations with Spiritual Social Activists.* Berkeley: Parallax Press, 1990.

Ingram, Paul O. "Nature's Jeweled Net: Kūkai's Ecological Buddhism." *Pacific World* 6 (1990):50–64.

Inoue, Shin'ichi. *Putting Buddhism to Work: A New Theory of Management and Business,* trans. Duncan Williams. Tokyo: Kodansha International, 1997.

Jaini, Padmanabh S. "Indian Perspectives on the Spirituality of Animals." In *Buddhist Philosophy and Culture: Essays in Honour of N. A. Jayawickrema,* ed. David J. Kalupahana and W. G. Weeraratne, 169–78. Colombo: N. A. Jayawickrema Felicitation Volume Committee, 1987.

Jayaprabha. "Ethics and Imagination." *Golden Drum: A Magazine for Western Buddhists,* August-October 1989, 10–11.

Johnson, Wendy. "Tree Planting at Green Gulch Farm." *Inquiring Mind* 7, no. 2 (spring 1991):15.

————. "The Tree at the Bottom of Time." *Tricycle* 5, no. 2 (winter 1995):98–99.

————. "Spring Weeds." *Tricycle* 5, no. 3 (spring 1996):92–93.

————. "Daughters of the Wind." *Tricycle* 6, no. 3 (spring 1997):90–91.

————. "Planting Paradise." *Tricycle* 6, no. 4 (summer 1997):85.

Jones, Ken. "Enlightened Ecological Engagement." *Buddhist Peace Fellowship Newsletter* 10, no. 3-4 (fall 1988):32.

———. *The Social Face of Buddhism: An Approach to Political and Social Activism.* London: Wisdom Publications, 1989.

———. "Getting Out of Our Own Light." In *Dharma Gaia: A Harvest of Essays in Buddhism and Ecology*, ed. Allan Hunt Badiner, 183–90. Berkeley: Parallax Press, 1990.

———. *Beyond Optimism: A Buddhist Political Ecology.* Oxford: Jon Carpenter, 1993.

Jung, Hwa Yol. "The Ecological Crisis: A Philosophic Perspective, East and West." *Bucknell Review* 20, no. 3 (winter 1972).

———. "Ecology, Zen, and Western Religious Thought." *Christian Century*, 15 November 1972, 1153–56.

Jung, Hwa Yol, and Petee Jung. "Gary Snyder's Ecopiety." *Environmental History Review* 41, no. 3 (1990):75–87.

Jurs, Cynthia. "Earth Treasure Vases: Eco-Buddhists Bring an Ancient Teaching from Tibet to Help Heal the Land." *Tricycle* 6, no. 4 (summer 1997).68–69.

Kabilsingh, Chatsumarn. *A Cry from the Forest: Buddhist Perception of Nature, A New Perspective for Conservation Education.* Bangkok: Wildlife Fund Thailand, 1987.

———. "How Buddhism Can Help Protect Nature." In *Tree of Life: Buddhism and Protection of Nature*, ed. Shann Davies, 7–15. Hong Kong: Buddhist Perception of Nature Project, 1987. Reprinted in *Vajradhatu Sun*, April-May 1988, 9, 20.

———. "Buddhist Monks and Forest Conservation." In *Radical Conservatism: Buddhism in the Contemporary World: Articles in Honour of Bhikkhu Buddhadasa's 84th Birthday Anniversary*, 301–10. Bangkok: Thai Inter-Religious Commission for Development/ International Network of Engaged Buddhists, 1990.

———. "Early Buddhist Views on Nature." In *Dharma Gaia: A Harvest of Essays in Buddhism and Ecology*, ed. Allan Hunt Badiner, 8–13. Berkeley: Parallax Press, 1990.

Kalupahana, David J. "Toward a Middle Path of Survival." *Environmental Ethics* 8, no. 4 (winter 1986):371–80. Reprinted in *Nature in Asian Traditions of Thought: Essays in Environmental Philosophy*, ed. J.

Baird Callicott and Roger T. Ames, 247–56 (Albany: State University of New York Press, 1989).

Kapleau, Philip. *To Cherish All Life: A Buddhist Case for Becoming Vegetarian.* San Francisco: Harper and Row, 1982.

———. "Animals and Buddhism." *Zen Bow Newsletter* 5, no. 2 (spring 1983):1–9.

Karunamaya. "The Whys and Hows of Becoming a Vegetarian." *Golden Drum: A Magazine for Western Buddhists,* August-October 1989, 12–13.

Kaye, Lincoln. "Of Cabbages and Cultures: Buddhist 'Greens' Aim to Oust Thailand's Hilltribes." *Far Eastern Economic Review,* December 13, 1990, 35–37.

Kaza, Stephanie. "Emptiness As a Basis for an Environmental Ethic." *Buddhist Peace Fellowship Newsletter,* spring 1990, 30–31.

———. "Toward a Buddhist Environmental Ethic." *Buddhism at the Crossroads* 6, no. 4 (fall 1990):22–25.

———. "Buddhism and Ecology: Suggested Reading." *Inquiring Mind* 7, no. 2 (spring 1991):20.

———. "Acting with Compassion: Buddhism, Feminism, and the Environmental Crisis." In *Ecofeminism and the Sacred,* ed. Carol J. Adams, 50–69. New York: Continuum, 1993.

———. *The Attentive Heart: Conversations with Trees.* New York: Ballantine Books, 1993.

———. "Conversation with Trees: Toward an Ecologically Grounded Spirituality." *ReVision* 15 (winter 1993):128–36.

Ketudat, S., et al. *The Middle Path for the Future of Thailand: Technology in Harmony with Culture and Environment.* Honolulu: Institute of Culture and Communication, East-West Center; Chiang Mai: Faculty of the Social Sciences, Chiang Mai University, 1990.

Keyser, Christine. "Endangered Tibet: Report from a Conference on Tibetan Ecology." *Vajradhatu Sun,* December 1990-January 1991, 1, 12.

Khoroche, Peter, trans. *Once the Buddha Was a Monkey: Arya Sura's Jatakamala.* Chicago: University of Chicago Press, 1989.

Komito, David. "Mādhyamika, Tantra, and 'Green Buddhism'." *Pacific World* 8 (1992).

Kraft, Kenneth. "The Greening of Buddhist Practice." *Zen Quarterly* 5, no. 4 (winter 1994):11–14. Reprinted in *This Sacred Earth: Religion, Nature, Environment*, ed. Roger S. Gottlieb, 484–98 (New York: Routledge, 1996).

Kraus, James W. "Gary Snyder's Biopoetics: A Study of the Poet as Ecologist." Ph.D. diss., University of Hawaii, 1986.

LaFleur, William R. "Saigyō and the Buddhist Value of Nature." Parts 1 and 2. *History of Religions* 13, no. 2 (November 1973):93–127; no. 3 (February 1974):227–47. Reprinted in *Nature in Asian Traditions of Thought: Essays in Environmental Philosophy*, ed. J. Baird Callicott and Roger T. Ames, 183–209 (Albany: State University of New York Press, 1989).

———. "Sattva—Enlightenment for Plants and Trees." In *Dharma Gaia: A Harvest of Essays in Buddhism and Ecology*, ed. Allan Hunt Badiner, 136–44. Berkeley: Parallax Press, 1990.

Lakanaricharan, Sureerat. "The State and Buddhist Philosophy in Resource Conflicts and Conservation in Northern Thailand." Ph.D. diss., University of California, Berkeley, 1995.

Langford, Donald Stewart. "The Primacy of Place in Gary Snyder's Ecological Vision." Ph.D. diss., Ohio State University, 1993.

Larson, Gerald James. "'Conceptual Resources' in South Asia for 'Environmental Ethics'." In *Nature in Asian Traditions of Thought: Essays in Environmental Philosophy*, ed. J. Baird Callicott and Roger T. Ames, 267–77. Albany: State University of New York Press, 1989.

Lesco, Phillip A. "To Do No Harm: A Buddhist View on Animal Use in Research." *Journal of Religion and Health* 27 (winter 1988):307–12.

Levitt, Peter. "An Intimate View." In *Dharma Gaia: A Harvest of Essays in Buddhism and Ecology*, ed. Allan Hunt Badiner, 93–96. Berkeley: Parallax Press, 1990.

———. "For the Trees." *Ten Directions*, spring-summer 1993, 34–35. Reprinted in *Turning Wheel*, spring 1994, 25–26.

Ling, T. O. "Buddhist Factors in Population Growth and Control: A Survey Conducted in Thailand and Ceylon." *Population Studies* 23, no. 1 (March 1969):53–60.

Lohmann, Larry. "Who Defends Biological Diversity? Conservation Strategies and the Case of Thailand." In *Biodiversity: Social and Ecological Perspectives*, ed. Vandana Shiva. Penang: World Rainforest Movement; London and Atlantic Highlands, N.J.: Zed Books, 1991.

———. "Green Orientalism." *Ecologist* 23, no. 6 (1993):202–4.

———. "Visitors to the Commons: Approaching Thailand's 'Environmental' Struggles from a Western Starting Point." In *Ecological Resistance Movements: The Global Emergence of Radical and Popular Environmentalism*, ed. Bron Raymond Taylor, 109–26. Albany: State University of New York Press, 1995.

Loori, John Daido. "Born As the Earth." *Mountain Record*, winter 1991, 2–10.

———. "Being Born As the Earth: Excerpts from a Spirited Dharma Combat with John Daido Loori." *Mountain Record*, winter 1992, 14–18.

———. "The Sacred Teachings of Wilderness: A Dharma Discourse on the Living Mandala of Mountains and Rivers." *Mountain Record*, winter 1992, 2–9.

———. "River Seeing the River." *Mountain Record*, spring 1996, 2–10.

Macy, Joanna. *Despair and Personal Power in the Nuclear Age*. Philadelphia: New Society Publishers, 1983.

———. *Dharma and Development: Religion as Resource in the Sarvodaya Self-Help Movement*. Rev. ed. West Hartford, Conn.: Kumarian Press, 1985.

———. "Interdependence in the Nuclear Age: An Interview with Joanna Macy by Stephan Bodian." *Karuna*, fall 1985, 8–9.

———. "In Indra's Net." In *The Path of Compassion: Writings on Socially Engaged Buddhism*, ed. Fred Eppsteiner, 170–81. Berkeley: Parallax Press, 1988.

———. "Sacred Waste." *Buddhist Peace Fellowship Newsletter* 10, no. 3-4 (fall 1988):22–23.

———. "Empowerment beyond Despair: A Talk by Joanna Macy on the Greening of the Self." *Vajradhatu Sun* 11, no. 4 (April-May 1989):1, 3, 14.

———. "Deep Ecology and Spiritual Practice." *One Earth*, autumn 1989, 18–21.

———. "Guardians of Gaia." *Yoga Journal*, November-December 1989, 53–55.

———. "The Ecological Self: Postmodern Ground for Right Action." In *Sacred Interconnections: Postmodern Spirituality, Political Economy, and Art*, ed. David Ray Griffin, 35–48. Albany: State University of New York Press, 1990.

———. "The Greening of the Self." In *Dharma Gaia: A Harvest of Essays in Buddhism and Ecology*, ed. Allan Hunt Badiner, 53–63. Berkeley: Parallax Press, 1990.

———. *World as Lover, World as Self.* Berkeley: Parallax Press, 1991.

———. "Schooling Our Intention." *Tricycle* 3, no. 2 (winter 1993):48–51.

Maezumi, Taizan. "The Buddha Seed Grows Consciously: The Precept of Non-killing." *Ten Directions*, spring 1985, 1, 4.

———. "A Half Dipper of Water." *Ten Directions*, spring-summer 1990, 11–12.

McClellan, John. "Nondual Ecology." *Tricyle* 3, no. 2 (winter 1993):58–65.

McDaniel, Jay B. "Revisioning God and the Self: Lessons from Buddhism." In *Liberating Life: Contemporary Approaches to Ecological Theology*, ed. Charles Birch, William Eakin, and Jay B. McDaniel, 228–57. Maryknoll, N.Y.: Orbis Books, 1990.

McDermott, James P. "Animals and Humans in Early Buddhism." *Indo-Iranian Journal* 32, no. 2 (1989):269–80.

Metzger, Deena. "The Buddha of the Beasts." *Creation*, May-June 1989, 25.

———. "Four Meditations." In *Dharma Gaia: A Harvest of Essays in Buddhism and Ecology*, ed. Allan Hunt Badiner, 209–12. Berkeley: Parallax Press, 1990.

Mininberg, Mark Sando. "Sitting with the Environment." *Mountain Record*, winter 1993, 44–47.

Miyakawa, Akira. "Man and Nature or in Nature?" *Dharma World* 21 (March-April 1994):47–49.

Naess, Arne. "Interview with Arne Naess." In *Deep Ecology: Living As If Nature Mattered*, ed. Bill Duvall and George Sessions, 74–76. Salt Lake City: Peregrine Smith Books, 1985.

———. "Self-Realization: An Ecological Approach to Being in the World," In *Thinking Like a Mountain: Towards a Council of All Beings*, ed. John Seed, Joanna Macy, and Arne Naess, 19–30. Philadelphia: New Society Publishers, 1988.

———. "Mountains and Mythology." In *The Sacred Mountains of Asia*, ed. John Einarsen, 89. Boston: Shambhala Press, 1995.

Nagabodhi. "Buddhism and Vegetarianism." *Golden Drum: A Magazine for Western Buddhists*, August-October 1989, 3.

———. "Buddhism and the Environment." *Golden Drum: A Magazine for Western Buddhists*, February-April 1990, 3.

Nash, Nancy. "The Buddhist Perception of Nature Project." In *Tree of Life: Buddhism and the Protection of Nature*, ed. Shann Davies, 31–33. Hong Kong: Buddhist Perception of Nature Project, 1987.

Natadecha-Sponsel, Poranee. "Buddhist Religion and Scientific Ecology as Convergent Perceptions of Nature." In *Essays on Perceiving Nature*, ed. Diana M. DeLuca, 113–18. Honolulu: Perceiving Nature Conference Committee, 1988.

———. "Nature and Culture in Thailand: The Implementation of Cultural Ecology and Environmental Education through the Application of Behavioral Sociology." Ph.D. diss., University of Hawaii, 1991.

Newbury, Roxy Keien. "The Green Container: Taking Care of the Garbage." *Mountain Record*, winter 1991, 51–53.

Nhat Hanh, Thich. *Being Peace*. Berkeley: Parallax Press, 1987.

———. *Interbeing: Commentaries on the Tiep Hien Precepts*. Berkeley: Parallax Press, 1987.

———. "The Individual, Society, and Nature." In *The Path of Compassion: Writings on Socially Engaged Buddhism*, ed. Fred Eppsteiner, 40–46. Berkeley: Parallax Press, 1988.

———. "The Last Tree." In *Dharma Gaia: A Harvest of Essays in Buddhism and Ecology*, ed. Allan Hunt Badiner, 217–21. Berkeley: Parallax Press, 1990.

———. "Seeing All Beings with the Eyes of Compassion." *Karuna: A Journal of Buddhist Meditation*, summer-fall 1990, 6–10.

———. *Peace Is Every Step: The Path of Mindfulness in Everyday Life*, ed. Arnold Kotler. New York: Bantam Books, 1991.

———. "Look Deep and Smile: The Thoughts and Experiences of a Vietnamese Monk." In *Buddhism and Ecology*, ed. Martine Batchelor and Kerry Brown, 100–109. London and New York: Cassell, 1992.

———. *Love in Action: Writings on Nonviolent Social Change*. Berkeley: Parallax Press, 1993.

Nhat Hanh, Thich, et al. *For a Future to Be Possible: Commentaries on the Five Wonderful Precepts*. Berkeley: Parallax Press, 1993.

———. *A Joyful Path: Community Transformation and Peace*. Berkeley: Parallax Press, 1994.

Nolan, Kathy Fusho. "The Great Earth." *Mountain Record*, spring 1996, 70–72.

Norberg-Hodge, Helena. *Ancient Futures: Learning from Ladakh*. San Francisco: Sierra Club Books, 1991.

———. "May a Hundred Plants Grow from One Seed: The Ecological Tradition of Ladakh Meets the Future." In *Buddhism and Ecology*, ed. Martine Batchelor and Kerry Brown, 41–54. London and New York: Cassell, 1992.

Ophuls, William. "Buddhist Politics." *Ecologist* 7, no. 3 (1977):82–86.

Pauling, Chris. "A Buddhist Life Is a Green Life." *Golden Drum: A Magazine for Western Buddhists*, February-April 1990, 5–7.

Payutto, Prayudh. *Buddhist Economics: A Middle Way for the Marketplace*. Bangkok: Buddhadhamma Foundation, 1994.

Pei, Shengji. "Managing for Biological Diversity in Temple Yards and Holy Hills: The Traditional Practices of the Xishuangbanna Dai Community, Southwestern China." In *Ethics, Religion, and Biodiversity: Relations between Conservation and Cultural Values*, ed. Lawrence S. Hamilton with Helen F. Takeuchi, 118–12. Cambridge: White Horse Press, 1993.

Pitt, Martin. "The Pebble and the Tide." In *Dharma Gaia: A Harvest of Essays in Buddhism and Ecology*, ed. Allan Hunt Badiner, 102–5. Berkeley: Parallax Press, 1990.

Perl, Jacob. "Ecology of Mind." *Primary Point* 7, no. 2 (summer 1990):4–6.

Pongsak, Ajahn. "In the Water There Were Fish and the Fields Were Full of Rice: Reawakening the Lost Harmony of Thailand." In *Buddhism and Ecology*, ed. Martine Batchelor and Kerry Brown, 87–99. London and New York: Cassell, 1992.

Randhawa, M. S. *The Cult of Trees and Tree Worship in Buddhist and Hindu Scripture*. New Delhi: All-Indian Arts and Crafts Society, 1964.

Raye, Bonnie del. "Buddhists Concerned for Animals." In *Turning the Wheel: American Women Creating the New Buddhism*, ed. Sandy Boucher, 289–94. San Francisco: Harper and Row, 1988.

Reed, Christopher. "Down to Earth." In *Dharma Gaia: A Harvest of Essays in Buddhism and Ecology*, ed. Allan Hunt Badiner, 233–35. Berkeley: Parallax Press, 1990.

Rissho Kosei-kai. *A Buddhist View for Inclusion in the Proposed 'Earth Charter' Presented to the Preparatory Committee of the United Nations Conference on Environment and Development (UNCED), December 15, 1991*. (Obtainable from 2-11-1 Wada, Suginami-ku, Tokyo 166, Japan.)

Roberts, Elizabeth. "Gaian Buddhism." In *Dharma Gaia: A Harvest of Essays in Buddhism and Ecology*, ed. Allan Hunt Badiner, 147–54. Berkeley: Parallax Press, 1990.

———, ed. *Earth Prayers*. San Francisco: Harper Collins, 1991.

Robinson, Peter. "Some Thoughts on Buddhism and the Ethics of Ecology." *Proceedings of the New Mexico-West Texas Philosophical Society* 7 (1972):71–78.

Rolston, Holmes, III. "Respect for Life: Can Zen Buddhism Help in Forming an Environmental Ethic?" *Zen Buddhism Today* 7 (September 1989):11–30.

Rudloe, Anne. "Pine Forest Teachings: Bringing Joy and Compassion to the Environmental Wars." *Primary Point* 7, no. 2 (summer 1990):14–15.

Ruegg, D. Seyfort. "*Ahimsā* and Vegetarianism in the History of Buddhism." In *Buddhist Studies in Honour of Walpola Rahula*, ed. Somaratna Balasooriya et al., 234–41. London: Gordon Fraser; Sri Lanka: Vimamsa, 1980.

Sagaramati. "Do Buddhists Eat Meat?" *Golden Drum: A Magazine for Western Buddhists*, August-October 1989, 6–7.

Sakya Trizin. *A Buddhist View on Befriending and Defending Animals*. Portland: Orgyan Chogye Chonzo Ling, 1989.

Sandell, Klas, ed. *Buddhist Perspectives on the Ecocrisis*. Sri Lanka: Buddhist Publication Society, 1987.

———. "Buddhist Philosophy as Inspiration to Ecodevelopment." In *Buddhist Perspectives on the Ecocrisis*, ed. Klas Sandell, 30–37. Sri Lanka: Buddhist Publication Society, 1987.

Sasaki, Joshu. "Who Pollutes the World." In *Zero: Contemporary Buddhist Life and Thought,* vol. 2, 151–57. Los Angeles: Zero Press, 1979.

Schelling, Andrew. "Jataka Mind: Cross-Species Compassion from Ancient India to Earth First! Activists." *Tricycle* 1, no. 1 (fall 1991):10–19.

Schmithausen, Lambert. *Buddhism and Nature: The Lecture Delivered on the Occasion of the EXPO 1990 (An Enlarged Version with Notes)*. Tokyo: International Institute for Buddhist Studies, 1991.

———. *Plants as Sentient Beings in Earliest Buddhism*. Faculty of Asian Studies, Australian National University, 1991.

———. *The Problem of the Sentience of Plants in Earliest Buddhism*. Tokyo: International Institute for Buddhist Studies, 1991.

———. "The Early Buddhist Tradition and Ecological Ethics." *Journal of Buddhist Ethics* 4 (1997):1–42.

Schneider, David Tensho. "Saving the Earth's Healing Resources." *Yoga Journal*, July-August 1992, 57–63.

Schumacher, E. F. "Buddhist Economics." In *Valuing the Earth: Economics, Ecology, Ethics*, ed. Herman E. Daly and Kenneth N. Townsend, 173–81. Cambridge, Mass.: MIT Press, 1993.

Seed, John. "Rainforest Man: An Interview by Stephen Bodian." *Yoga Journal*, November-December 1989, 48–51, 106–8.

———. "Wake the Dead!" In *Dharma Gaia: A Harvest of Essays in Buddhism and Ecology*, ed. Allan Hunt Badiner, 222–26. Berkeley: Parallax Press, 1990.

———. "The Rainforest as Teacher: An Interview with John Seed." *Inquiring Mind* 8, no. 2 (spring 1992):1, 6–7.

Seed, John, Joanna Macy, and Arne Naess, eds. *Thinking Like a Mountain: Towards a Council of All Beings*. Philadelphia: New Society Publishers, 1988.

Sendzimir, Jan. "Satellite Eyes and Chemical Noses." *Primary Point* 7, no. 2 (summer 1990):15, 16, 18.

Seung Sahn. "Not Just a Human World." *Primary Point* 7, no. 2 (summer 1990):3–4.

Shaner, David Edward. "The Japanese Experience of Nature." In *Nature in Asian Traditions of Thought: Essays in Environmental Philosophy*, ed. J. Baird Callicott and Roger T. Ames, 163–82. Albany: State University of New York Press, 1989.

Shaner, David Edward, and R. Shannon Duval. "Conservation Ethics and the Japanese Intellectual Tradition." *Conservation Ethics* 11 (fall 1989):197–214.

Shaw, Miranda. "Nature in Dōgen's Philosophy and Poetry." *Journal of the International Association of Buddhist Studies* 8, no. 2 (1985):111–32.

Shepard, Philip T. "Turning On to the Environment without Turning Off Other People." *Buddhism at the Crossroads* 6, no. 4 (fall 1990):18–21.

Shimizu, Yoshiaki. "Multiple Commemorations: *The Vegetable Nehan* of Itō Jakuchū." In *Flowing Traces: Buddhism in the Literary and Visual Arts of Japan*, ed. James H. Sanford et al., 201–33. Princeton: Princeton University Press, 1992.

Shively, Donald H. "Buddhahood for the Nonsentient: A Theme in Nō Plays." *Harvard Journal of Asiatic Studies* 20, no. 1-2 (June 1957):135–61.

Sivaraksa, Sulak. "Rural Poverty and Development in Thailand, Indonesia, and the Phillipines." *Ecologist* 15, no. 5-6 (1985):266–68.

———. *Siamese Resurgence: A Thai Buddhist Voice on Asia and a World of Change*. Bangkok: Asian Cultural Forum on Development, 1985.

———. *A Socially Engaged Buddhism*. Bangkok: Thai Inter-Religious Commission for Development, 1988.

———. "A Buddhist Perception of a Desirable Society." In *Ethics of Environment and Development: Global Challenge, International Response*, ed. J. Ronald Engel and Joan Gibb Engel, 213–21. Tucson: University of Arizona Press, 1990.

———. "True Development." In *Dharma Gaia: A Harvest of Essays in Buddhism and Ecology*, ed. Allan Hunt Badiner, 169–77. Berkeley: Parallax Press, 1990.

———. "Building Trust through Economic and Social Development and Ecological Balance: A Buddhist Perspective." In *Radical Conservatism: Buddhism in the Contemporary World: Articles in Honour of Bhikkhu Buddhadasa's 84th Birthday Anniversary*, 179–98. Bangkok: Thai Inter-Religious Commission for Development/ International Network of Engaged Buddhists, 1990.

———. *Seeds of Peace*. Berkeley: Parallax Press, 1992.

———. "How Societies Can Practice the Precepts." In *For a Future to Be Possible: Commentaries on the Five Wonderful Precepts*, ed. Thich Nhat Hanh, 110–14. Berkeley: Parallax Press, 1993.

Skolimowski, Henryk. *Eco-Philosophy: Designing New Tactics for Living*. Salem, N.H.: Marion Boyars, 1981.

———. "Eco-Philosophy and Buddhism: A Personal Journey." *Buddhism at the Crossroads* 6, no. 4 (fall 1990):26–29.

———. *A Sacred Place to Dwell: Living with Reverence upon the Earth*. Rockport, Mass.: Element, 1993.

Snyder, Gary. *Earth House Hold: Technical Notes and Queries to Fellow Dharma Revolutionaries*. New York: New Directions Books, 1957.

———. *Turtle Island*. New York: New Directions Books, 1974.

———. *Riprap and Cold Mountain Poems*. San Francisco: Four Seasons Foundation, 1976.

———. *Six Sections from Mountains and Rivers without End*. San Francisco: Four Seasons Foundation, 1976.

———. *The Real Work: Interviews and Talks, 1964–1979*. New York: New Directions, 1980.

——. "Buddhism and the Possibilities of a Planetary Culture." In *Deep Ecology: Living As If Nature Mattered*, ed. Bill Devall and George Sessions, 251–53; Salt Lake City: Peregrine Smith Books, 1985. Reprinted in *The Path of Compassion: Writings on Socially Engaged Buddhism*, ed. Fred Eppsteiner, 82–85 (Berkeley: Parallax Press, 1988).

——. "The Etiquette of Freedom." *Sierra*, September-October 1989, 75–77, 113–16.

——. *The Practice of the Wild*. San Francisco: North Point Press, 1990.

——. *No Nature: New and Selected Poems*. New York and San Francisco: Pantheon Books, 1992.

——. "Indra's Net As Our Own." In *For a Future to Be Possible: Commentaries on the Five Wonderful Precepts*, ed. Thich Nhat Hanh, 127–35. Berkeley: Parallax Press, 1993.

——. "A Village Council of All Beings: Ecology, Place, and Awakening of Compassion." *Turning Wheel*, spring 1994, 12–15.

——. "Exhortations for Baby Tigers: The End of the Cold War and the End of Nature." *Shambhala Sun* 4, no. 2 (November 1995):31–33. Reprinted in *A Place in Space: Ethics, Aesthetics, and Watershed* (Washington, D.C.: Counterpoint, 1995).

——. *A Place in Space: Ethics, Aesthetics, and Watersheds*. Washington, D.C.: Counterpoint, 1995.

——. "Walking the Great Ridge Omine on the Diamond-Womb Trail." In *The Sacred Mountains of Asia*, ed. John Einarsen, 71–77. Boston: Shambhala Press, 1995.

——. "Nets of Beads, Webs of Cells." *Mountain Record* 14, no. 3 (spring 1996):50–54.

Sōtōshū Shūmuchō. *International Symposium: The Future of the Earth and Zen Buddhism*. Tokyo: Sōtōshū Shūmuchō, 1991.

Sponberg, Alan. Review of *Dharma Gaia: A Harvest of Essays in Buddhism and Ecology*, edited by Allan Hunt Badiner. *Environmental Ethics* 14 (fall 1992):279–82.

——. "Green Buddhism and the Hierarchy of Compassion." *Western Buddhist Review* 1 (December 1994):131–55.

Sponsel, Leslie E. "Cultural Ecology and Environmental Education." *Journal of Environmental Education* 19, no. 1 (1987):31–42.

Sponsel, Leslie E., and Poranee Natadecha-Sponsel."Buddhism, Ecology, and Forests in Thailand: Past, Present, and Future," In *Changing Tropical Forests: Historical Perspectives on Today's Challenges in Asia, Australasia and Oceania: Workshop Meeting, Canberra, 16–18 May 1988*, ed. John Dargavel, et al., 305–25. Canberra: Australian National University Centre for Resource and Environmental Studies, 1988.

———. "Nonviolent Ecology: The Possibilities of Buddhism." In *Buddhism and Nonviolent Global Problem-Solving: Ulan Bator Explorations*, ed. Glenn D. Paige and Sarah Gilliatt, 139–50. Honolulu: Center for Global Nonviolence Planning Project, Spark M. Matsunaga Institute for Peace, University of Hawaii, 1991.

———. "The Relevance of Buddhism for the Development of an Environmental Ethic for the Conservation of Biodiversity." In *Ethics, Religion, and Biodiversity: Relations between Conservation and Cultural Values*, ed. Lawrence S. Hamilton with Helen F. Takeuchi, 75–97. Cambridge: White Horse Press, 1993.

———. "The Role of Buddhism for Creating a More Sustainable Society in Thailand." In *Counting the Costs: Economic Growth and Environmental Change in Thailand*, ed. Jonathan Rigg, 27–46. Singapore: Institute of Southeast Asian Studies, 1995.

Spretnak, Charlene. *The Spiritual Dimension of Green Politics*. Santa Fe, N.M.: Bear and Co., 1986.

———. "Dhamma at the Precinct Level." In *The Path of Compassion: Writings on Socially Engaged Buddhism*, ed. Fred Eppsteiner, 199–202. Berkeley: Parallax Press, 1988.

———. "Green Politics and Beyond." In *Turning the Wheel: American Women Creating the New Buddhism*, ed. Sandy Boucher, 284–88. San Francisco: Harper and Row, 1988.

———. *States of Grace: The Recovery of Meaning in the Postmodern Age*. San Francisco: Harper San Francisco, 1991.

Stone, David. "How Shall We Live? Deep Ecology Week at the Naropa Institute." *Vajradhatu Sun*, August-September 1990, 13–14.

Story, Francis. *The Place of Animals in Buddhism.* Kandy, Sri Lanka: Bodhi Leaves, Buddhist Publication Society, 1964.

Tanahashi, Kazuaki. "Garbage First." *Turning Wheel*, winter 1994, 39.

Taylor, J. L. *Forest Monks and the Nation-State: An Anthropological and Historical Study in Northeastern Thailand.* Singapore: Institute of Southeast Asian Studies, 1993.

―――. "Social Activism and Resistance on the Thai Frontier: The Case of Phra Parajak Khuttajitto." *Bulletin of Concerned Asian Scholars* 25, no. 2 (1993):3–16.

Thompson, James Soshin. "The Mind of Interbeing." *Ten Directions*, spring-summer 1990, 16–17.

―――. "Returning to the Source: Radical Confidence for Environmentalists." *Ten Directions*, spring-summer 1993, 31–33.

―――. "Radical Confidence: What is Missing from Eco-Activism." *Tricycle* 3, no. 2 (winter 1993):40–45.

Thurman, Robert. "Buddhist Views of Nature: Variations on the Theme of Mother-Father Harmony." In *On Nature*, ed. Leroy S. Rouner, 96–112. Notre Dame, Ind.: University of Notre Dame Press, 1984.

Timmerman, Peter. "It Is Dark Outside: Western Buddhism from the Enlightenment to the Global Crisis." In *Buddhism and Ecology*, ed. Martine Batchelor and Kerry Brown, 65–77. London and New York: Cassell, 1992.

Titmuss, Christopher. "Interactivity." In *The Path of Compassion: Writings on Socially Engaged Buddhism*, ed. Fred Eppsteiner, 182–89. Berkeley: Parallax Press, 1988.

―――. "On the Green Credo." *Tricycle* 3, no. 2 (winter 1993):55–57.

―――. *The Green Buddha.* Devon, U.K.: Insights Books, 1995.

Tou-hui, Fok. "Where Is the Green Movement Going." *The Light of Dharma* 81 (February 1989).

Treace, Bonnie Myotai. "Home: Born As the Earth Training." *Mountain Record*, winter 1991, 36–41.

Visalo, Phra Phaisan. "The Forest Monastery and Its Relevance to Modern Thai Society." In *Radical Conservatism: Buddhism in the Contemporary World: Articles in Honour of Bhikkhu Buddhadasa's 84th*

Birthday Anniversary, 288–300. Bangkok: Thai Inter-Religious Commission for Development/ International Network of Engaged Buddhists, 1990.

Venturini, Riccardo. "A Buddhist View on Ecological Balance." *Dharma World* 17 (March-April 1990):19–23.

Wallis, Nick. "Buddhism and the Environment." *Golden Drum: A Magazine for Western Buddhists*, August-October 1989, 4–5.

Wasi, Prawase. "Alternative Buddhist Agriculture." In *Radical Conservatism: Buddhism in the Contemporary World: Articles in Honour of Bhikkhu Buddhadasa's 84th Birthday Anniversary*, 172–78. Bangkok: Thai Inter-Religious Commission for Development/ International Network of Engaged Buddhists, 1990.

Waskow, Arthur. "What Is Eco-Kosher." In *For a Future to Be Possible: Commentaries on the Five Wonderful Precepts*, ed. Thich Nhat Hanh, 115–21. Berkeley: Parallax Press, 1993.

Watanabe, Manabu. "Religious Symbolism in Saigyo's Verses: Contribution to Discussions of His Views on Nature and Religion." *History of Religions* 26, no. 4 (May 1987):382–400.

Williams, Duncan. "The Interface of Buddhism and Environmentalism in North America." B.A. thesis, Reed College, 1991.

Wise, Nina. "Rock Body Tree Limb." In *Dharma Gaia: A Harvest of Essays in Buddhism and Ecology*, ed. Allan Hunt Badiner, 99–101. Berkeley: Parallax Press, 1990.

———. "Full, As in Good." *Turning Wheel*, spring 1994, 26–27.

World Wildlife Fund International. *The Assisi Declarations: Messages on Man and Nature from Buddhism, Christianity, Hinduism, Islam and Judaism*. Geneva: World Wildlife Fund International, 1986.

Yamaoka, Seigen H. *A Buddhist View of the Environment*. San Francisco: Buddhist Churches of America, 1991.

Yamauchi, Jeffrey Scott. "The Greening of American Zen: An Historical Overview and a Specific Application." M.A. thesis, Prescott College, 1996.

Yokoyama, W. S. "Circling the Mountain: Observations on the Japanese Way of Life." In *Buddhism and Ecology*, ed. Martine Batchelor and Kerry Brown, 55–64. London and New York: Cassell, 1992.

Notes on Contributors

David Landis Barnhill received his Ph.D. from Stanford University in religious studies, with a minor in Japanese literature. He is currently associate professor of intercultural studies and chair of the Religious Studies Department at Guilford College in North Carolina. He has published articles on the Japanese poet Bashō as well as other aspects of Japanese religion and literature. He also is co-chair of the Religion and Ecology Group of the American Academy of Religion.

Christopher Key Chapple is professor of theological studies and director of Asian and Pacific studies at Loyola Marymount University in Los Angeles. He is the author of *Karma and Creativity* (State University of New York Press, 1986) and *Nonviolence to Animals, Earth, and Self in Asian Traditions* (State University of New York Press, 1993); co-translator of the *Yoga Sutras* (Sri Satguru Publications, 1990); and editor of several books, including *Ecological Prospects: Scientific, Religious, and Aesthetic Perspectives* (State University of New York Press, 1994).

Malcolm David Eckel is associate professor of religion in the Department of Religion, College of Arts and Sciences, Boston University. He is the author of *To See the Buddha: A Philosopher's Quest for the Meaning of Emptiness* (Harper San Francisco, 1992) and the Buddhism editor for *The HarperCollins Dictionary of Religion*, general editor, Jonathan Z. Smith (Harper San Francisco, 1995).

Rita M. Gross, professor of comparative studies in religion at the University of Wisconsin, Eau Claire, is the author of *Buddhism after Patriarchy: A Feminist History, Analysis, and Reconstruction of Buddhism* (State University of New York Press, 1993) and *Feminism and Religion— An Introduction* (Beacon Press, 1996).

Ruben L. F. Habito, professor of world religions and spirituality, Perkins School of Theology, Southern Methodist University, and resident teacher,

Maria Kannon Zen Center, Dallas, Texas, also taught at Sophia University, Tokyo, from 1978 to 1989. A dharma heir of Yamada Koun Rōshi of the Sanbō Kyōdan Zen tradition, he has written *Healing Breath: Zen Spirituality for a Wounded Earth* (Orbis Books, 1993) and other works in Japanese and English.

Ian Harris is reader in religious studies at the University College of St. Martin, Lancaster, England, author of *The Continuity of Madhyamaka and Yogācāra in Early Mahāyāna Buddhism* (E. J. Brill, 1991), and editor (with S. Mews, P. Morris, and J. Shepherd) of *Contemporary Religions: A World Guide* (Longman, 1992). He studied at the Universities of Cambridge and Lancaster, receiving a doctorate in Buddhist philosophy from the latter, and has written a number of articles on Buddhism and ecological ethics. He is a founding member of the U.K. Buddhist Studies Association and editor of the *Bulletin of the British Association for the Study of Religions*. His current research is focused on Buddhism and politics in the twentieth century.

Paul O. Ingram is professor of religion at Pacific Lutheran University and is currently president of the Society for Buddhist-Christian Studies. He is the author of *The Modern Buddhist-Christian Dialogue* (E. Mellen Press, 1988) and co-editor (with Frederick J. Streng) of *Buddhist-Christian Dialogue: Essays in Mutual Transformation* (University of Hawaii Press, 1986). His most recent book is *Wrestling with the Ox: A Theology of Religious Experience* (Continuum Publications, 1997).

Stephanie Kaza is associate professor of environmental studies at the University of Vermont, where she teaches religion and ecology, environmental philosophy, and nature writing. She is a long-time Sōtō Zen practitioner affiliated with Green Gulch Zen Center. Her book, *The Attentive Heart: Conversations with Trees* (Fawcett Columbine, 1993), is a collection of meditative essays on West Coast trees.

Kenneth Kraft is currently chairman of the Religion Studies Department at Lehigh University. He received his Ph.D. in East Asian studies from Princeton University. He is the author of *Eloquent Zen* (University of Hawaii Press, 1992) and the editor of *Inner Peace, World Peace* (State University of New York Press, 1992) and *Zen: Tradition and Transition* (Grove Press, 1988). His work on engaged Buddhism has led to an interest in the ethical and cultural significance of nuclear waste.

Lewis Lancaster is professor of East Asian languages and Buddhist studies at the University of California, Berkeley, and is currently in charge of the Ph.D. program in the Group in Buddhist Studies on that campus. He has recently written "The Sources for the Koryo Buddhist Canon: A Search for Textual Witnesses" and "The History of the Study of Twentieth Century Forgeries of Dunhuang Manuscripts" and is editor of *Religion and Society in Contemporary Korea* (Institute of East Asian Studies, University of California at Berkeley, 1997). He has been active in the world of computers, organizing the Electronic Buddhist Text Initiative, a consortium of more than forty groups around the world dealing with Buddhism and the new technology. A CD-ROM containing Sanskrit Buddhist texts is underway.

John Daido Loori is the resident teacher and spiritual leader of Zen Mountain Monastery in upstate New York. He has completed formal training in rigorous kôan Zen and in the subtle teachings of Master Dôgen's Zen. Drawing on his background as scientist, artist, naturalist, parent, and Zen priest, Abbot Loori speaks to Western students from the perspective of shared background. His books include *The Eight Gates of Zen* (Dharma Communications, 1992), *Two Arrows Meeting in Mid-Air: The Zen Koan* (Charles E. Tuttle, 1994), and *The Heart of Being: Moral and Ethical Teachings of Zen Buddhism* (Charles E. Tuttle, 1996).

Poranee Natadecha-Sponsel, M.A. in philosophy, Ohio University, Ed.D. in educational foundations, University of Hawaii, is an adjunct professor of philosophy and academic officer for the dean of graduate and professional studies at Chaminade University of Honolulu, where she teaches courses on philosophy, religion, and gender. She has taught in the Women's Studies Program at the University of Hawaii, and before that she taught for many years in the Department of Philosophy and Religion at Kasetsart University in Bangkok. She has collaborated with Leslie E. Sponsel in field research and resulting publications on various aspects of the relationship between Buddhism and ecology. She has also conducted research on linkages between women, economic development, and environment in Thailand. She serves on the Executive Board of the Hawaii Association of International Buddhists. She is co-editor with Leslie E. Sponsel of *Ecology, Ethnicity, and Religion in Thailand* (under review).

Steve Odin received his Ph.D. from the State University of New York, Stony Brook, and is an associate professor in the Department of Philosophy at the University of Hawaii at Manoa, where he teaches Japanese philosophy, comparative philosophy, and American philosophy. He is the author of *Process Metaphysics and Hua-yen Buddhism: A Critical Study of Cumulative Penetration vs. Interpenetration* (State University of New York Press, 1982) and *The Social Self in Zen and American Pragmatism* (State University of New York Press, 1996).

Graham Parkes is professor of philosophy at the University of Hawaii and a former senior fellow at the Harvard University Center for the Study of World Religions. He is the editor of *Heidegger and Asian Thought* (University of Hawaii Press, 1987) and *Nietzsche and Asian Thought* (University of Chicago Press, 1991), translator of Nishitani Keiji's *The Self-Overcoming of Nihilism* (State University of New York Press, 1990), and author of *Composing the Soul: Reaches of Nietzsche's Psychology* (University of Chicago Press, 1994).

Steven C. Rockefeller is professor of religion at Middlebury College in Vermont, where he formerly served as dean of the college. He received his master of divinity degree from Union Theological Seminary in New York City and his Ph.D. in the philosophy of religion from Columbia University. He is the author of *John Dewey: Religious Faith and Democratic Humanism* (Columbia University Press, 1991) and the co-editor of *The Christ and the Bodhisattva* (State University of New York Press, 1987) and *Spirit and Nature: Why the Environment Is a Religious Issue* (Beacon Press, 1992).

Alan Sponberg taught Buddhist studies at Princeton and Stanford for eleven years before moving to develop the Asian Humanities Program at the University of Montana, where he is currently professor of Asian philosophy and religion. His research interests focus on the cross-cultural transformations of Buddhism, both historical and contemporary, and he has published on the development of early Mahāyāna Buddhism as well as on contemporary Buddhist revival movements in China, India, and the West. He has practiced Buddhism for more than twenty years and is, as Dh. Saramati, an ordained member of the Western Buddhism order.

Leslie E. Sponsel earned the B.A. in geology from Indiana University and the M.A. and Ph.D. in anthropology from Cornell University. He is a professor at the University of Hawaii, where he directs the ecological anthropology concentration and teaches courses on human ecology in general and on tropical forest ecosystems in particular, spiritual ecology, peace studies, and human rights. Since 1986 he has been visiting southern Thailand almost yearly to collaborate with colleagues at Prince of Songkhla University in exploring various aspects of the relationship between religion and ecology. During the summers of 1994 and 1995, he held a Fulbright grant and conducted research on the role of sacred trees and sacred forests in the conservation of biodiversity in southern Thailand. He edited *Indigenous Peoples and the Future of Amazonia: An Ecological Anthropology of an Endangered World* (University of Arizona Press, 1996); co-edited with Thomas Gregor *The Anthropology of Peace and Non-violence* (Lynne Rienner Publisher, 1994); co-edited with Thomas Headland and Robert Bailey *Tropical Deforestation: The Human Dimension* (Columbia University Press, 1996); and co-edited with Poranee Natadecha-Sponsel *Ecology, Ethnicity, and Religion in Thailand* (under review).

Donald K. Swearer is the Charles and Harriet Cox McDowell Professor of Religion at Swarthmore College, where he teaches courses in Asian and comparative religions. He was the Numata Visiting Professor of Buddhist Studies at the University of Hawaii in 1993 and a Guggenheim Fellow in 1994. His recent publications include *The Buddhist World of Southeast Asia* (State University of New York Press, 1995) and *The Legend of Queen Cama* (State University of New York Press, 1998).

Mary Evelyn Tucker is an associate professor of religion at Bucknell University in Lewisburg, Pennsylvania, where she offers courses in world religions, Asian religions, and religion and ecology. She received her Ph.D. from Columbia University in the history of religions, specializing in Confucianism in Japan. She has published *Moral and Spiritual Cultivation in Japanese Neo-Confucianism* (State University of New York Press, 1989) and is co-editor with her husband, John Grim, of *Worldviews and Ecology* (Bucknell University Press, 1993). She and John Grim are currently directing a series of ten conferences on religions of the world and ecology at the Harvard University Center for the Study of World Religions. They are also editors for a series on ecology and justice from Orbis Press. She is a committee member of the United National Environmental Programme for the Environmental Sabbath and vice-president of the American Teilhard Association.

Duncan Ryūken Williams is a Ph.D. candidate in religion at Harvard University specializing in Japanese religious history. He has been a visiting lecturer at Brown University, Trinity College, and Sophia University, Tokyo. He is the translator of Shinichi Inoue's *Putting Buddhism to Work* (Kodansha, 1997).

Jeff Yamauchi is the resident naturalist at Zen Mountain Center and founder of Earth Witness Foundation, which is dedicated to integrating Buddhism and environmentalism through education. He received his master's degree in environmental studies from Prescott College, Arizona, in 1996. He is currently planning an extended solo trip along the John Muir Trail in the California Sierra Nevada.

Index

Abhidharma, visions of hell by, 382
Abhinavagupta, 395
Abortion, 15–16
Abraham, religions of, 358
Abram, David, 225
    on Gary Snyder's worldview,
        199–202
Absorption, meditative, 356
Abstention, from eating meat or
    fish, 152, 153, 155, 162n
Abuse
    of nature, 180
    within sangha, 55
Accountability, for the environmen-
    tal crisis, 279–280
Achan Pongsak Techathamamoo
    activism of, 55
    on protection of nature, 47
Action, 78
Activism
    anti-nuclear, 270–273, 282–283
    of Phra Prajak Kuttajitto, 34–35,
        55
    responses to, 55–56
Advaita Vedānta tradition, 336
Aesthetic order, 80–81
Aggression, 300
Agreement, as a maṇḍala, 78
Ahiṃsā, 49
    environmental protection and,
        317–318
    violence and, 137
Aidagara, 94
Ainu, view of nature of, 194
Almon, Bert, on Gary Snyder, 194,
    202–203

Alms, 34
Almspersons, 53
Altruism, 357–358, 370–371
Amarillo, Texas, plutonium at, 284
Ambiguity, of Buddhist religious
    concepts, 396
American Buddhism, ecology and,
    219–245
American society
    consumption by, 289n
    violence in and alienation of, 210
American Zen Buddhism, at Zen
    Mountain Center, 249–250
Ames, Roger T.
    on aesthetic order, 80–81
    on Christianity, 73
    on environmental philosophy, 89,
        93
Ānanda, jātaka about, 141
Ananda Kanchanapan, 23
Anapanasati Sutta, 235
Anāthapiṇḍika, support of Bud-
    dhists by, 9
Anātman doctrine, 353, 358
Anattā, 49
Ancestral worship, 151
Anger
    precept against, 182
    in Zen Buddhism, 166–167
Animal consciousness, 132
"Animal day," 386
Animal liberation movement, 319
Animals
    cognition by, 132
    communication with, 195–196
    compassionate and wise, 135

endangered, 252–253*t*
foolish, 136–137
in Gary Snyder's poetry, 195, 196
at Green Gulch Zen Center, 221
in hierarchy of life-forms, 356
human consciousness and, 131–
    132
Japanese rites to release, 149–157
in jātaka narratives, 131–144,
    145–146*t*
karma of, 277
killing and eating of, 196–197
protection of, 385–386
Animal sacrifice, in Buddhism,
    138–140
Animal stories, as metaphors, 132
Annen, 390
Anthropocentrism
    Buddhist, 340–342
    Judeo-Christian, 327–328
Antistructure, 51
Anukampā, 26, 42*n*
Anurak, 26, 27. See also Care
Anurakkhā, 26. See also Care
Anurak thamachāt, 27–28, 37
Anuttara-samyaksambodhi, 178
Apocryphal jātaka, 134
Apple Canyon
    fire management in, 256–258
    land use at, 254–256
    Zen Mountain Center in, 250–251
Aquino, Benigno, 58
Aramaki, Noritoshi, on the life of
    Śākyamuni, 11–12
Arhat-hood, stages of, 356
Aristotelian-Thomistic teleology,
    Christianity as, 73
Arjuna, 335
Art
    anti-nuclear activism in, 272
    Japanese, 390–393
Arya Sūra, jātaka retold by, 134
Āsanna-paccathika, 351
Asceticism, 47
    need for modern, 15–16

Ascetics, classification of, 36
Ashikaga shogunate, *hōjō-e*
    ceremony and, 153–154
Asia, Buddhism in, 4
Aśoka, edicts of, 386
Asoke tree (*Saraca indica*), 52
Assertive action, 182
Atta, 29
Augustine, 85
Australia, Buddhism in, 4
Autumn equinox, *ekos* at, 228
Avataṃsaka, 357
Avataṃsaka Buddhism, 100
    Indra's net in, 189–190
Avīci, 382
Avidyā, 384
Awakening, in Zen Buddhism, 167–
    169
Aware, in Japanese notions of
    nature, 99
Awareness
    environmental, 255
    intelligence as one form of, 200
    in reinhabitory ethic, 226–227
Axiological cosmology, of Hua-yen
    Buddhism, 98

Baker, Richard, 221
Bakhtin, Mikhail, on interspecies
    community, 202–204
Banana plant, 393–394
Banyan tree (*Ficus bengalensis*), 52
Bashō, *haiku* poetry of, 99, 197,
    330–331, 339, 393–394
Basho, 94
Bashosei, 389
"Basic Buddhism," 353–359
Basketmaking workshop, at Zen
    Mountain Center, 262
Basney, Lionel, on Gary Snyder,
    211
Batchelor, Stephen, 344
Beatniks, 327–328
Beauty, 395
Beavers, 182–183

Bechert, Heinz, on Buddhist history, 381
Beginners' Mind Temple, 221
Being, 344
"Being-time," 213
Benefit, 151
of Zen practice, 166
Bennett, John, on ecologically appropriate societies, 48–50
Berque, Augustin, on Japanese culture, 389
Berry, Thomas, 144
on Hsiang-yen's *kōan*, 287
on species protection, 131
Betweenness, 94
Bhagavad Gītā, concept of nature in, 335–336, 338
Bhajanaloka, in nature, 381
Bhāvana-mārga, 357
Bhuddabhadra, 390
Bhūmi, 357
Bija, 78
Biocentric equality, environmental problems with, 120–121
Biocentrism
Buddhadāsa's philosophy as, 30, 39–40
interdependent co-origination and, 343
Phra Prayudh's philosophy and, 36–37
Biodegradable materials, 180
Biodiversity, 65*n*
in international law, 318, 319–320
protection of, 318–319
Biological impact report, for Zen Mountain Center, 251–254, 255–256, 263
Bioregional community
practice of, 207–208
the world as, 191–194
Biosphere
ecology of, 188–191
impact of nuclear ecology on, 269–270

in international law, 318
overpopulation and excessive consumption and, 291
Biosphere cultures, 226
Birch, Charles, on life, 82–83
Bird hell, 382
Birth
Buddhist views of, 296–297
interdependence and, 296
necessity of, 294
as a positive occasion, 293
public policy toward, 294
Birth control, 15–16
encouragement of, 308
Blame, 184
Bodhi, 178
Bodhi Bhikkhu, on the environmental ethic, 47–48
Bodhicitta, "enlightened gene" as, 301–307
Bodhiñāna, 34
Bodhisattva, 76, 358, 394
activism against radioactive waste in, 271
jātaka about, 138–139
nature and, 10–11
reproduction and, 301–307
Bodhisattva stages, 357
Bodhisattva vows, 271, 284
Bodhi tree (*Ficus religiosa*), 52
center of the cosmos under, 345
Bodhyanga, 356
Body
Hindu attitudes toward, 335–336
honoring of, 180
human, 345–346
living, 81–82
as one of three mysteries, 77, 78–79, 115
"Body of bliss," 75–76
Böhme, Jakob, naturalist philosophy of, 116
Bohr, Niels, 80
Bommyōkyō, as basis for *hōjō-e* rites, 150–151

Bonding, sexuality and, 308
Book of Serenity, The, 169–170
Books, dissemination of environ-
    mentalist thought through,
    124–125
Bot, 52
Brahmadatta, King, jātaka about,
    141
Brahmajāla Sūtra, as basis for
    hōjō-e rites, 150–151
Brahmin, fowler, and partridge,
    jātaka about, 138
Brahmin and Jaina monk in
    Banaras, jātaka about, 137–138
Brahmins, King of Kosala, and
    animal sacrifice, jātaka about,
    139–140
Brazil, Buddhism in, 4
Breathing, 235
    mindfulness and, 273
Brown, Edward Espe, 221
Buddha, the, 75
    attempt on the life of, 134
    enlightenment of, 345
    forests and, 33–34
    historical images of, 76
    hōjō-e rite and, 150–151
    in Indian fine art, 394
    in jātaka narratives, 133–135,
        135–144
    life of, 11–12
    mercantile support of, 9–10
    as one of the Three Treasures,
        177–178
    Phra Prayudh on, 37
    teachings of, 47–48
    the world as, 385
Buddhadāsa Bhikkhu, 21
    career of, 24
    criticisms of, 25
    Dhammapiṭaka and, 30–33
    ecological hermeneutic of, 24–25,
        36–37
    Ian Harris on, 39–40

    on nature as *dhamma*, 24–30
    scholarship of, 41n, 42n
Buddhadharma, 373
"Buddha eye," 114
Buddhaghosa, 351
    jātaka translations by, 133–134
Buddhahood, 103, 358
    for nonsentient beings, 117, 122–
        124
    for sentient beings, 113, 114,
        296–297
Buddha-nature, 13–14, 346
    of all things, 118
    ecological problems with univer-
        sality of, 121–122
    intrinsic value as, 320
    of living things, 113–114, 116–
        117, 191
    of non-being, 122
    of plants, 332, 393
    of radioactive waste, 122–123, 124
    of tubercle bacillus, 122
Buddha Treasure, 178
Buddhavacanaṃ, 39
"Buddha way," 123
Buddhi, 336
Buddhism
    and aid to the poor and op-
        pressed, 6–8
    animal awareness in, 133–135
    anti-ecological aspects of, 55–56
    asceticism of, 15–16
    celibacy and, 306
    decline in Thai adherence to, 45
    developmental dimension of,
        353–359, 360–364, 368–371
    ecological differences between
        Christian tradition and, 79–80
    ecology and, 3, 5, 156–157, 162n,
        351, 387–388, 389–394
    ecology in contemporary Thailand
        and, 21–40
    environmental philosophy and,
        89, 351, 377–396

feminism and, 305–306
fertility control and, 291–311
forests and, 33–34, 35–36
Gaia theory and, 104–105
global ethics and, 313–322
hierarchical aspects of, 353, 366–368, 368–371, 371–374
history of, 4, 381, 354–358, 385–387
hōjō-e ceremony and, 156–157
intrinsic value in, 320
on intrinsic value of species, 320
karma in, 275, 276
mercantile activity and, 9–10, 12
mindfulness in, 273–274
nondual forms of, 353, 375n
nuclear ecology and, 269–287
number of participants in, 4, 53
Phra Prayudh on, 36–37
and practices of Buddhists, 56–59
Protestant, 8
relational dimension of, 353–359, 360–364
relationships among types of, 354–359
reproduction and, 291–311
reverence for life in, 320–321
samsāric cosmogony of, 382–383
Sinhala, 8
spiritual quest of, 38
transplantation from India to China and Japan, 113–114
views of nature in, 10–12, 13–14, 14–15, 33–34, 327–346, 340
wealth and, 9–10
Western influence on, 8
Buddhist birth stories. *See also* Jātaka narratives
animals and environment in, 131–144
animals in, 145–146t
"Buddhist Economics," 100, 344
Buddhist environmentalism, 37–40
challenges for, 280–284
conflicts between Dharma work

and, 281–282
eco-karma in, 277–280
in teachings of Kūkai, 75–79
in the United States, 219–245
Buddhist Perception of Nature Project, 397n
Buddhist Precepts, environmental relevance of, 177–184
Buddhist tradition, 3–4, 72
animal sacrifice in, 138–140
animals in, 133
enlightenment in, 345–346
environmental crisis and, 269
hōjō-e in Japanese, 149–150
karma in, 275
limitations of, 205–208
toward nature, 333–335
Buddhist worldview
interdependence in, 295–301
organic principles of, 75–79
Western worldview versus, 46, 56–59, 59–60
Bun, 53
Burning hell, 382
Business meetings, of monks, 36
Butsudō, 123

Cahuilla Indians, at Zen Mountain Center, 254
California, Jen-yen in, 7
California Organic Farming Association, Green Gulch center in, 229–230
Callicott, J. Baird
on Christianity, 73
on environmental philosophy, 89
on metaphysical implications of ecology, 93
Calocedrus decurrens, at Zen Mountain Center, 254
Care, for nature and the environment, 26–30
Carpooling, at Spirit Rock center, 236

Catholicism, Greek philosophy versus, 115
Cause and effect
    Mystic Law of, 96–97
    in overpopulation, 299
Celibacy, Buddhism and, 306
Cemeteries, 52
Centrism, 344–345
Ceramic industry, 262
Cetanā, 275
Chai Podhisita, 60
Chaiya, Thailand, Buddhadāsa Bhikkhu at, 24–25
Chan-jan, 390
Chanoyu ceremony, 99
Chanting, in gratitude, 227–228
Charity, 6
Chatsumarn Kabilsingh, 33–34
Chattel, women as, 295
Chedi, 52
Chen-yen, 75
Ch'i force, 104
Chigasakishi Motomura, early *hōjō-e* ceremony at, 152–153
Chih-i, 390
Childlessness
    Buddhism and, 301
    prejudices against, 305
Children
    cultural relativism regarding, 294–295
    exploitation of, 295
    interdependence and, 296
    as parental carbon-copies, 304
    proper care for, 293, 300
China
    arrival of Buddhism in, 12–13, 113
    dismal environmental record of, 333
    views of nature in, 15
Chinese, Sanskrit and, 379
Chinkonsai ceremony, 160n
Choice, 84
Chokushi, 153
Christian creationism, 377

Christianity
    Buddhist aesthetic order versus logical order of, 80–82
    God in, 358
    teachings about nature in, 72–73
Christian tradition. *See also* Judeo-Christian tradition
    ecological differences between Buddhism and, 79–80, 83–85
    environmental destructiveness of, 83–84, 111–112
    nature in, 380
    original sin in, 83
Chuang-tzu, influence on Buddhism of, 123–124
Church, the, on aiding the poor and oppressed, 6
Chuzhoi, 202
Citta, 76
Citta Thera, 34
Cit wāng, 26
Clarity, 180
Climate, 94–95
Cobb, John B., Jr., on life, 82–83
Cochida, Doug, on the human body, 345–346
Cognition, by animals, 132
Cold Mountain poet, poetry of, 331
Collective cultural perceptions, 5
    of aiding the poor and oppressed, 5–8
    of cooking, 14
    of mercantile activity, 9
Colors, five Buddhist, 77–78
Comfort, spiritual enlightenment and, 298–299
Commerce, 9
Communication, 78
    sexuality and, 307–311
    between species, 195–196
Communionism, 194–195
Community. *See also* Monastic community
    bioregional, 191–194, 207–208

Gary Snyder on, 187–188
interspecies, 202–204
land as, 91
nature and, 90, 187–188, 213
spirit of, 51–52, 208–213, 240
Community relations
at Green Gulch center, 230
at Spirit Rock center, 237
Compassion, 357–358
hierarchy of, 366–368, 368–371,
376n
Compost piles, at Green Gulch
center, 231
Concentration, 115
Conscience, social, 91
Consciousness, 142
animals and, 144
evolution of, 362–363, 367
as one of the Six Great Elements,
76–78
Conservation aesthetic, of Aldo
Leopold, 91–92, 100–101
Conservationists, environmental
philosophy and, 90–91
Consumption. *See* Material con-
sumption
Control, of nature, 181
Convention on Biological Diversity,
on intrinsic value of species, 319
Cook, Francis H., on organismic
interrelatedness, 97–98
Cooked foods, 14
Coomaraswamy, Ananda K., on
Indian religious art, 394–395
Cooperation, among natural entities,
29–30
Corruption, within *sangha*, 55
Cosmic ecology, in Hua-yen
Buddhism, 98
Cosmological thought, 377, 382
Council of Elders, agenda of, 55
Cowell, E. B., 133–134
Creation, God and, 115
Creationism, 377

Creativity, 81
Cuckoo and king of Banaras, jātaka
about, 136
Cūla Thera, 34
Cultivation, Gary Snyder on, 362
Cultural artifacts, 211
Cultural relativism, regarding
women, 294–295
Cultural transmission, 210–213

Daigohonzon maṇḍala, 96
Daikon, 390
Dainichi-kyō, 76
Dainichi Nyorai, 114, 115–116
the Buddha as, 75–76
dharma of, 78–79
as life force, 82–83
self-enjoyment of, 117
Dalai Lama, 15, 38, 40, 219, 346
on Buddhist practice, 343
on the Golden Rule, 316–317
Harvard lectures by, 340–341
on radioactive waste disposal,
271
on spirit and nature, 328–330,
338, 343–344
Darsana-mārga, 357
Dasmann, Ray, 226
Dean, Tim, on the Other, 204
Death
human responses to, 71
interfusion of life with, 122
and life in nature, 188–191
Dedication chants, at Green Gulch
center, 227–228
De Divisione Naturae (Erigena),
115
"Deep ecology," 112–113, 214–
215n, 344
biocentric equality in, 120–121
bioregionalism as, 192
"Deep self," 49
Defilement, 33
precept against, 183

Deforestation, Buddhism and, 47–48
Degradation, and lack of spiritual
   enlightenment, 298–299
Deities, fertility of, 308–309
Delusion, enlightenment versus,
   170–171
Department of Environmental
   Conservation, 182–183
Dependent co-origination, 76
Dependent origination doctrine, 151
Deprivation, lack of spiritual
   enlightenment and, 298–299
Descartes, René, 74
de Silva, Lily, 45
Desire, 83
   spiritual enlightenment and, 298–
   299
Devadatta, 134
   jātaka about, 135, 136–137
Devall, Bill, 127*n*
   on biocentric equality, 120–121
Development, limiting of, 258
Developmental dimension of
   Buddhism, 353–359, 360–364
   reaffirmation of, 368–371
Devolutionary cosmogony, 383
Dhamma, nature as, 24–30, 33–34.
   See also Dharma
Dhammadhātu, 29–30
Dhammajāti, 26, 381. *See also* Nature
Dhammapada, 342
Dhammapiṭaka, 21
   Buddhadāsa and, 30–33
   ecological hermeneutic of, 36–37
   Ian Harris on, 39–40
   on nature and the pursuit of
   enlightenment, 30–37
   nature management advocated by,
   32–33
   Phra Prajak Kuttajitto and, 34–35
   scholarship of, 30–31, 41*n*
   as title, 41*n*
Dhamma Saphā, booklets of, 42*n*
Dhamma Study Group, 26

Dharma, 48, 120, 333, 352, 354,
   369, 372, 373
   adherence of monks to, 52
   of Dainichi, 78–79, 115–116
   dimensions of, 359
   environmental wisdom of, 53–54
   hōjō-e ceremony and, 152
   of noxious bacteria and radioactive waste, 121–122
   as one of the Three Treasures,
   177–178
   at Spirit Rock center, 238–239
   Tao and, 328
   Thai ecocrisis and, 47–48
   understanding of, 345
   Western assimilation of, 360
Dharma Bums (Kerouac), 213
Dharmadhātu, 381, 384–385
Dharmakāya, 76, 114, 115–116,
   116–117, 118, 127*n*, 385
   "sick" buildings, microbes, and
   toxic waste as part of, 120–121
Dharmakīrti, poetry of, 338–339
Dharma-maṇḍala, 78
*Dharma*s, realm of, 384–385
Dharmatā, 381
Dharma Treasure, 178
Dharma work, anti-nuclear activism
   as, 282–283
Dhira Phantumvanit, on Thailand's
   ecocrisis, 45
Dhutaṅga, 36, 47
Dhyāna, 356
Dialogue, Mikhail Bakhtin on, 203
Dīgha Nikāya, 355–356
Dillard, Annie
   on choice, 84
   on necessity, 84–85
Disease, in shamanistic cultures, 204
"Distant enemy," 351
Diversity, 193, 194. *See also*
   Biodiversity
Dobson, Andrew, on ecologically
   appropriate societies, 48–50

Dōgen, 101–102, 113, 119
  on delusion and enlightenment, 170–171
  environmental reconciliation of teachings of, 121–124
  philosophy of, 116–118
  on plants as Buddha-nature, 393
  on the zero-point, 168
Domination, in hierarchies, 365–366
Dominionistic approach to nature, 333–334
Dōshi, in *hōjō-e* ceremony, 152
Dōtoku, 123
Draft International Covenant on Environment and Development environmental protection in, 318
"Dreamtime," 213
Drengson, Philip, on ecologically appropriate societies, 48–50
Drugoi, 202
Dualism, 375*n*
Dukkha, 25
Dummedha Jātaka, 138–139
Dūra-paccathika, 351
Duty, 344
  acting out of, 370–371
Dysteleological nature of the universe, 383–384
Dzogchen Buddhism, 353

Earth
  as bioregional community, 191–194
  eco-karma of, 278
  reality of, 345–346
  seeing from the perspective of, 170–171
  sensitivity to the pain of, 172–173
  withdrawal of humans from, 287
Earth Charter, 324*n*
  Buddhism and, 313–322
  popular approval of, 322
Earth Council, 322
Earth Day ceremonies, at Green Gulch center, 233–234

"Earth Witness Foundation" program, at Zen Mountain Center, 259–260
East, dismal environmental record of the, 333–334
East Asian Buddhism, environmental philosophy and, 89
EcoBuddhism, 37–40, 378, 395–396, 397*n*
Ecofeminists, 56
Eco-karma, 277–280
Eco-koans, 286–287
Ecological community, 188–191
Ecological conscience, 91
Ecological culture
  at Green Gulch center, 231–232
  at Spirit Rock center, 237–238
  transmission of, 240
Ecological living, reinhabitation as, 225–227
Ecological monitoring, at Buddhist centers, 241
Ecological role models, Buddhist centers as, 243–245
Ecological sensitivity, building Zen centers with, 255
Ecological sustainability, 239–240
  at Zen Mountain Center, 260–261
Ecological worldview, relational fields in, 92–95
Ecology
  American Buddhism and, 219–245
  Buddhism and, 3, 5, 353, 387–388, 389–394
  field model of nature in, 92–95
  interrelatedness and, 373–374
  jātaka narratives and, 140–142
  kōans concerning, 284–287
  life in, 82–83
  mercantile activities and, 9
  metaphysics of, 93
  nuclear, 269–287
  shamanism and, 199–202
  subfields of, 270

Thai Buddhism and, 21–40
Zen Buddhism and, 165–173
Economic systems
  Asian, 387
  Western, 386–387
Ecosystem, Indra's net as, 189–190
Ecosystem cultures, 226
Ecosystemic macrocosm, 93
Education
  at Green Gulch center, 232
  at Spirit Rock center, 238
Egalitarianism, Buddhism and, 352
Ego
  collective, 305
  in reproduction and consumption,
    302–303
"Eightfold path," 356
Einstein, Albert, 80
Ekos, 228
Elders
  early prominence of, 356
  trees as, 195
Electric power generation, at Zen
  Mountain Center, 260–261
Elements, Six Great, 76–77
Elephanta temple, 336, 337
Elephants and quail, jātaka about,
  136–137
"Elite" Buddhism, 197
Emotional sensitivity, 99
Emotions, 395
Empathy, 26
Emptiness, 26, 94, 337, 345, 347*n*
  Dalai Lama on, 329
  form and, 329, 338
  interdependent co-origination and,
    342–343
  metaphysic of, 103
Endangered species, at Zen Moun-
  tain Center, 251–254, 252–253*t*
Energy sources
  at Green Gulch center, 233–234
  in reinhabitory ethic, 227
  at Spirit Rock center, 238–239

Engi, of nature, 122
"Enlightened gene," transmission
  of, 301–307
Enlightenment, 34, 297, 331, 332,
  357–358
  attainment of, 76, 345–346
  delusion versus, 170–171
  fusion of self and natural world
    in, 331
  "gene" of, 301–307
  limiting reproduction and, 306–307
  self and, 341–342
  seven limbs of, 356
  sexuality and, 310–311
  as solving the environmental
    crisis, 370
  in Zen Buddhism, 167–169, 178
Enlightenment, the, 383
Ensō, 272
Environment
  consumption and population and,
    292
  designing a harmonious, 104
  hōjō-e ceremony and, 154–156
  in jātaka narratives, 131–144
  killing of, 179
Environmental crisis, 21
  Buddhadāsa Bhikkhu's concern
    about, 26
  Buddhism and, 269
  Buddhist writings on, 37–40
  causes of, 23–24, 327–328
  Christianity and, 72–73
  determining accountability for,
    279–280
  Earth Charter and, 313–322
  enlightenment as solving, 370
  as fostered by incorrect Western
    worldview, 31–32
  Greco-Roman philosophy and,
    73–74
  Japan's role in, 119
  overpopulation and excessive
    consumption in, 291–311

problems of Buddhism and, 56–59, 359
relevance of Zen precepts to, 177–184
religion and, 63n, 377–378
in Thailand, 45–50, 59–60
Western worldview and, 111–112
Zen Buddhism and, 165–167, 172–173, 250
Environmental ethics, 89–90, 112, 370, 371–374
Buddhism and, 377–396, 399n
nuclear industry in, 278–279
reinhabitation and, 225–227
of Watsuji Tetsurō, 94–95
at Zen Mountain Center, 249–264
Environmental philosophy, 89
of Aldo Leopold, 89–92, 104–105
of Buddhism, 377–396
Gaia concept in, 104–105
of Japanese Buddhism, 92–104
Environmental programs, at Zen Mountain Center, 249–250, 259–261, 263–264
Equilibrium society. *See* Green society
Equitable distribution of resources, 292
Erigena, John Scotus, philosophy of, 115
Errors, seeing others', 180–181
Eshō funi principle, 96–97
"Eskimos," animal icons of, 197
Esoteric Buddhist teachings, of Kūkai, 75–79, 79–80, 114
Ethical norms, 3–4
and the poor and oppressed, 5–8
Ethical principles, universal, 321–322
Ethics. *See* Environmental ethics; Global ethics; Golden Rule; Land ethic
Eucalyptus, poisons in, 242
Eucharist, 199
Europe, Buddhism in, 4

Evil
Golden Rule and, 316
not creating, 178
Evolution of consciousness, 362–363, 367
Excrement, place of, 382
Existence
of the Buddha and sentient beings, 114
Buddhist metaphor for, 79
Dōgen on, 117–118
Experience, 83
Expressive symbols, Buddhahood revealed through, 103
Extension, 78

Fa-hsien, 390
Family
importance of, 206, 207
perpetuation of lineage of, 301–302, 306–307
Fang sheng ch'ih, 386
Faraday, Michael, 80
Farming, at Green Gulch center, 229–230, 231–232
Fatal disease, Buddhist attitudes toward, 122
Faults, seeing others', 180–181
Fear, and lack of spiritual enlightenment, 298–299
Feminism
Buddhism and, 305–306
patriarchal stereotyping and, 309–310
Feng shui, 104
Ferry, Luc, 126n
Fertility, of mythological women, 308–309
Fertility control
Buddhism and, 291–311
Ficus bengalensis, 52
Ficus religiosa, 52
Field model of nature, 92–95
Ainu and, 194

Film, dissemination of environmentalist thought through,
   124–125
Fine arts
   Buddhism and, 391–394
   in India, 394–395
Fire management
   at Buddhist centers, 242
   at Zen Mountain Center, 256–258
Fish
   eating of, 71, 152
   in *hōjō-e* ceremony, 153, 155
Fish ponds, 386
Five Great Elements, reality as
   composed of, 114
Five spiritual faculties, 356
Flower Garland Sūtra, 171
Food, of monks, 52
Food practices
   at Buddhist centers, 243
   at Green Gulch center, 231
   at Zen Mountain Center, 261
Food-web community, 188–191
Force field, 94
Forest, restoration of, 54
Forests. *See also* Trees
   Buddhism and, 33–34, 35–36, 36,
      47, 337
   in Gary Snyder's poetry, 195–196
   at Green Gulch center, 223–224, 229
   jātaka about, 140
   mental freedom in, 33
   monasteries in, 53
   restoration of, 65*n*
   at Spirit Rock center, 223–224
   at Zen Mountain Center, 256–257
Form
   Dalai Lama on, 329
   Emptiness and, 329, 338
Four Foundations of Mindfulness,
   235
Four Iron Cauldrons, 139
Four levels of meditative absorption, 356

Four Maṇḍalas, in Shingon training,
   77–78
Fowler, Brahmin, and partridge,
   jātaka about, 138
Franciscan worldview, 72
Freedom, mental, 33
Fresh–rotten food, 14
Friendliness, 342
Friendship, 244–245
Fūdo, 94–95
Fugu fish, eating of, 71
Fujiwara no Sanesuke, 156
Fujiwara Teika, *waka* poetry of, 99
"Fundamental nature," 338
Futagoyama, in *The Lotus Sūtra,*
   332–333, 337

Gabe (Generic American Buddhist
   Environmentalist), challenges
   confronting, 280–284
Gadamer, Hans Georg, on inter-
   preting texts, 379
Gaia theory of nature, 103–104
Gananath Obeyesekere, 8
Gandhi, Mohandas K., 58
Gangetic plain, life on, 11–12
Gardening, at Zen Mountain Center,
   261
Garden of Eden, nature as, 10, 14
Gāthā, 134
Gati, 356
Gay life-styles, validation of, 305
Geertz, Clifford, 4
Geidō, 99, 101, 103
   origins of, 99
Generosity, 181–182
Genjōkōan, 102
Genshin, visions of hell by, 382
Gere, Richard, 347*n*
Giving, 179, 181–182
Global community, Earth Charter
   and, 313–322
Global ethics, Buddhism and, 313–
   322

Global interdependence
  Golden Rule and, 317
  growth of, 313–314
  in international law, 318
Global thinking, 193
Gocaragāma, 34
God
  in Christianity, 73
  humanity and, 73
  individuals' relationship with, 112
  life and, 75
  as life force, 82–83
  the natural world as, 115
  as nature, 85
  in religions of Abraham, 358
Godhead, 126–127*n*
Gods, 356
Golden Gate National Recreation
    Area (GGNRA), Green Gulch
    center relations with, 230
Golden Rule, 321
  as basis of all ethics, 316–317
Golden stag, jātaka about, 135
Goldstein, Joseph, founding of
    Spirit Rock Meditation Center
    by, 222
Gombrich, Richard, 11
Good, 34
  actualizing for others, 179
  Golden Rule and, 316
  practicing, 178–179
Gorbachev, Mikhail, Earth Charter
    and, 315
Go-Uda, Emperor, 158*n*
Governance, at Buddhist centers,
    240–241
Grace, 371, 373
Grāhyagrāhakakalpanā, 384
Grapard, Allan G., 337
  on enlightenment, 332
  on Japanese love of nature, 340
Graphs, of Buddhist concepts, 354,
    365, 367
Grass, jātaka about, 141

Gratitude, 32
  jātaka about, 135
  in reinhabitory ethic, 226–227
Gratitude to the land
  at Green Gulch center, 227–228
  at Spirit Rock center, 235
"Great Luminous One," Dainichi as,
    76
Great Maṇḍalas, 77–78
"Great Sun," the Buddha as, 75
Greco-Roman philosophy
  Catholicism versus, 115
  ecological differences between
      Buddhism and, 79–80
  environmental crisis and, 73–74
Greed, 46
  Buddhist challenges to, 23–24
  destructiveness of, 29, 172–173
  mercantile activities and, 9
  overcoming, 26 27
  precept against, 181 182
  in Zen Buddhism, 166–167
"Green Buddhism," 33–34. *See also*
    Environmental ethics
  critique of, 371–374
  hierarchy of compassion and,
      351–374
  at Zen Mountain Center, 263–264
Green Cross International, 315
Green Dragon Temple, 221
Green Gulch Zen Center
  activism against radioactive waste
      at, 271
  as ecological role model, 243–245
  environmental practices of, 220
  evaluation of, 227–234
  future challenges at, 241–243
  land history of, 220–221
  locality of, 223–225, 229–230
  points of tension at, 239–241
Green society, monastic community
    as, 48–50, 53–54, 54–56
Greenwashing, 180
Gross, Rita, 4

Gross national product (GNP), 22, 23
Groves, sacred, 52
Grumbine, Edward, 225
Gūjin, 123
Gypsy moths, 182

Habitats, at Zen Mountain Center, 251
Habito, Ruben, criticism of Buddhism by, 55
Hachiman, 152–153
Haibutsu kishaku policy, 159*n*
Haiku poetry, 99, 101, 330–331. *See also* Poetry
Hakamaya, Noriaki, 37
Hamilton, Michael P., 253*n*
Han cultural area, arrival of Buddhism in, 12–13
Han Shan, 213, 217*n*
Happiness, 32, 33
Harada-Yasutani lineage, 165, 174*n*
Hargrove, Eugene C., land ethic of, 92
Harming others, 316–317, 317–318
Harmony, 346
	actualization of, 182
	environmental, 104
	Three Pure Precepts and, 178–179
Harris, Ian, 21
	criticisms of ecoBuddhism by, 37–40
Hazard management, at Buddhist centers, 242–243
Hearn, Lafcadio, on Japanese love of nature, 330
Heart Sūtra, The, 329
	on Emptiness and form, 338
	at Green Gulch center, 228
Hebrew Bible, the poor and oppressed in, 6
Heidegger, Martin, 384, 385
Hell, Buddhist visions of, 382–383
Hell-beings, 356
Hell of no interval, 382

Hell of repetition, 382
Hell where everything is cooked, 382
Helping others, 316–317, 317–318
Hermitage, at Spirit Rock center, 238–239
"Hermit strand," 337
Hierarchy, 351–374
	in Buddhist philosophy, 351–353
	in Christian worldview, 80
	of compassion, 366–368, 368–371, 376*n*
	of oppression, 364–366, 376*n*
Hijō jōbutsu, 390
Hīnayāna Buddhism, 316
	rise of, 355
Hindu tradition
	limitations of, 205
	views of nature in, 335–339
Hippies, 327–328
"Historical Buddhas," 76
Hito, 94
Hōi, 123
	of humans and tubercle bacilli, 122
Hōjō-e ceremony, 149–157, 159*n*, 160*n*
	described, 149–150, 151–152
	earliest recorded, 152–153
	environmental issues concerning, 154–156, 156–157
	environmentally deleterious preparations for, 155–156
	at Hachiman shrines, 152–153
	at Iwashimizu Hachiman Shrine, 152, 153–154
	origins of, 160*n*
	textual basis of, 150–151
Hōjō-e hōsoku, source of *hōjō-e* ceremony in, 151
Hōjōgawa, release of fish and clams into, 152
Hōjō-ike/enketsuchi, release of carp into, 152

Holiness, of nature, 346
Holism, 193
  in emerging Western ecological
    models, 74–75
Homoerotic activity, 307–308
Honesty, 183–184
Hopi, 205
Hoshinji, 221
Hospital care, for the poor and
  oppressed, 7
Hosshin, 114
Hosshin seppō theories, 102, 115–
  116, 117
Hossō school, 152
Hostility, jātaka about, 136–137
Housing, for Zen Mountain Center
  visitors, 258–259
Hsiang-yen, 287
Hua-yen Buddhism
  aesthetics of nature in, 99–100
  cosmic ecology in, 98
  Indra's net in, 189–190
Hui-kuo, 75–76, 87n
Hui-yuan, 390
Hūṃ, 114
Human behavior, animal stories as
  explaining, 134–135
Humanistic approach to nature, 333
Humanity
  Buddhist philosophy toward,
    340–341
  God and, 73
  nature and, 384–385
Human nature, 94
  and excessive fertility and
    consumption, 302–303
Human rights
  in international law, 321
  reproduction and consumption as,
    292–295
  rights of other living things
    versus, 111–112
Humans. *See also* Population
  animals and, 131–132

in hierarchy of life-forms, 356
as predators, 71–72
relationship of land to, 90–91
relationship of nature to, 94–95,
  191–194
Humility, in reinhabitory ethic,
  226–227
Hunger, lack of spiritual enlightenment and, 298–299
Hungry ghosts, 356
Hunter's Point jail project, 230
Hunting
  Gary Snyder on, 198
  jātaka about, 135
Hwalin, Jen-yen's hospital in, 7
Hwa Yol Jung, on communionism,
  194–195

Iba Iseki, early *hōjō-e* ceremony at,
  152–153
Ichinen sanzen principle, 96–97
Ienaga Saburō, on Japanese art,
  391–393
Ignorance, 384
Ikeda Daisaku, on Nichiren Bud-
  dhism, 97
Ill-spoken words, jātaka about, 136
Impermanence, 353, 358
  of nature, 99
Impermanence-Buddha-nature, 102
India, 13
  concept of nature in, 335–339
  fine arts in, 394–395
  rise of Buddhism in, 12–13, 385–
    386
  Vajrayāna Buddhism from, 310–
    311
Indian Buddhist literature, nature in,
  3sdaaaaa38–339
Individuality, 94
Indra's net, 98, 171
  bioregionalism and, 191–194
  ecological community as, 188–191
Indriya, 356

Indulgences, sale of, 6
Industrialization, 112
Inquiring Mind newsletter, 222
Insight meditation practice, 222
Insight Meditation Society, 222
Integrity, 183–184, 203
Intellect, Hindu attitudes toward,
    336
Intelligence, as one form of
    awareness, 200
Intention, 78
    karma and, 275, 277
Interdependence. *See also* Field
    model of nature
    in Buddhist thought, 295–301
    of nature, 28–30, 74–75, 90, 97–
    98, 122, 198, 202–204,
    344–345
"Interdependence Day," at Spirit
    Rock center, 238
Interdependent co-origination, 342–
    343
Interior life movements, in emerging
    Western ecological models, 75
International law, environmental
    protection in, 318
"Interpenetration of part and part,"
    98
"Interpenetration of part and
    whole," 97–98
Interrelatedness
    of all existence, 369–370
    in Buddhism, 371–374
    in hierarchies, 364–365, 366–368
Interspecies community, 202–204
Intimacy of things, 183
Intrinsic value
    in Buddhism, 320
    of people, 321
Inua, 198
Inupiaq people, animal icons of,
    197
Irrationality, in Buddhism, 199
Islam, 334–335
    God in, 358

Itō Jakuchū, 390–393
Itō Seirō, 155, 158n, 159n, 160n
Iwashimizu Hachiman Shrine, 149–
    150, 158n, 159n, 160n, 161n
    hōjō-e ceremony at, 152, 153–154
Iwashimizusai, 159n
Iworu, 194
Izutsu Toshihiko, on Zen Buddhism,
    93–94

Jaina monk and Brahmin in
    Banaras, jātaka about, 137–138
Jaini, Padmanabh S., 133
Jainism, animals in, 133
Japan, 71
    animal-release rites in medieval,
    149–157
    arrival of Buddhism in, 113–114
    dismal environmental record of,
    119, 333–334
    Kūkai in, 87n
    military forces in, 161n
    state rituals of, 153–154, 154–156
    view of nature in, 327–328, 330
Japanese art and literature, termi-
    nology of, 99
Japanese Buddhism
    aesthetic concept of nature in,
    99–101
    concept of nature in, 89–105,
    339–340
    environmental ethics in, 92–103,
    388–389
    field model of nature in, 92–95
    nature and self in, 95–97
    nature in, 124–125, 197–198,
    330–331
    salvific function of nature in,
    101–103
    Shintō and, 102–103
Japanese literary tradition, nature
    in, 101–102
Jātakamālā (Arya Sūra), 134
Jātakamālā, nature and the, 11
Jātaka narratives, 39

animals in, 133–135, 145–146*t*
apocryphal, 134
ecology and, 140–142
morals in, 142–144
vegetarianism in, 137–138
Javasakuṇa Jātaka, 135
Jen-yen
aid to the poor and oppressed by,
7–8
mercantile support of, 10
Jetavana, jātaka about, 139–140
Jiji muge doctrine, 98
Jinen, 95–96, 102
Jinen honi doctrine, 102
Jōkei, 153
Jorgenson, John, 13
Jōriki, benefits of, 167–169
Joy, sexuality and, 310–311
Judaism, God in, 358
Judeo-Christian tradition. *See also*
Christian tradition
anthropocentrism of, 327–328
Jūji enjishō, 152–153

Kālacakra, maṇḍala of, 395
Kalyana mitta, 244
Kamakura shogunate, *hōjō-e*
ceremony and, 153–154, 160–
161*n*
Kami, 152
in Shinto, 113–114
Kamo Festival, 153
Kamo no Chōmei, 209–210
Kampuchea, Buddhism in, 59
Kansenji official, 158*n*
Karma, 353
technology and the understanding
of, 274–277
transference of, 277
waste and, 280–281
Karma-maṇḍala, 78
Karme Chöling Tibetan Center, 244
Karmic accountability, for the
environmental crisis, 279–280
Karmic cosmology, 200–201

Karmic worldview, 276
Karunā, 367
Kasuga Festival, 153
Kataññū, 32
Kegon Buddhism
aesthetics of nature in, 99–100
ecological worldview of, 97–98
field model of nature in, 94
Indra's net in, 189–190
Keishu, 150
Kellert, Stephen R., on Eastern
environmentalism, 333–334,
339–340
Kenshō-godō, benefits of, 167–169
Kerouac, Jack, 213
Kessai, 153, 162*n*. *See also* Absten-
tion
Keyes, Charles F., on Buddhist
history, 381
Khoroche, Peter, 134
Khuddaka Nikāya, 32
Khunying Suthawan Sathirathai, on
Thailand's ecocrisis, 45
Ki force, 104
Kilesa, 33
Killing
of animals, 149–150
jātaka about, 135
Kim, Hee-jin, 128*n*
King, Martin Luther, Jr., 58
King of Banaras and cuckoo, jātaka
about, 136
King of Death, 32
King of Kosala, Brahmins, and
animal sacrifice, jātaka about,
139–140
Kitkitdizze, 213
Gary Snyder on life at, 193
Kōans
described, 285–286
ecological, 284–287
in Zen Buddhism, 168–170, 171,
172, 174*n*
Kohák, Erazim, on moral sense of
nature, 346

Kokālika Jātaka, 136
Kongocho-kyō, 76
Kongōmyōkyō, 152
   as basis for *hōjō-e* rites, 150–151
Kornfeld, Jack, founding of Spirit
   Rock Meditation Center by,
   222
Kotthamachāt, 32
Kraft, Kenneth, 58
Kraus, James W., 215*n*
Kṛṣṇa, 335, 338
Kū, 94
Kūkai, 102–103, 390
   aesthetic order of, 81–82
   Buddhist environmental paradigm
      of, 75–79
   Christianity versus teachings of,
      79–80, 83, 85
   environmental reconciliation of
      teachings of, 121–124
   teachings of, 72, 84, 113–116
   titles of, 87*n*
   universalism of, 120–121
Küng, Hans
   on the Golden Rule, 316
   on religion and global ethics, 314
Kunstadter, Peter, 45
Kusa grass, jātaka about, 141
Kusala, 34
Kusanjāli Jātaka, 141
Kuti, 52
Kyoto School, field model of nature
   of, 94

LaFleur, William R., 126*n*, 332
   on Bashō's poetry, 331
   on nature in Buddhism, 102–103
Lake Hemet fire, 257
Land
   gaining control of, 155–156
   society's relationship to, 89–91
Land aesthetic, 91
Land ethic, 92, 319
   of Aldo Leopold, 89–91, 100–
      101, 225–226

   at Zen Mountain Center, 256
Land restoration, at Green Gulch
   center, 229
Landscaping, at Green Gulch center,
   233–234
Land stewardship
   at Buddhist centers, 240–241,
      241–242
   by Green Gulch center, 229–230
   by Spirit Rock center, 235–237
   by Zen Mountain Center, 256–
      259, 263–264
Language
   Buddhist concepts of nature and,
      378–379
   in international legal documents,
      319–320, 321
Laṅkāvatāra Sūtra, on kinship of
   humans and animals, 143
Laos, Buddhism in, 59
Lao-tzu, influence on Buddhism of,
   123–124
Latter Day of the Law, 96
Laṭukika Jātaka, 136–137
Law, decline of, 209
Laypersons, 52
   in Buddhist tradition, 342
   in *hōjō-e* ceremony, 152
   rites of intensification of, 52–53
Le Guin, Ursula, 203
Leopold, Aldo, 95, 98, 106*n*, 225–
   226, 319
   environmental philosophy of, 89–
      92
   field model of nature of, 92–95
   Gaia theory and, 104–105
   on rightness, 188
   Zen Buddhism and, 100–101
Lesbian life-styles, validation of,
   305
Lévi-Strauss, Claude, 14
Lies
   precept against, 180
Life
   affirmation of, 179

Buddhist attitude toward, 135
cellular, 82–83
conditions enhancing, 297–298
and death in nature, 188–191
God and, 75
interfusion of death with, 122
Nichiren Daishonin's view of,
96–97
quality of, 300–301
reverence for, 320–321
stealing of, 179
in Zen Buddhism, 177
Life force, 75
Life-forms
hierarchy of, 356
taxonomy of, 358–359
Life-style diversity, intolerance of,
305
Liminality
of monastic community, 49, 50–54
in rites of passage, 51
Lion and woodpecker, jātaka about,
135
Lions and tigers, jātaka about, 141–
142
Li Po, poetry of, 339
Li-shih wu-ai doctrine, 97
Living body, 81–82
"Living in the present moment,"
Zen Buddhism and, 167, 171
Living things
karma of, 277–278
protection of, 318–319
seeing from the perspective of,
170–171
Local thinking, 193
Logic, *kōans* and, 285–286
Logical order, 81
Lohakumbi Jātaka, 139–140
Loori, John Daido
on the environmental crisis, 285
Zen films of, 125
Lotus Sūtra, The, 96–97, 337, 390
on plants, 332
on volcanos, 332–333

Love, in nature, 189, 190–191
"Love mode," 376*n*
Loving-kindness, 32–33
Loving kindness meditation, at
Spirit Rock center, 235
Lun, 123
Lung-mei currents, 104

Maccurāja, 32
Macy, Joanna, 38, 344
on anti-nuclear activism, 282
on radioactive waste disposal, 271
on the self, 370–371
Madhyamaka tradition
Emptiness in, 342–343
nature in, 338
Mae chii, 56
Maezumi Rōshi, Hakuyu Taizan,
founding of Zen Mountain
Center by, 250
Magic
in Buddhism, 199
in nature, 200–201
Mahākassapa, 34
Mahā-maṇḍala, 77–78
Mahāparinirvāṇa Sūtra, 390
Mahāratnakūṭa Sūtra, 134
Mahāvairocana Tathāgata, 114
the Buddha as, 75, 79
Mahāyāna Buddhism, 103, 113,
316, 343
Gary Snyder and, 205, 206
hierarchy in, 356–358
magic and, 200
reproduction and, 301–307
rise of, 355
Mahāyāna "three-body" theory, in
Kūkai's buddhology, 75–76
Mai hen kae tua, 26
Maintained Three Treasures, 178
Maio-lo, on life and environment,
96–97
Male heirs, 301–302
Mallika, Queen, jātaka about, 139–140
Manas, 336

Maṇḍalas, 77–78
  ecosystems as, 194
  of Kālacakra, 395
Mandela, Nelson, 58
Mantras, 77, 115
Manzanita Village, 244
Mappō, 96, 209
Māra, 32
Mārga, 355
Marginality, in rites of passage, 51
Marin Conservation Association, 223
Marin County
  Buddhist centers in, 220, 223
  land-management issues in, 241–242
Marin County Water District, 237
Marxism, 9
  view of Buddhists under, 56
Masayuki Taira, 155, 162n
"Master Banana Plant," 393
Masturbation, 307–308
Matao Noda, 94
Material consumption
  countering excessive, 291, 299–301
  decreasing, 260
  environment and population and, 292
  religious criticism of excessive, 293, 303
Materialism, 112
  among Buddhists, 57
  deleterious effects of, 22–24, 45, 46
Material well-being, spiritual enlightenment and, 298–299
Maternity, women and, 301–302
Maxwell, James Clerk, 80
McDaniel, Jay, 324n
McDermott, James P., 133
Meat
  eating of, 137–138, 152
  in *hōjō-e* ceremony, 153
Media, the, on the environmental crisis in Thailand, 23–24
"Medicinal Herbs," 390

Medieval period of Japanese history, 99
Meditation, 222. *See also* Seated meditation
  at Spirit Rock center, 235
Meditative absorption, 356
Meetings, 52
Mehden, Fred R. von der, 56
  on Thai Buddhism, 57–58
Meiji government
  Buddhist persecution by, 389
  hōjō-e ceremony and, 159n
Memorial ceremonies, at Green Gulch center, 233–234
Men, in Buddhist philosophy, 341
Mental freedom, 33
Mercantile activities, greed in, 9
Mercantile activity
  Buddhism and, 9–10, 12
  collective cultural perception of, 9
Merchant laymen, in Buddhism, 10
Merit, 53
Merton, Thomas, 24
Metaphors, animal stories as, 132
Metaphysics
  of ecology, 93
  nature in, 380
Mettā, 32–33, 49
Metta Sutta, 235
Micchadiṭṭhi, 32
Middlebury College, Dalai Lama at, 328–330
Middle Way, 311
  making life worthwhile through, 297–301
Mikkyō, 75
Mikkyō Buddhism, 103
Mikoshi, 152, 158–159n
Minamoto clan, deities of, 154
Mind
  clouding of, 180
  Hindu attitudes toward, 336
  as one of the Six Great Elements, 76–77

as one of three mysteries, 77, 78–79, 115
purification of, 329
in reinhabitory ethic, 227
Mindfulness
benefits of, 167–169
in Buddhist practice, 273–274
Mizukara, onozukara and, 95–97
Moderation
Buddhist resources for, 295–301
at Green Gulch center, 234
in reproduction and consumption, 294, 302–303
Modern economic culture
benefits of, 22
origin and deleterious effects of, 22–24
Modes of cooperation, in nature, 90
Mokṣa, 384
Mokuseki bushshō, 114
Molesworth, Charles, on cultural transmission, 211–212
Monasteries, 35–36
animal preserves at, 386
ecologically inappropriate deeds of, 386
forest, 53
Monastic community. *See also* Thai *saṅgha*
celibacy in, 64*n*
ecologically appropriate attributes of, 49
Gary Snyder on, 187–188
as a green society, 48–50
layout of, 52–53
liminal attributes of, 49, 50–54
limitations of, 54–56
as a retreat, 57
social stratification in, 64*n*
Monastic life, 337
Monji symbols, 103
Monkhood, 64–65*n*
entering and leaving, 51
Monk of wisdom, 24

Monks
dwelling places for, 34–36, 37
laypersons and, 342
responses to activism of, 55–56
responsibilities of, 36, 47–48
rules of conduct for, 51–52
"Monks of convenience," 386
Mono no aware worldview, environmental problems with, 119–120
Moon, poetic images of, 338–339
Moralistic approach to nature, 333
Morality, 206–207
"Mother-goddesses," female deities as, 308–309
Motherhood
female sexuality and, 308–309
nurturing and, 309–310
Mountains
near Green Gulch and Spirit Rock centers, 223
seeing from the perspective of, 170–171
in Zen Buddhism, 166–167
Mountains and Rivers Temple, 244
Mount Diablo, 223
Mount Tamalpais, 223
Mount Tamalpais State Park, 237
Mt. Tremper Zen Center, 244
Mu, 94
Mu busshō, 122
Mudrās, 77, 115
Muir Beach, 220, 221
Muir Woods National Monument, 220
Mujō, of nature, 99
Mujō beings, 102
Mujō-busshō, 102
Mujōdō no taigen, benefits of, 168–169
Mujō-seppō theory, 117
Mu no basho, 94
Muromachi shogunate, *hōjō-e* ceremony and, 153
Murota, Y., on Japanese view of nature, 327

Murphy, Patrick D., 198
  on community and ecology, 202–204
  on cultural transmission, 212
Music, 84
Myanmar, Buddhism in, 59
Myōhō renge kyō, 96–97
"Mysterious depths," 99
Mystic Law of cause/effect, 96–97
Mythological and shamanistic community, Gary Snyder and, 194–204
Mythological women, fertility of, 308–309
Mythology
  fertility in, 308–309
  nature and, 194–199
  power of, 198–199

Naess, Arne, 127n, 344, 370–371
  "deep ecology" of, 214–215n
  ecological philosophy of, 93
  Spinozan foundations of, 387
Nāgārjuna, 316
  on Emptiness, 342–343
Nakano Hatayoshi, on early *hōjō-e* ceremonies, 153
"*Namu myōhō renge kyō*," 97
Nan-ch'uan, 286–287
Nash, Roderick, on religion and nature, 327
Nāthaputta, jātaka about, 137–138
National Buddhist Association, animal-protection radio broadcasts by, 386
Native American cultures. *See also* Shamans
  Gary Snyder's worldview and, 194–195, 196–197, 198, 199, 205
  social conflict in, 204
Nature
  affirmation of, 179
  beauty of, 91–92
  Buddhist views of, 10–12, 13–14,
    14–15, 33–34, 46, 327–346,
    337–338, 378–379, 382–384
  caring for, 26–30
  Chinese views of, 12–13
  Christian teachings about, 72–73
  communion with, 194–195
  community and, 187–188, 213
  "conquest" of, 388–389
  controlling, 181
  Dalai Lama on, 328–330
  dhamma as, 24–30
  ecological definitions of, 379–381
  elevating self relative to, 181
  enlightenment and, 331
  in environmental philosophy, 89
  field model of, 92–95
  finding faults and errors in, 180–181
  Gaia theory of, 103–104
  giving to and receiving from, 181–182
  God and, 73, 85, 115
  happiness and, 33
  Hindu attitudes toward, 335–336
  honoring the body of, 180
  human consciousness and, 142
  humanistic, moralistic, negativistic, and dominionistic approaches to, 333–334
  human responses to, 71–72
  Ian Harris on, 38–40
  incorrect Western worldview of, 31–32, 327
  Indian views of, 335–339
  interdependence of, 28–30, 74–75, 90, 97–98, 122, 198, 202–204
  in Japanese Buddhism, 95–97, 99–101, 124–125
  in Japanese literary tradition, 101–102
  Japanese views of, 89–105, 119–120, 327–328, 330, 333–334, 339–340

law of, 32
love in, 189, 190–191
magic in, 200–201
management of, 32–33
mythology and, 194–199
order in, 81–82
perfection of, 180–181
philological issues of, 395–396
relationship of humans to, 94–95
revising human attitudes toward,
    111–112
salvific function of, 101–103
as sentient being, 101–102
soteriological function of, 102–
    103
stealing of, 179
terminological meanings of, 379–
    380
Western culture and, 74–75
wild aspects of, 337
Nature Conservancy, The, 222
Nature preserve, Zen Mountain
    Center as, 259
Nāṭyaśāstra, 395
"Near enemy," 351
Necessity, 84–85
Negativistic approach to nature, 333
Neighborhood values, 207–208
Neoplatonic theology, 115
Nevada Nuclear Test Site, anti-
    nuclear activism at, 272–273,
    283–284
New Buddhism, 389
Nibbāna, 33
Nichiren Buddhism, 97
Nichiren Daishonin, on oneness of
    life and its environment, 96–97
Nichiren Shōshū sect, 96
Nidāna, 356
Nihilism, 385–386
Nihonjinron thinkers, 389
Ningen, human nature as, 94
Nirmāṇakāya, 76, 116
Nirvāṇa, 384, 384, 389–390

ambiguity of, 396
quest for, 342
representation of, 337–338
saṃsāra and, 103, 119–120
stopping and, 332
Nirvāṇa Sūtra, on sentient beings,
    116–117
Nishida Kitarō
    field model of nature of, 94
    on Subject/Object antithesis, 384–
        385
Noble Truths, first, 7
Noh drama, 99, 101
Non-attachment, 42n
Nondual forms of Buddhism, 353,
    375n
Non-egalitarianism, of Buddhism,
    352
"Non-killing precept," 152, 154,
    161n.
    Fujiwara no Sanesuke and, 156
Nonreproductive life-styles,
    validation of, 305
Nonreproductive sexuality, dis-
    couragement of, 307–308
Nonsentient beings
    Buddhahood of all, 117, 122–124,
        390
    protection of, 318–319
Nonseparateness. *See also* Separa-
    tion
    in Zen Buddhism, 169–170
Non-substantiality, 353
Nonviolence, 52
    social change through, 58
No-self, ideal of, 342, 344
Noss, Reed, 225
Nothingness, 94
Novices, 53
Nuclear ecology, 288n
    Buddhism and, 269–287
    described, 269–270
    kōans for, 284–287
    mindfulness and, 273–274

Nuclear Guardianship Project, 271
Nuclear waste. *See* Radioactive
  waste
Nuns, Buddhist, 56
Nurturing, motherhood and, 309–
  310
Nutrition, importance of, 300–301

Oak Tree Canyon, at Spirit Rock
  center, 237
Obedience, 52
Obligation, 188
  to reproduce, 301–302
Oda, Mayumi, anti-nuclear activist
  art of, 272
Ōjōyōshū, 382
Okada Sōji, 160*n*
Olcott, Henry Steel, on Sri Lanka, 8
Olson, Grant A., 30
Omine Akira, on nature in Japanese
  literary tradition, 101–102
One-Bodied Three Treasures, 178
Oneness, 197
  of nature, 188–191
Onozukara, mizukara and, 95–97
Oppressed, the
  aid to, 5–8
  nature as, 14–15
Oppression, hierarchy of, 364–366,
  376*n*
Order, in nature, 81–82
Organic balance, in emerging
  Western ecological models, 75
Oriental despotism, 387
Original sin, 83
Oryoki meals, at Green Gulch
  center, 232
Other, the
  David Abram on, 201–202
  Mikhail Bakhtin on, 202–204
  nature as, 187–188
"Other-power," 102
"Outside world," in Zen practice,
  165–167

Overcrowding, 300–301
Overpopulation
  environmental crisis and, 291,
    292
  lack of spiritual enlightenment
    and, 298–299

Pakati state, 27
Pali scriptures, 25
Pali *suttas*, 34
Pali texts and traditions
  Dhammapiṭaka's use of, 30–31
  jātaka as, 133–134
Panpsychism, 72
Pansa, 48
Paradise, in Hindu tradition, 337
Pāramitā, 357
Parenthood
  Buddhism and, 305–306
  nurturing and, 309–310
  in religious symbolism, 311
Parinirvāṇa scene, 391
Parking
  placing limits on, 241
  at Spirit Rock center, 236
Parliament of the World's Religions,
  316
Participation, 188
Partridge, fowler, and Brahmin,
  jātaka about, 138
"Passing through," 195
Paṭicca-samuppāda, 295, 344
Pāṭimokkha, 52
Patriarchal stereotyping, 309–310
Patronizing attitude, in aid to the
  poor and oppressed, 6
Paul, Sherman, on community, 213
Peace, Buddhism versus, 281–282
Peace and quiet, nature as fostering,
  24–25
People. *See* Humans
Perfection, 180–181
Perfection of Wisdom literature, 357
Periphuseōn (Erigena), 115

Pernetarian society. *See* Green society
Personalists, 341
Petee Jung, on communionism, 194–195
Phenomena, Buddhist interpretation of, 76–77, 113–114, 115–116, 117–118
Photovoltaic (PV) system, at Zen Mountain Center, 260–261
Phra paññā, 24
Phra Phaisan Visalo
  on forest monasteries, 53
  on the monastic community, 50
Phra Prajak Kuttajitto, activism of, 34–35, 55
Phra Prayudh Payutto. *See* Dhammapiṭaka
Physics
  Buddhism and, 80
  in international law, 318
  nuclear ecology and, 270
Physis, 95–96
Pilgrimage, 238
Pindapata, 52
Pine Springs Ranch, 251
"Pink cloud," awakening as, 169
Pinus coulteri, at Zen Mountain Center, 254
Pinus Jeffreyi, at Zen Mountain Center, 254
Place of excrement, 382
Plants
  Buddha-nature of, 332, 390–391
  in dedication chants, 228
  endangered, 253t
  in Gary Snyder's poetry, 195, 196
  at Green Gulch Zen Center, 221
  karma of, 278
Platform Sūtra of the Sixth Patriarch, The, 285
Plato, 74
  realm of Ideas of, 81
Plumwood, Valerie, 225

Plutonium, 272
  at Amarillo, Texas, 284
  toxicity of, 276
Poetry. *See also* Bashō; *Haiku* poetry
  of Cold Mountain, 331
  of Dharmakīrti, 338–339
  of Gary Snyder, 190, 194–199, 208–209, 211–212
  of India, 338–339
  of Li Po, 339
  of Saigyō, 102–103, 197, 331–332
  of Śāntideva, 284–285, 343
Politics
  within *saṅgha*, 55
  the voice of nature in, 196
Pollution, Buddhist attitudes toward, 122
Poor, the
  aid to, 5–8
  nature as, 14–15
Population, environmental crisis and, 291–311
Pottery workshop, at Zen Mountain Center, 262
Poverty, overpopulation and, 298–299
"Power mode," 376*n*
Powers and potencies, 123
  in hierarchies, 365–366, 366–368
Practice. *See also* Food practice; Insight meditation practice; Sōtō Zen practices; Work as practice; Zen practice
  community of, 205–208
Prajñāpāramitā literature, views of nature in, 10–11
Prakṛti, 335–336, 380–381
Pratītya-samutpāda, 76, 295, 342, 353, 381
Precious Garland, The (Nāgārjuna), on the Golden Rule, 316
Predators
  humans as, 71–72
  jātaka about, 141–142

Pretas, 356
Prip-Møller, Johannes, on Chinese Buddhist monasteries, 386
Pronatalism, religion and, 292–295
Protestant Buddhism, 8
Protestantism, 199
Prothero, Stephen, 8
Pudgala, 341
Pudgalavādin, 341
Pure Land theory, 102
Purification, of the mind, 329
Puruṣa, 335, 336

Quail and elephants, jātaka about, 136–137
Queen, Christopher, 8
Quercus kelloggii, at Zen Mountain Center, 254
Quezada, Juan, on pottery technique, 262

Radiation ecology, 270
Radioactive waste
    Buddha-nature of, 120–121, 121–122, 122–123, 124
    disposal of, 270–271
    impermanence of, 121–122
    karmic implications of, 274–277
    *kōan*s concerning, 286–287
    mindfulness and, 273–274
    in nuclear ecology, 269
    Zen Buddhism and, 281–284
Radiobiology, 270
Radish motif, 390–391
Rahūla, jātaka about, 138
Rainy season retreat, 48, 57
Rasas, 395, 402*n*
Raw–cooked food, 14
Reality. *See also* Universe
    Buddhist view of, 76
    in Western culture, 74
    in Zen Buddhism, 93–94, 177
Reality embodiment, 114
Realized Three Treasures, 178

Realm of Ideas, 81
Rebirth, Buddhism and, 296–297
Recycling, 383
Reforestation, 54, 65*n*
Rehanek, Woody, on Gary Snyder's poetry, 197
Reincarnation, 143
Reincorporation, in rites of passage, 51
Reinhabitation, 191–192, 239
    as ethical ecological living, 225–227
Reinhabitory peoples, 226
Relational dimension of Buddhism, 353–359, 360–364
Relational fields, in ecological worldview, 92–95
Relationships
    differences between Christian and Buddhist concepts of, 79–80
    sexuality as leading to spiritual and dharmic, 311
Relativism, regarding women, 294–295
Religion
    Buddhism as first world, 4
    cosmological thinking in, 377
    in environmental issues, 63*n*
    excessive consumption and, 293, 299–301
    global interdependence and, 314
    reproduction and, 293, 301–307
    sexuality and, 307, 310–311
Religion journal, 38
Religious fundamentalism, 23
Reproduction. *See also* Sexuality
    Buddhism and, 291–311
    emotionality and greed in, 304
    environmental crisis and, 292
    motivations for, 301–307
    pressures toward, 304–305
    religion and, 292–295
    as religious duty, 293
    selfishness and, 302–303
    sexuality and, 307–311

Resources
  equitable distribution of, 292
  interdependence and consumption
    of, 296
  reproduction and consumption of,
    296
Responsibility, 183–184, 287, 346
  for being wrong, 285
  for the environmental crisis, 279–
    280, 285
  at Green Gulch center, 228–232
  karma and evasion of, 275
  in reinhabitory ethic, 227
  at Spirit Rock center, 235–237
Responsive rapport, between all
  things, 119
Restraint, at Green Gulch center,
  234
Retreat centers, ecological concerns
  about, 244
Retreats, 48, 57
  at Green Gulch center, 234
  at Zen Mountain Center, 254–
    255, 262–263
Revelation, as a *maṇḍala*, 78
Revitalization movements, 57–58
Reynolds, Frank E., on Buddhist
  cosmological thinking, 382
Ṛg Veda, animals in, 133
Rice, offerings of, 160–161*n*
Rightness
  Aldo Leopold on, 188
  in nature and aesthetics, 91–92
Rights of living things, human
  rights versus, 111–112
Riji muge doctrine, 97–98
Rio Declaration, 314–315
Rio Earth Summit, Earth Charter
  and, 314–315
Rio + 5 review, 315
Risk, ethics of, 279
Risshinben, 128*n*
Ritchie, Donald, on Japanese
  attitude toward nature, 334

Rites of intensification, of lay
  individuals, 52–53
Rites of passage, 51
Rivers
  seeing from the perspective of,
    170–171
  in Zen Buddhism, 166–167
Riyaku shinkō, 151
Rochester Zen Center, 244
Rockefeller, Steven C., on environ-
  mental ethics, 342
Romantic movement, as source of
  land ethic, 89–90
Round, Graham, 54
Round, Philip, 54
Ruegg, D. Seyfort, on eating meat,
  137
Rukkhadhamma Jātaka, 140
Rūpic cosmogony, 383
"Rūsui chōja shijin," 150–151
Ryder, Japhy, 213
Ryoan-ji, Zen garden at, 330

Saccadhamma, interdependence of
  nature as, 29
Sacralizing the landscape, at Green
  Gulch center, 233–234
Sacredness, in earth covenants and
  charters, 320–321
Saddharmasmṛtyupasthāna Sūtra
  (Abhidharma)
  Buddhist visions of hell in, 382
Sahakorn, the world as, 28
Saichō, 102–103, 114, 390
Saigyō, 391–393 poetry of, 102–
  103, 197, 331–332, 333, 339
Saito, Yuriko, on Japanese environ-
  mentalism, 119–120
Śākyamuni
  asceticism of, 15–16
  as "historical Buddha," 76
  life of, 11–12
Sala, 52
Salt, offerings of, 160–161*n*

Salvation, 102–103
Salvific function of nature, 101–103
Samādhi, 115
 power of, 167
Sāmaññaphala Sutta, 355–356
Samaya-body of the Buddha,
 natural elements as, 114
Samaya-maṇḍala, 78
Sambhogakāya, 75–76, 116
Saṃjiva, 382
Sāṃkhya tradition, 336
Saṃsāra, 78, 380–381, 384
 ambiguity of, 396
 nirvāṇa and, 103, 119–120
Saṃsāric cosmogony, of Buddhism,
 382–383
Samyaksambodhi, 178–179
Saṃyutta Nikāya, Buddha's
 teachings in, 356
Sanbō Kyōdan community, Zen
 Buddhism of, 165
Sanbō Kyōdan lineage, kōans in,
 168–169
Sanchokusai, 153
San Geronimo Valley Planning
 Group, 222
 Spirit Rock center and, 236, 237
Saṅgha, 33–34
 Gary Snyder on, 187–188
 limitations of, 205–208
 as one of the Three Treasures,
 177–178
 at Zen Mountain Center, 260
Sanghakamma, 36
Saṅgha Treasure, 178
Sangopyen, 24
Sanitsuda Ekachai, 58
San Jacinto Mountains, 254
Sankeisha, 152
Sanmitsu, 77, 115
Sanshin, in Kūkai's buddhology,
 75–76
Sanskrit, Chinese and, 379
Śāntideva, poetry of, 284–285, 343

Santikaro, on activism, 55
Saraca indica, 52
Satipatthana Sutta, 235
Satori, in Zen Buddhism, 101
Sattvaloka, in nature, 381
Scharf, Robert, on nihonjinron
 thinkers, 389
Schmithausen, Lambert, 37
 on jungle and forest, 337
 on plant karma, 278
Schoolyard Garden, 230
Schumacher, E. F., 100, 344
Science, Buddhism and, 80–81
Scripturalism, 4
Seated meditation, benefits of, 167–
 169
"Secret teaching," 75
Seed syllables, 78
Seiganji temple, 390
Self, 29
 awakening to one's true, 167–169
 elevation of, 181
 enlightenment and, 331, 341–342
 in ethics, 94–95, 370–371
 extinction of, 49
 in Hindu traditions, 336–337
 in Japanese Buddhism, 95–97
 in reinhabitory ethic, 226–227
 seeing from the perspective of,
 170–171
 in Zen Buddhism, 167–171, 172
Self-discovery, of Zen practitioners,
 165–166
Selfish ignorance, in Zen Buddhism,
 166–167
Selfishness
 destructiveness of, 29, 172–173
 overcoming, 26–27
 reproduction and, 302–303
Selflessness, and caring for nature,
 26–30
Self-perpetuation, reproduction as,
 304–305
Self-realization, 370–371

Self-reliance, 342
Self-sufficiency, at Zen Mountain
    Center, 260–261
Semi-feudalism, 387
Seminature, Japanese love of, 334
Sen no Rikyū, 99
Sentient beings, 101–102
    Buddhahood for all, 113, 114,
        296–297
    human rights and, 320–321
    in nature, 381
    Nirvāṇa Sūtra on, 116–117
    protection of, 318–319
"Sentient landscape," nature as, 201
Separation, 25. *See also* Non-
        separateness
    in rites of passage, 51
    Zen Buddhism and, 177
Sessho kindan ideology, 149, 154
Sesshū, 99, 394–395
Sessions, George, 127*n*
    on intuition of biocentric equality,
        120–121
Seven limbs of enlightenment, 356
Seventh Day Adventists, in Apple
    Canyon, 251
Seventh National Development
    Plan, in Thailand, 23–24
Sexuality, 180. *See also* Repro-
        duction
    bonding and, 308
    Buddhism and, 293
    communication and, 307–311
    negative feelings toward, 307–
        308
    religion and, 310–311
    reproduction and, 307–311
    sacredness of, 310–311
Shamanism, ecology and, 199–202
Shamanistic and mythological
    community, Gary Snyder and,
    194–204
Shamans, 199–202
    as religious practitioners, 205

Shambhala Center, 244
Shame, and aid to the poor and
    oppressed, 6
Shih-shih wu-ai doctrine, 98
Shimbutsu bunri policy, 159*n*
Shimizu, Yoshiaki, on Itō Jakuchū's
    art, 391
Shin, 76
    of nonsentient beings, 117
    in Shintō, 113–114
Shin bukkyō, 389
Shingon doctrine
    meditation practices of, 116
    worldview of, 75–79
Shinjin, 154
Shinran, Pure Land theory of, 102
Shintō
    Buddhism in Japan and, 102–103,
        113–114
    hōjō-e ceremony and, 149–150,
        152–153
    nature in, 197–198
Shitsu-u, 123
Shizen, 95–96, 102
Shōbōgenzō (Dōgen), 102, 393
Shōen, *hōjō-e* ceremony and, 155–
    156
Shōji, 122
Shōjin kessai, 153, 162*n*
Shrader-Frechette, Kristen, on risk,
    279
Shrine deity, 152
Shuji, 78
Shujo, 101–102
Silence, at Green Gulch center, 234
Simplification, at Green Gulch
    center, 234
Sinhala Buddhism, 8
Six *bodhisattva* virtues, 357
Six Great Elements, 83
    reality as composed of, 76–78
Six realms, of karmic cosmology,
    200–201
Skandha, 341

Skillful means, countering ecolog-
ical problems with, 172–173
Slavery, 295
Smith, April, 242
Smith, Patricia Clark, on social
conflict among Native Ameri-
can women, 204
Snyder, Gary, 187–213, 220, 235
    on bioregional community, 191–194
    on community spirit, 208–213
    on food webs, 188–191
    on "Green Buddhism," 362–363
    Mikhail Bakhtin and Patrick
    Murphy on, 202–204
    on mythological community, 194–
    204
    on nature as community, 187–188
    as "nature poet," 205
    poetry of, 190, 194–199, 208–
    209, 210–212, 214*n*, 215*n*, 272
    on radioactive waste disposal,
    270–271
    on reinhabitory ethic, 226–227
    on sanctity of nature, 346
    on shamanistic community, 194–
    204
    on spiritual practice, 205–208
    on Tao and *Dharma*, 328
Social conscience, 91
Society, as aiding the poor and
oppressed, 5–8
Soils, at Zen Mountain Center, 250–
251
Sokoji Temple, 220
Sokushin Ze-Butso (Dōgen Zenji),
174*n*
Solar power, at Zen Mountain
Center, 260–261
Somneuk Natho, 54, 58
Sōmoku, in teachings of Kūkai, 114
Sōmoku jōbutsu doctrine, 391, 393
Soper, Kate, on nature in envi-
ronmentalist discussions,
379–380

Sorting of things, 123
Sōseki Natsume, 101
Sosen kaikō, 151
Soteriological function of nature,
102–103
Sōtō Zen practices, 117–118, 233
Soul
    Hindu attitudes toward, 335–336
    of a living thing, 113
Species, intrinsic value of, 319–320,
320–321
Species protection, 131
Speech, as one of three mysteries,
77, 78–79, 115, 116
Spirit, 198
    Dalai Lama on, 328–330
    Hindu attitudes toward, 335–336
Spirit Rock Design Committee, 236
Spirit Rock Meditation Center
    as ecological role model, 243–245
    environmental practices of, 220
    evaluation of, 235–239
    future challenges at, 241–243
    land history of, 222–223
    locality of, 222–225
    points of tension at, 239–241
Spiritual Darwinism, 362–363
Spiritual training, 362
Spiritual quest, of Buddhism, 38
Spretnak, Charlene, on long-term
thinking, 283
Spring equinox, *ekos* at, 228
Śramaṇic dietary laws, animals in,
133, 137
Sri Lanka, Buddhist revival on, 8
State rituals of Japan, 153–154,
154–156
Stealing, precept against, 179
Sthiramati, on the external world,
385
Stockholm Declaration, 315
Stopping, 331–332
Strong, Maurice, Earth Charter and,
315

Stupas, 52
Subatomic microcosm, 93
Subject-Object antithesis
  humanity-nature relationship as, 384
  in Zen Buddhism, 168–171
Substantial objects, in ecological
  worldview, 93
"Suchness," 79
Suffering, 7, 25, 32–33, 52
  Earth Charter and, 318
  self-centered desire and, 304
  spiritual enlightenment and, 298–
  299
Sukha, 32, 33
Sukhāvatī, 337
Sukhāvatīyūha, 337
Sulak Sivaraksa, on peace at any
  price, 281–282
Sumie paintings, 99, 101
Summer solstice, *ekos* at, 228
Suññatā, 26
Śūnyavāda doctrine, 357
Supernatural, the. *See also* Magic
  Thomas Aquinas and, 380
Superstition, in Buddhism, 199
Sustainability. *See* Ecological
  sustainability
Sūtras, 357
  in Buddhist teachings, 118
Suvaṇṇamiga Jātaka, 135
Suvarṇaprabhāsa Sūtra, as basis for
  hōjō-e rites, 150–151
Suzuki, D. T.
  on aesthetics of nature, 99–100
  on "conquest of nature," 388–389
  on Zen Buddhism, 360
Suzuki Rōshi, Shunryu, founding of
  Green Gulch Zen Center by,
  220–221
Svabhāva, 380–381
Swearer, Donald, on environmental
  problems, 58
Symbioses, 90
Sympathy, 26, 42*n*

Taiwan, Jen-yen's hospital in, 7, 8
Tambiah, Stanley, 8
Tanahashi, Kazuaki, anti-nuclear
  activist art of, 272
T'ang period, Buddhism during, 386
Taṇhā, 83
"Tantric sex," 310
Tantric worldview, 385
Tao, *Dharma* and the, 328
Taoism
  biocentric equality in, 120–121
  environmental philosophy and, 89, 93
  Gaia theory and, 104–105
Taoist thinkers, influence on
  Buddhism of, 123–124
Tao of art, 99
Tapana, 382
Tariki, 102
Tassajara Mountain Center, 221
Tathāgata embryo, 390
Tathāgatagarbha, 303, 390
Tax-avoidance, through monkhood,
  386
Te, 123
Techno-karma, 277
Technology
  benefits of, 22–23
  excesses of, 378
  global community and, 313–314
  and the understanding of karma,
  274–277
  vision versus, 72
Teilhard de Chardin, Pierre, 362
Television, dissemination of
  environmentalist thought
  through, 124–125
Telovāda Jātaka, 137–138
Temples
  biodiversity around, 65*n*
  in monastic communities, 52–53
  retreats to, 48
Ten *bodhisattva* stages, 357
Tendai Buddhism, 114
  concept of life in, 96–97

Thai Buddhism, 21, 59–60
Thailand
    Buddhism in contemporary, 21,
        23–24, 47, 397*n*
    environmental crisis in, 45–50,
        59–60
    monastic communities in, 50–54
    number of villages, temples,
        monks, and novices in, 53
Thai Pali *tipiṭaka*, 30
Thai *saṅgha*, 21, 50. *See also*
        Monastic community
    Dhammapiṭaka at, 30–31
    as green society, 54–56
Thamachāt, 26, 27, 27–28. *See also*
        Nature
Thāyashimsat heaven, 151
Theophany, natural creatures as, 115
Theosophical Society, Sri Lankan, 8
Theras, early prominence of, 356
Theravāda Buddhism, 40
    jātaka in, 134
    rise of, 355
    in Thailand, 47
Theravāda cosmology, 383–384
Thich Nhat Hanh, 219, 235
    on nuclear waste, 274
    on peace at any price, 281–282
    on radioactive waste disposal, 271
Thomas Aquinas, the supernatural
        and, 380
Thoreau, Henry David, 311
Thought, 83
"Three-body" theory, in Kūkai's
        buddhology, 75–76
"Threefold learning," 369
"Threefold teaching," 356
Three Jewels. *See* Three Treasures,
        the
Three Mysteries, 77, 78–79, 83, 115
Three Pure Precepts, 178–179
Three Treasures, the, 177–178
    gratitude for, 227–228
"Three world-levels," 356

Tibet, 328, 347*n*, 397*n*
    as radioactive waste disposal site,
        271
    Vajrayāna Buddhism in, 311
T'ien tao, 123
T'ien-t'ai school, 113
Tigers and lions, jātaka about, 141–
        142
Titans, 356
Tittira Jātaka, 138
Tools, 211
Tōriten, 151
"Total-being," 123
Totem animals, 197
Tōtōkōzen-in Ichigonkannondō, 152
Toxic waste, Zen Buddhism and, 281
Traffic management, at Spirit Rock
        center, 236
Transhuman nature, 203–204
Trash removal, at Green Gulch
        center, 231
Trees. *See also* Forests
    in Gary Snyder's poetry, 195
    jātaka about, 140, 141–142
    memorialized at Green Gulch
        center, 234
    non-native, 242
    at Zen Mountain Center, 254
Trikāya, in Kūkai's buddhology,
        75–76
Triloka, 356
Trisikṣā, 369
Truth
    manifesting, 180
    of Zen Buddhism, 287
"Truth word," 75
Tsurugaoka Hachiman Shrine, 160*n*
Tubercle bacillus
    Buddha-nature of, 122
    Taoist/Zen Buddhist role of, 120–
        121
Turner, Victor, on liminality, 51
Tu Weiming, 104
    on the Golden Rule, 317

Twelve positive causes and conditions, 356

Ujigami, 154
Uji philosophy, 102
Undeveloped land, at Buddhist centers, 239–240
Union, jātaka about, 140
United Nations, Earth Charter and, 313
United Nations Conference on Environment and Development (UNCED), Earth Charter and, 314–315
UN Conference on the Human Environment, 315
United Nations World Commission on Environment and Development, Earth Charter and, 314–315
United States
aid to the poor and oppressed in, 6
Buddhism in, 4
Jen-yen in, 7
views of nature in, 333–334
Universal Declaration of Human Rights, Earth Charter and, 313
Universalism, in Buddhist thought, 118
Universe. *See also* Cosmological thought; Reality
Buddhist conception of, 383, 384–385
differences between Christian and Buddhist concepts of, 79–80
as Indra's net, 189–190
seeing from the perspective of, 170–171
self-maintenance of, 181
in Zen Buddhism, 177, 178
Unji gi, 114
Upāya, 77, 172, 354
Uruvelā, 34
Usa Hachiman Shrine, first *hōjō-e* ceremony at, 152–153, 160*n*

Vairocana, 385
Vajrayāna Buddhism, 307–311
rise of, 355
Vanavaccha Thera, 34
van Gennep, Arnold, on rites of passage, 51
Van Gogh, Vincent, 330
Vassa, 48
Vasubandhu, on Subject/Object antithesis, 384
Vegetable Nirvāṇa (Itō Jakuchū), 390–393, 392*illus.*
Vegetables, Buddhahood of, 390–391
Vegetarianism, 243
at Green Gulch center, 231
in jātaka narratives, 137–138
at Spirit Rock center, 237
Vegetation, in teachings of Kūkai, 114
Venturini, Riccardo, on Buddhist traditions, 340
Verticality. *See* Hierarchy
Vessava, King, jātaka about, 140
Vihara, 52
Vijñānavāda school, on Subject/Object antithesis, 384
Vimalakīrtinirdeśa Sūtra, 394
Vinaya, 52
Vinaya rules, 36
Violence, 137
from overcrowding, 300
Vipassanā, 222, 238
"Virtue ethic," 370
Virtues, 357
of Three Treasures, 177–178
Visākhā Pūja sermon, 25
Vision, technology versus, 72
Visualization, 116
Vital power, 104
Viveka, 25, 336
"Voice of the Watershed" walks, at Green Gulch center, 232
Voluntary services, 7–8

Vyaddha Jātaka, 141–142
Vyāsa, on violence, 137

Wabi, 99
Waka poetry, 99
Walking meditation
  at Green Gulch center, 232
  at Spirit Rock center, 238
Walpola Rahula, 8
Walters, Derek, on *feng shui*, 104
Waste. *See also* Radioactive waste
  Zen Buddhist attitude toward,
    280–281
Waste management, at Buddhist
    centers, 242–243
Waste recycling
  at Buddhist centers, 243
  at Green Gulch center, 231
  at Spirit Rock center, 237
Watanabe, Masao, on Japanese view
    of nature, 328, 330
Water conservation
  at Buddhist centers, 242–243
  at Green Gulch center, 231–232
Wats, 48
  in monastic communities, 52
Wat Suan Mōkh
  Buddhadāsa Bhikkhu at, 25–26
  founding of, 24
Watsuji Tetsurō, on the human/
    nature relationship, 94–95
Watt, W. Montgomery, on early
    Islam, 334–335
Watts, Alan, on Zen Buddhism, 360
Way of Heaven, in Taoism, 123
Wealth, 9–10
  worldly increases in, 22–23
Welch, Lew, Gary Snyder and, 195
West, the
  Buddhism and, 206–207
  Christianity and, 73
  environmental ethic of, 72–73
  environmentally damaging
    enterprises of, 386–387

incorrect worldview of, 31–32
influence on Buddhism of, 5, 6–7,
    8, 45
mercantile activities of, 9
views of nature in, 10
Western Buddhism, 199
Western thought, environmentalists
    in, 124–125
Western worldview
  Buddhist worldview versus, 46,
    56–59, 59–60, 79–80
  environmental crisis and, 111–112
  hierarchy in, 80
"Whale banquets," 156
Wheelwright, George, 220, 221
Wheelwright, Hope, 221
Whistler, James McNeill, 330
White, Lynn, Jr.
  on Christianity, 72–73
  on Judeo-Christian anthropo-
    centrism, 327–328
Whitehead, Alfred North, 107*n*
  field model of nature and, 93
  on the living body, 81–82
Whitman, Walt, 144
  on animals, 131–132
"Wild," the, 311, 337
Wilderness retreat, at Zen Mountain
    Center, 262–263
"Wild mind," in reinhabitory ethic,
    227
Winter solstice, *ekos* at, 228
Wisdom, 357–358, 367
  in reinhabitory ethic, 227
Women
  cultural relativism regarding,
    294–295
  discrimination against, 56
  maternity and, 301–302
  motherhood and, 308–309
  nurturing and, 309–310
  relationships with men, 310–311
Woodpecker and lion, jātaka about,
    135

Work as practice, at Green Gulch
center, 233
Workshops. *See* Environmental
workshops
World Charter for Nature, on
intrinsic value of species, 319
World Commission on Environment
and Development, on environmental protection, 317–318
Wrongness, responsibility for, 285
Wu Ti, Emperor, 386

Xing, 128*n*

Yab-yum icon, 310–311
Yakṣas, in Indian fine art, 394
Yasai Nehan (Itō Jakuchū), 390–
393, 392*illus.*
Yin/yang forces, 104
Yoga, 336
Yogācāra doctrine, 357
Yogācāra school, 152
on Subject/Object antithesis, 384
Yoga Sūtras, 137
Yoga tradition, 336
Yojō, 99
Yōkyoku, 394
Yoshimasa, 153
Yoshimitsu, 153
Yoshimochi, 153
Yoshinori, 153
Yuasa Yasuo, on the human/nature
relationship, 94–95
Yūgen, 99, 101
Yūjō beings, 102

Zazen, 240
benefits of, 167–169, 171, 172
Zeami, 99
Zen Buddhism, 13, 124–125, 353
Aldo Leopold and, 100–101
biocentric equality in, 120–121
criticism of universality of, 119–120
D. T. Suzuki and, 388–389

ecology and, 165–173
field model of nature in, 93–94
Kegon infrastructure of, 97–98,
99–100
practices in, 117–118
rise of, 355
waste in, 280–281
Zen Center, founding of Green
Gulch Zen Center by, 220–221
Zen circle, 272
Zendo, 240
at Green Gulch center, 228, 234
Zen garden, at Ryoan-ji, 330
Zen Mountain Center
biological survey of, 251–254
endangered flora and fauna at,
252–253*t*
environmental ethics at, 263–264
environmental program at, 249–
250, 259–261, 263–264
environmental workshops at,
261–263
fire management at, 256–258
flora and fauna at, 251
forester at, 256–257
land stewardship by, 256–259
land use at, 254–256
locality of, 249, 250–254
residents at, 258
soils at, 250–251
Zen practice, 174*n*
ecological action and, 172–173
Gary Snyder and, 205
human body in, 345–346
pitfalls in, 165–167
three fruits of, 167–171, 172
Zen Precepts, environmental
relevance of, 177–184
Zen *satori*, 101
Zen training, at Zen Mountain
Center, 262
Zero-point, in Zen Buddhism, 168,
169–170
Zhao, Teaching Master, 169–170